ENERGY EFFICIENCY – A BRIDGE TO LOW CARBON ECONOMY

Edited by **Zoran Morvaj**

Energy Efficiency – A Bridge to Low Carbon Economy
Edited by Zoran Morvaj

Published by InTech

Janeza Trdine 9, 51000 Rijeka, Croatia

Publishing Process Manager Daria Nahtigal

Technical Editor Teodora Smiljanic

Cover Designer InTech Design Team

Additional hard copies can be obtained from orders@intechopen.com

Energy Efficiency – A Bridge to Low Carbon Economy, Edited by Zoran Morvaj
p. cm.
ISBN 978-953-51-0340-0

For EU product safety concerns:
IN TECH d.o.o., Prolaz Marije Krucifikse Kozulić 3, 51000 Rijeka, Croatia,
info@intechopen.com or visit our website at intechopen.com.

Meet the editor

Dr Z. Morvaj, holds PhD in electrical engineering, and has more than 20 years of teaching experience at universities in Croatia, UK and Thailand. During early nineties he worked for an UK energy and environmental consultancy firm, managing the projects for international development banks and aid organizations, policy development and capacity building projects to governments and performance improvement consultancy projects for industry. He moved to Thailand to manage local office just before the economic crises in South East Asia in 1997. He developed, designed, and managed implementation of a number of energy and environmental management projects for large industrial international companies from Japan, USA, UK and Thailand, which often transcended into a cost management projects because the theme-of-day at that time was cost cutting. Besides, he has practical experience in supporting local green development, related change management, training needs assessment, and setting up adequate institutional and financial support. He is an author of a number of professional papers, handbook and manuals, and author of a book on the energy and environmental subjects published by renowned international publishers and translated into Chines. Recently he developed and managed a large national multimillion dollar energy efficiency project in Croatia. Currently he manages energy and green development projects throughout South-east Europe and Central Asia for major international development organizations.

Contents

Preface

Energy efficiency is finally a common sense term. Nowadays almost everyone knows that using energy more efficiently saves money, reduces the emissions of greenhouse gasses which cause climate change phenomena and lowers dependence on imported fossil fuels like oil, gas and coal.

When we consider energy supply, energy efficiency is again the natural first step. By eliminating wasteful consumption and losses in the supply chain, we are actually increasing capacity of existing systems by creating so called 'negawatts', i.e. enabling supply of more customers without additional investments into energy generation and distribution capacities.

Therefore, whether we consider supply or demand side of an energy system, energy efficiency is always the first thing to do.

However, after this step one should think off what follows? We are living in a fossil age at the peak of its strength. Almost 90% of all primary energy used nowadays comes from fossil fuels and nuclear. This is due to phenomenal increase in use of fossil fuels as a consequence of rapid development of emerging economies. Competition for securing resources for fuelling further economic development is increasing, price of fuels will increase and geopolitical conflicts will become more likely as the availability of fossils fuels would gradually decline.

All of these will make stable energy supply at predictable prices less and less likely. We can see nowadays volatility of oil prices as a consequence of internal or external struggles in the Middle East. There are threats to close mayor oil supply routes, new energy partnerships are emerging, big oil companies are positioning themselves for maintaining their leading position no matter what, renewables are on the rise although not without hick ups, electric mobility is becoming more than eccentric dream, climate change is finally accepted as the reality and an international agreement on facing these challenges is emerging.

Evidently we are living in a rapidly changing word facing a multitude of challenges caused by these processes of endless change, technological as well as geopolitical. One consequence is growing complexity which has huge impact on society which requires adequate policy response and timely implemented actions. The policy space that climate change happily occupied for 5 to 10 years has temporarily been superseded by other issues, like economic growth and bank recapitalization. We read so often there is lack of leadership, lack of money, and so many challenges that are confronting us at the same time. The bandwidth of political leaders is restricted, and short term approach focusing mostly on the mandate at hand increases vulnerability of national economies which are dependent on fossil fuel imports.

Small nations and small economies will be first to suffer if caught unprepared in the midst of the struggle for resources among the large players. Here it is where energy efficiency has a potential to lead toward the natural second step – transition from fossil age into a bio-age!

This book aims to contribute to an increasing policy debate on transition from fossil fuel based economies toward new low carbon bio-age. The book has 4 sections:

Section I	Policy issues
Section II	Energy efficiency on demand side
Section III	Energy efficiency in buildings
Section IV	Energy efficiency on supply side

Section I presents contemporary work on the EE and RES policies focusing on several specific issues. Chapter 1 discusses smart energy cities in the context of transition towards low carbon economy. Chapter 2 elaborates on urban complexity, efficiency and resilience of the cites with implications on climate change mitigation and adaptation. Chapter 3 evaluates energy efficiency strategies in Austria in the context of the EU energy service directive. Chapter 4 approaches smart grids and energy efficiency from the perspective of customers, while Chapter 5 looks into the issues of energy consumption inequality and effects on human development.

Section II addresses the energy efficiency issues on the demand side of energy systems. Chapter 6 presents a method for energy efficiency improvement of electro-hydraulic working machines. Chapter 7 analyses energy efficiency of agricultural production.

Section III looks into energy efficiency of buildings which are almost universally the largest single category of energy end-users. Chapter 8 presents how to optimize an energy management in buildings through visualization software, while Chapter 9 considers how to succeed in implementing energy efficiency in buildings. Chapter 10 presents in details the results of a research project focused on improving energy efficiency of air conditioners using green roof plants.

Section IV deals with the EE issues on the supply side of energy systems. Chapter 11 presents criteria for assessment of energy carriers' systems sustainability. Chapter 12 discusses energy efficient design of auxiliary systems in fossil fuel power plants. Chapter 13 describes energy efficiency initiatives for Saudi Arabia both on supply and demand side. Chapter 14 provides cost comparison for electricity generation from renewable energy sources for different technologies. Chapter 15 proposes efficient control strategy for decentralized power supply systems. Chapter 16 discusses energy efficiency and electrical power generation and gives a view on energy governance issues.

Someone said that the only thing more harmefull then fossil fuel is fossilized thinking. It is my sincere hope that some of chapters in this book will influnce you to take a fresh look at the transtion to low carbon society and the role that energy efficeicny can play in that process.

Dr. Zoran Morvaj,
United Nations Development Programme, New York
USA

Part 1

Policy Issues

Smart Energy Cities - Transition Towards a Low Carbon Society

Zoran Morvaj[1], Luka Lugarić[2] and Boran Morvaj[2]
[1]United Nations Development Programme, New York
[2]University of Zagreb, Zagreb
[1]USA
[2]Croatia

1. Introduction

We are living in a fossil age. More than 90% of energy nowadays comes from fossil fuels. Fossil age still has some 100 years to go [1], but should we wait until the last moment before we make a switch? Population is increasing, urbanization is increasing, price of oil will be increasing, and eventually it will run out. The economies that delay transition to low carbon society, especially if dependent on import of fossil fuels, are risking major upheavals.

The transition policies should be crafted now - and implementation should follow without delay.

This of course entails a major shift in economies, and consequently there will be winners and losers. The losers in this shift of focus would be the existing pro-status-quo groups, lobbying to postpone changes. The winners may not even exist yet, which is why the ongoing political debates are unbalanced because the losers know they will lose and fight back now, but future winners still don't put up equally strong arguments.

The way out is by finding a long term roadmap, starting with national policies based on local resources which could drive the transition away from imported fossil fuels. Authors believe that this is a correct approach to a low carbon future, and should start in the cities - the places where most people live and use energy for everyday life and business needs.

A multitude of policy and technology developments have emerged in the last 10-15 years addressing sustainable development of cities, mitigation effects of climate change and creating better living conditions for citizens. Large cities are using their vast resources to search for their own development roadmap. However, a systematic approach does not exist yet and cities develop their plans individually.

Small nations and developing economies will be first to suffer if caught unprepared in the midst of the fast developing struggle for resources among the large players. Here it is where smart energy cities have a potential to lead the transition - from fossil age into a bio-age!

This chapter proposes a way for transition to sustainable energy development focusing on cities as implementing changes actors. The concept is created through the integration of practical experience from on-going projects and research results towards development of energy resilient economies.

2. Definition of key terms and concepts

2.1 Pillars of the low carbon society

Throughout history, economic transformations occur when new communication technology converges with new energy systems [2]. New forms of communication and new sources of energy are cornerstones of managing complex civilizational challenges ahead. The fusion of Internet, information and communication technologies (ICT) and renewable energy sources (RES) enables development of nations toward a low carbon society, the focus of this chapter.

As outlined in Figure 1, there are 5 basic pillars of the low carbon society:

1. **Energy efficiency**: all energy losses must be either eliminated or minimized in accordance with best available technologies;
2. **Renewable energy sources**: solar, wind, hydro, geothermal, biomass, ocean waves and tides—their falling costs make them increasingly competitive;
3. **Buildings as active consumers**: Buildings that generate most of their energy needs from locally available renewable energy sources;
4. **Electro mobility:** Electric vehicles, once deployed on a large scale will serve both as means of transportation but also as energy storage units throughout the city;
5. **Developing smart energy cities:** An integrated effort of improving social, economic, environmental systems in cities, with energy infrastructure transformed first, as an enabler of further developments.

When these five pillars come together, they make up an indivisible sustainable development platform—an emergent system, whose properties and functions are qualitatively different than the sum of its parts.

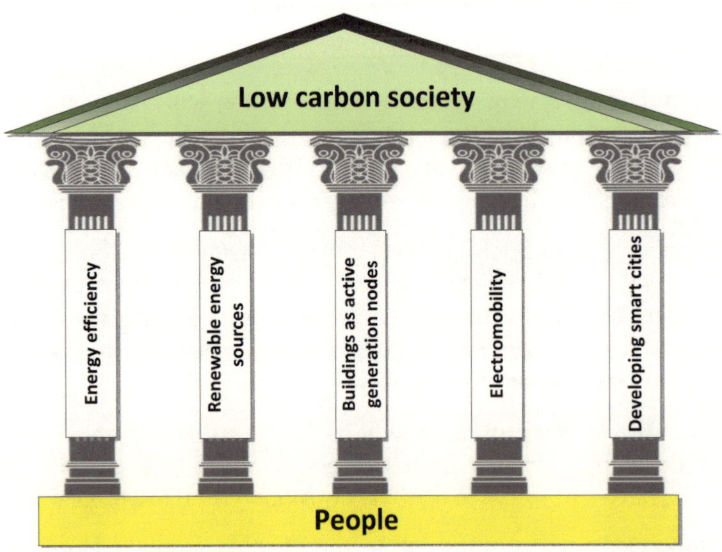

Fig. 1. Pillars of the low carbon society

Interconnectedness between the pillars creates cross-industry relationships, a system called distributed energy generation in which millions of existing and new businesses and homeowners become energy players to the advantage of final beneficiaries - the citizens.

The citizens - people as shown by Figure 1, are the foundation of the approach. Transition towards low carbon society must be consensual, involves change of behaviour and life style, thus people participation is essential.

2.2 Smart energy city

The United Nations estimates that already over 50% of the global population lives in cities [3]. Cities occupy only 2% of the Earth's surface but are the point of use of 75% of all resources required for everyday life and generate 75% of all waste [4]. Crucially, they produce 80% of global greenhouse gas emissions. Energy use is responsible for approximately 75% of these emissions, and 30-40% of that energy is used in buildings [5]. Sustainable future of the civilization depends to a great extent on changes in patterns of energy use and supply in cities.

Taking all this into account, for a city to become a smart energy city, it needs to evolve and address a multitude of technological and economic challenges in providing energy for basic needs of their citizens.

A smart energy city satisfies all energy needs of its citizens and goes beyond to provide innovative ways to increase the quality of life of its citizens in all areas. This is achieved by:

- Achieving the highest energy efficiency standards;
- Relying on local resources to provide for energy needs;
- Making all energy users active members of the local energy system;
- Developing smart homes and smart grids for demand management;
- Promoting electromobility;
- Using information to make insightful decisions on energy purchases or generation;
- Getting foresight to resolve problems proactively;
- Efficiently coordinating resources for effective operation of infrastructure systems.

An overview of key technologies and concepts which together comprise a smart energy city is shown in Figure 2.

2.3 Smart grids

The basic energy infrastructure of a smart energy city is the smart grid.

A smart grid implies integration of generation, transmission and distribution operations, monitoring and control functions, and suppliers and consumers through exchange of information in real time. Some of the widely quoted features are still under development while some have been implemented [6].

Buildings are the basic components of smart grids. The smart grid vision assumes all buildings will have a small renewable energy source installed and in case of increase of demand it can act as a small power plant, both externally to the grid and internally for its own consumption. Levels of observation at the new power grid, along with pertinent features are shown in Figure 3.

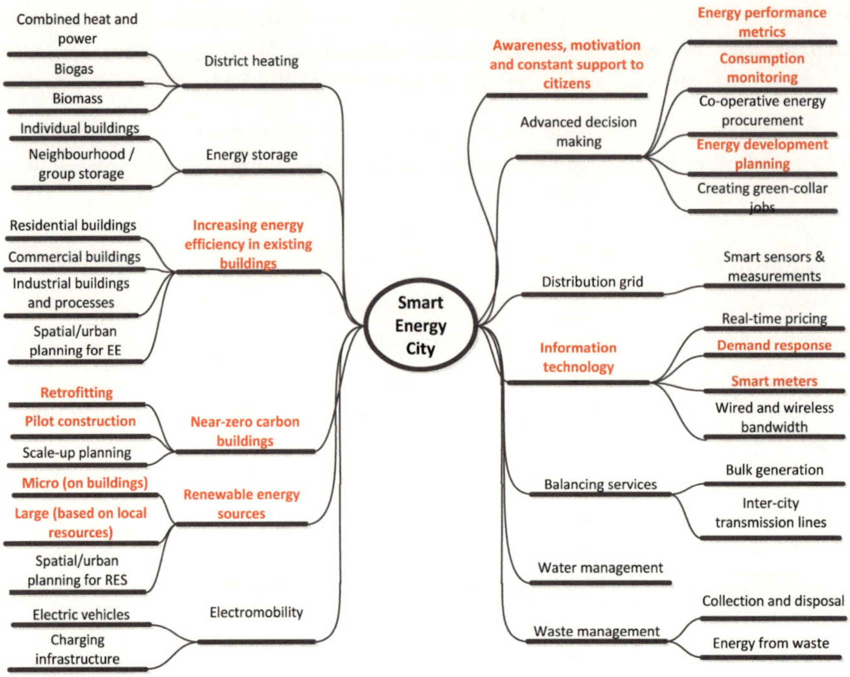

Fig. 2. Key concepts and technologies of a Smart Energy City

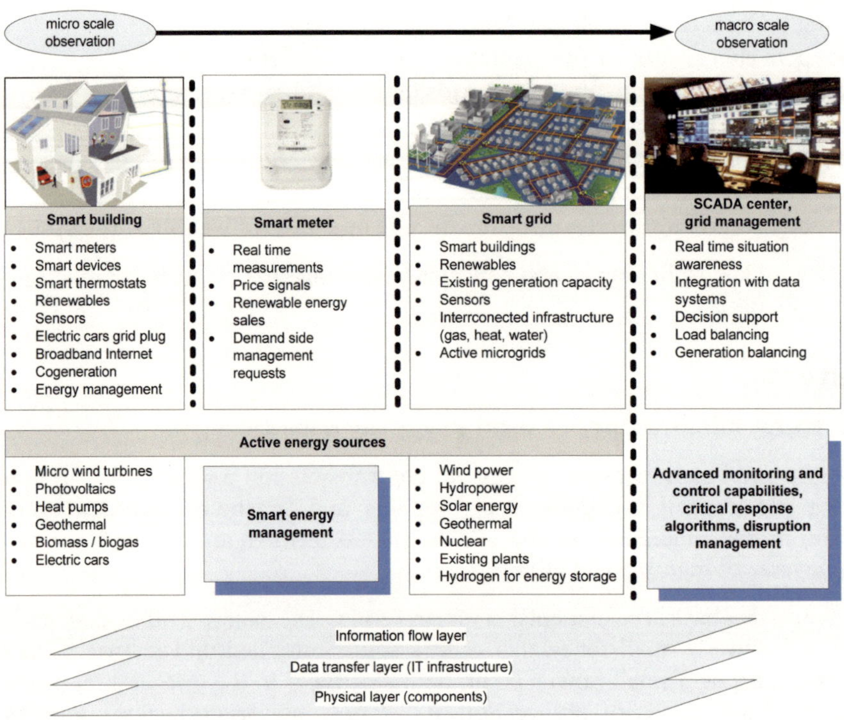

Fig. 3. Overview of the smart grid [7]

Vital to creation of smart cities is advancing infrastructural systems by using knowledge and technology in networking smart buildings.

2.4 Smart buildings

The definition of the term smart building has been used for more than two decades, and has been constantly evolving. In the 1980s "smart" was a building with implemented passive energy efficiency measures. In 1990s it was buildings with central, computer operated energy management systems. Today it includes all previous meanings with the addition of smart meters, networked appliances, advance energy management systems and renewable energy sources.

Smart buildings communicate with its surroundings (i.e. the energy distribution networks), and can adapt to conditions in the network, which building energy management systems can monitor and receive signals from. Smart buildings communicate between themselves, exchanging both information and energy, thus creating active microgrids . In general, the key components of a smart building are [8]:

- Local energy generation - producing energy either to be used within the building or injected to the grid;
- Sensors - monitoring of selected parameters and submit data to actuators;
- Actuators - which perform physical actions (i.e. open or close window shutters, turn on appliance, etc.)
- Controllers - monitoring inputs from sensors, managing units and devices based on programmed rules set by user;
- Central unit - used for programming and coordination of units in the system;
- Interface - the human-machine interface to the building automation system
- Network - communication between the units (RF, Bluetooth, wire);
- Smart meters - two-way, near or real-time communication between customer and utility company.

Capabilities and features of a model smart building are illustrated in Figure 4.

A smart building acts as a grid node as an energy producer through installed renewables or as an active participant in demand response management. Demand response (DR) programs can be classified into three groups [9]:

- **Incentive-Based:** represents a contract between utility and customer to ensure demand reductions from customers at critical times. This DR program gives participating customers incentives to reduce load during the agreed period which may be fixed or time-varying. Examples of the programs in this group are Direct Load Control and Interruptible & Curtailable Load.
- **Rate-Based:** a voluntary program where the customer pays a higher price during the peak hours and lower price during the off-peak hours. The price can vary in real time or a day in advance.
- **Demand Reduction Bids:** refers to relatively large customers to reduce their consumption. In this program customers send a demand reduction bid, containing demand reduction capacity and the price asked for, to the utility.

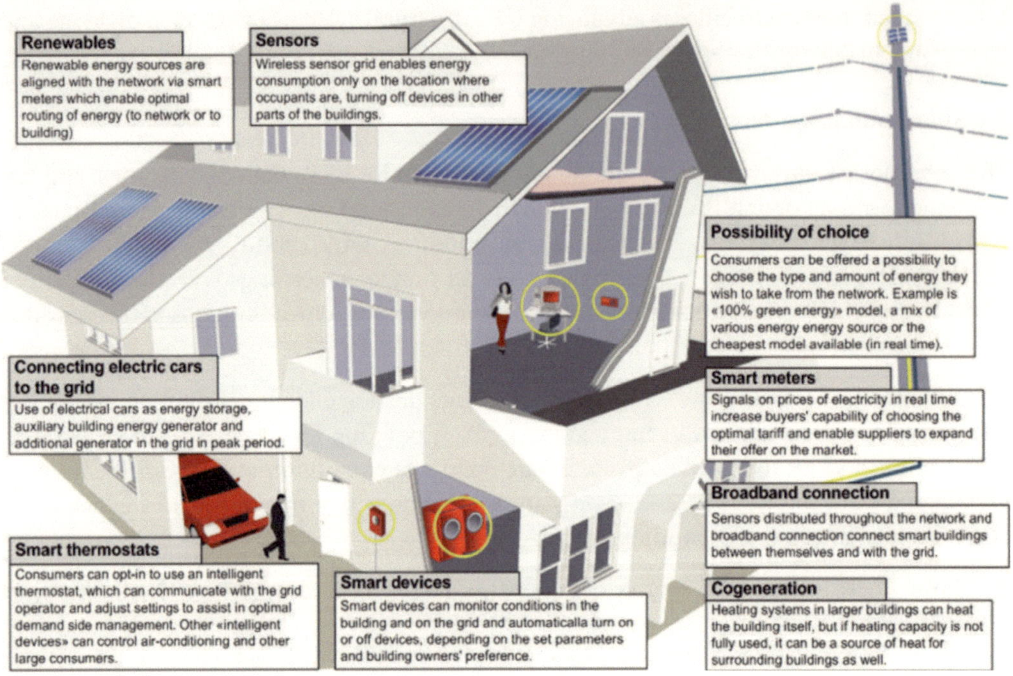

Fig. 4. Features of a smart building [6]

In an example given in [10], a demand response program based both on the price signal's value response and direct load control from the utility is considered. The imbalance of supply and demand is interpreted as the result of increased or decreased consumption and increased or decreased output of renewable energy resources. In case of shortage of supply, the price signal's value increases and buildings participating in the DR programme respond by turning off controllable load(s).

Algorithms for reducing energy consumption and regaining energy capacities are shown in Figure 5a and Figure 5b.

2.5 Energy management in cities

Energy management in cities can be defined as a continuous process aiming to [11, 12]:

- Avoid excessive and unnecessary use of energy through regulation and policy measures that stimulate behavioural changes;
- Reduce energy losses by implementing energy efficiency improvement measures and new technologies;
- Monitor energy consumption of all major users based on direct measurements of energy use (buildings, street lighting, water supply, public transport, etc.);
- Manage energy consumption by analysing energy consumption data and improving operational and maintenance practices.

To ensure continuity of energy efficiency improvements, energy consumption has to be managed as any other activity - an energy management system (EMS) must be implemented.

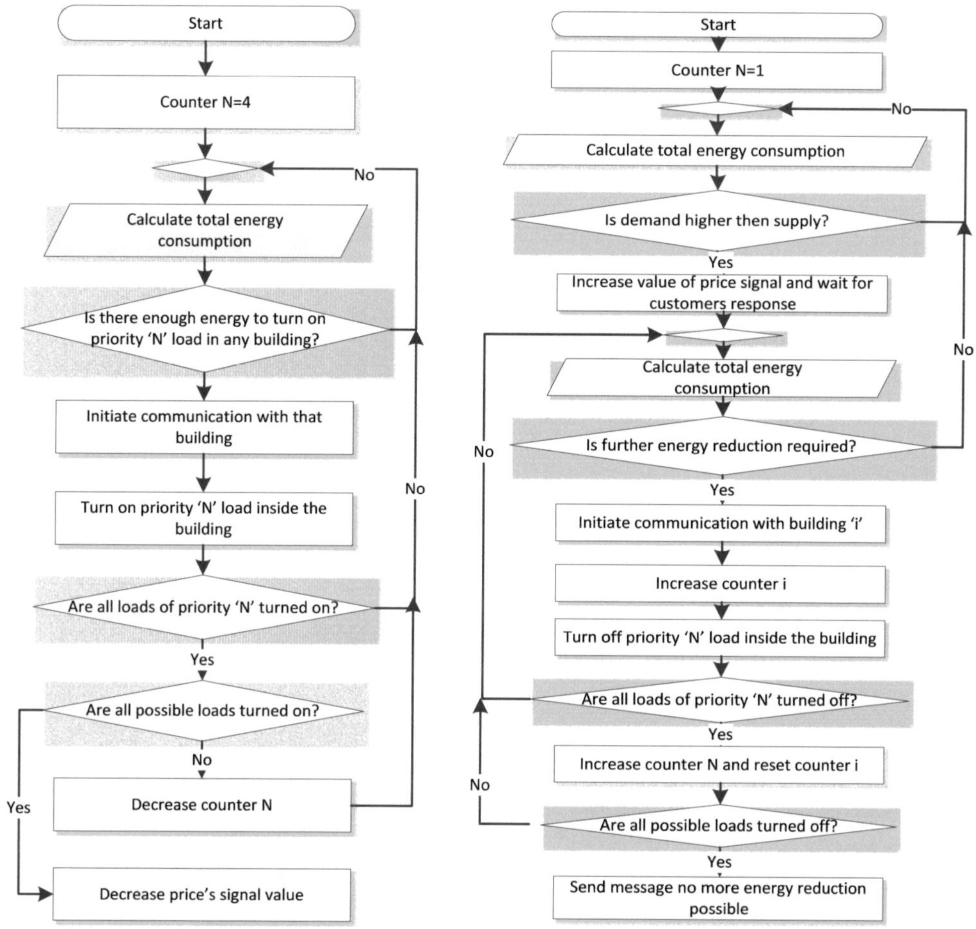

(a) (b)

Fig. 5. Algorithms for reducing (a) and regaining energy (b) in a model (from [10])

Essentially, energy management can be defined as a framework for ensuring continuous improvement in efficiency of energy use. It is supported by a body of knowledge and supported by measurements and ICT technology [13]. It does not only consider techno-economic features of energy consumption but makes energy efficiency an on-going social process calling for changes in behaviour and life style.

The energy management system (EMS) is a specific set of knowledge and skills based on organizational structure incorporating the following elements:

- Motivated and trained people with assigned responsibilities;
- Energy efficiency monitoring procedures inclusive of:
  - establishing baseline consumption;
  - defining consumption indicators;
  - setting improvement targets;
- Continuous measuring of energy use and improvement of efficiency until the best practice is reached.

The basic EMS concept and its key elements are shown in Figure 6.

A city's energy management team is responsible for regular analysis of collected data individually per building and aggregated analysis for all public buildings. The process of regular energy use measurement and analysis, as shown in Figure 7, provides relevant indicators that are needed for identification of measures that will lead to improved energy performances of buildings.

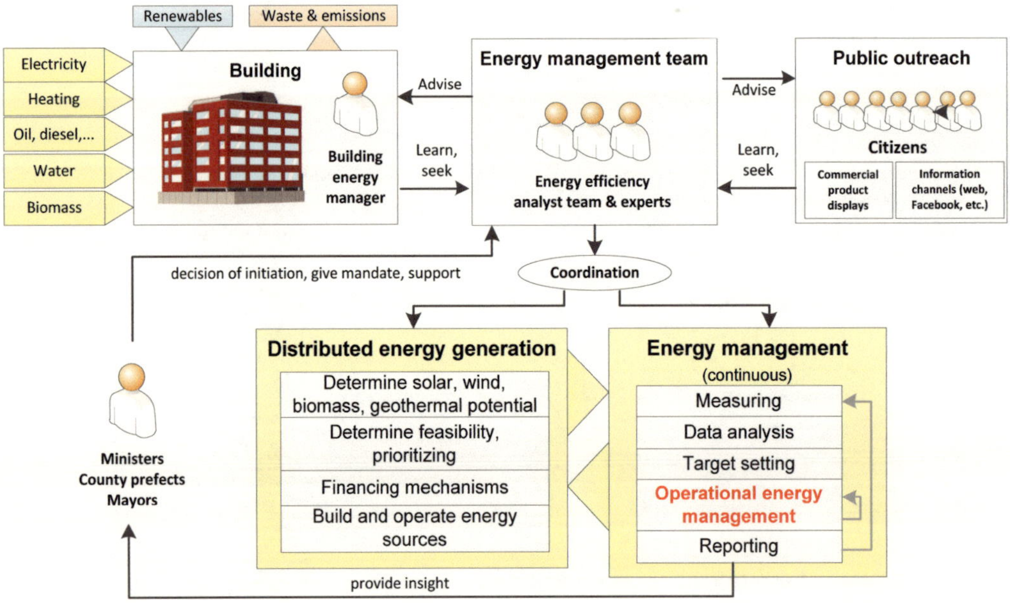

Fig. 6. Basic EMS concept in cities

2.6 Behaviour change

As said already, people are the foundation for introducing smart energy practices in cities because they will need to adopt their habits and behaviour to new realities of sustainable ways of energy use and supply.

The process of learning-while-doing and transfer of that knowledge from EE teams to the citizens and provision of essential information feedback from the implementation level back to the policy makers on national level in order to initiate policy adjustment is illustrated in Figure 8. The information feedback provided through EE teams is essential for accurate and objective analysis and evaluation of progress achieved and identification of needs for adjustment and adaptation of EE policies being implemented.

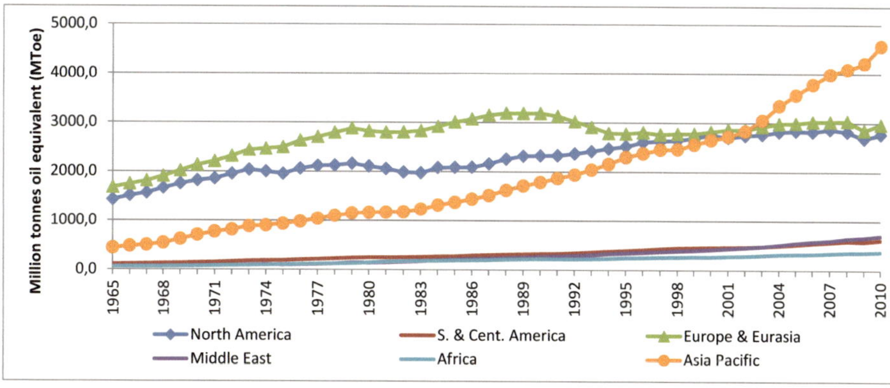

Fig. 10. Global primary energy consumption by geographic regions

Tables 1 and 2 show an imbalance between locations where the oil and gas resources are found and extracted and where the major demand for these occurs. As a consequence, there are is a multibillion dollar international energy commodity market, sensitive to speculations, political manoeuvring, artificial intermittent shortages and gluts, conflicts and wars.

Most of the recent conflicts are caused by the desire to secure access to fossil fuels.

Producers	Mt	% of world total
Russian Federation	502	12,6
Saudi Arabia	471	11,9
United States	336	8,5
Islamic Rep. of Iran	227	5,7
Peoples Rep. of China	200	5,0
Canada	159	4,0
Venezuela	149	3,8
Mexico	144	3,6
Nigeria	130	3,3
United Arab Emirates	129	3,2
Rest of the world	**1.526**	**38,4**
World	**3.973**	**100,0**

(2010 data)

Net exporters	Mt
Saudi Arabia	313
Russian Federation	247
Islamic Rep. of Iran	124
Nigeria	114
United Arab Emirates	100
Iraq	94
Angola	89
Norway	87
Venezuela	85
Kuwait	68
Others	**574**
Total	**1.895**

(2009 data)

Net importers	**Mt**
United States	510
People's Rep. of China	199
Japan	179
India	159
Korea	115
Germany	98
Italy	80
France	72
Netherlands	57
Spain	56
Others	**477**
Total	**2.002**

(2009 data)

Table 1. Global top crude oil producers, net exporters and importers

Producers	Mt	% of world total
Russian Federation	637	19,4
United States	613	18,7
Canada	160	4,9
Islamic Rep. of Iran	145	4,4
Qatar	121	3,7
Norway	107	3,3
Peoples Rep. of China	97	3,0
Netherlands	89	2,7
Indonesia	88	2,7
Saudi Arabia	82	2,5
Rest of the world	1.143	34,7
World	3.282	100,0

Net exporters	Mt
Russian Federation	169
Norway	101
Qatar	97
Canada	72
Algeria	55
Indonesia	42
Netherlands	34
Malaysia	25
Turkmenistan	24
Nigeria	24
Others	165
Total	808

Net importers	**Mt**
Japan	99
Germany	83
Italy	75
United States	74
France	46
Korea	43
Turkey	37
United Kingdom	37
Ukraine	37
Spain	36
Others	253
Total	820

Table 2. Global top natural gas producers, net exporters and importers

Taking a longer term view, we are definitely facing two converging trends:

1. The consumption will continue to grow in most of the economies, including these that are net exporters today (Tables 1 and 2);
2. The reserves of fossil fuels will gradually shrink.

As a consequence, net export capacity will shrink as well, making oil and gas more scares, thus more costly, and thus even more potent tool for political blackmailing. Key players in these games would be big economies who are still net importers (Tables 1 and 2). Small economies and particularly developing ones should seek not to be a part of these future struggles for resources.

But what are the alternatives?

3.2 Technologies

A view on current rising trend in utilization of renewable energy sources is given in Table 3. According to a 2011 projection by the International Energy Agency, solar power generators may produce most of the world's electricity within 50 years, dramatically reducing the emissions of greenhouse gases that harm the environment [17].Renewable energy sources, although still only 1/10 of the global primary energy supply, are on the rise (Figure 11). At the same time the costs of these new technologies were rapidly falling (Figure 12).

Indicator		2008	2009	2010
Global new investment in renewable energy (annual)	billion USD	130	160	211
Renewables power capacity (existing, not including hydro)	GW	200	250	312
Renewables power capacity (existing, including hydro)	GW	1.150	1.230	1,320
Hydropower capacity (existing)	GW	950	980	1,010
Wind power capacity (existing)	GW	121	159	198
Solar PV capacity (existing)	GW	16	23	40
Solar PV cell production (annual)	GW	6.9	11	24
Solar hot water capacity (existing)	GW-A	130	160	185
Ethanol production (annual)	billion litres	67	76	86
Biodiesel production (annual)	billion litres	12	17	19
Countries with policy targets	#	79	89	98

Table 3. Selected global indicators of renewable energy sources [18]

The levelled costs of all RES technologies are approaching (some already are there) so called grid parity with conventional power plants based on fossil fuels (Table 4).

Levelled cost is often cited as a convenient summary measure of the overall competiveness of different generating technologies. Levelled cost represents the present value of the total cost of building and operating a generating plant over an assumed financial life and duty cycle, converted to equal annual payments and expressed in terms of real dollars to remove the impact of inflation. Levelled cost reflects overnight capital cost, fuel cost, fixed and variable O&M cost, financing costs, and an assumed utilization rate for each plant type [19].

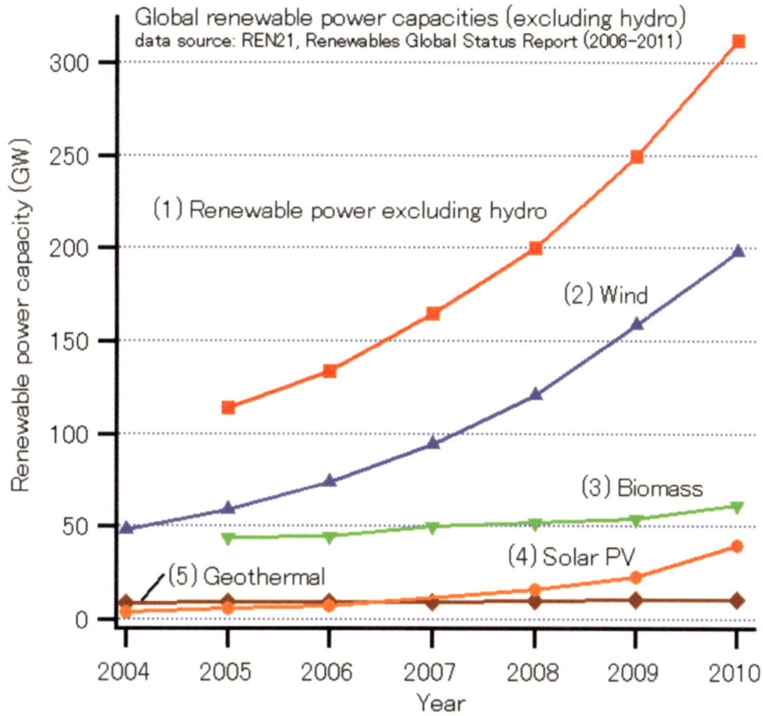

Fig. 11. Global renewable power capacity excluding hydro [18]

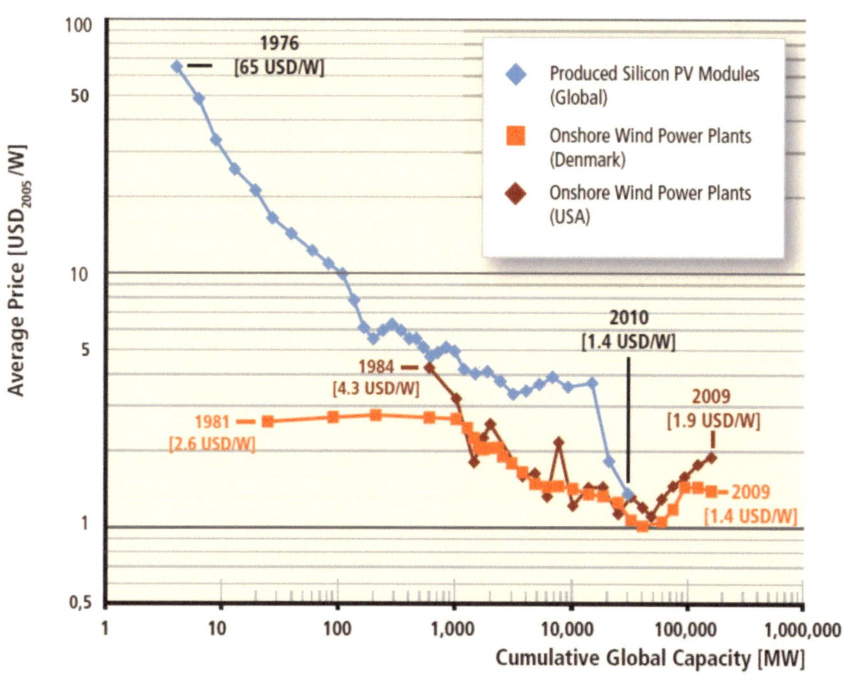

Fig. 12. Experience curves for PV modules and wind power plants [16]

Plant Type	Capacity Factor (%)	U.S. Average Levelized Costs (2009 $/megaw atthour) for plants entering service in 2016				
		Levelized Capital Cost	Fixed O&M	Variable O&M (including fuel)	Transmission Investment	Total System Levelized Cost
Conventional Coal	85	65,3	3,9	24,3	1,2	94,8
Advanced Coal	85	74,6	7,9	25,7	1,2	109,4
Advanced Coal w ith CCS	85	92,7	9,2	33,1	1,2	136,2
Natural Gas-fired						
Conventional Combined Cycle	87	17,5	1,9	45,6	1,2	66,1
Advanced Combined Cycle	87	17,9	1,9	42,1	1,2	63,1
Advanced CC w ith CCS	87	34,6	3,9	49,6	1,2	89,3
Conventional Combustion Turbine	30	45,8	3,7	71,5	3,5	124,5
Advanced Combustion Turbine	30	31,6	5,5	62,9	3,5	103,5
Advanced Nuclear	90	90,1	11,1	11,7	1,0	113,9
Wind	34	83,9	9,6	-	3,5	97,0
Wind - Offshore	34	209,3	28,1	-	5,9	243,2
Solar PV[1]	25	194,6	12,1	-	4,0	210,7
Solar Thermal	18	259,4	46,6	-	5,8	311,8
Geothermal	92	79,3	11,9	9,5	1,0	101,7
Biomass	83	55,3	13,7	42,3	1,3	112,5
Hydro	52	74,5	3,8	6,3	1,9	86,4

Table 4. Estimated Levelled Cost of New Generation Resources [19]

For technologies such as solar and wind generation that have no fuel costs and relatively small O&M costs, levelled cost changes rougly in proportion to the estimated overnight capital cost of generation capacity. For technologies with significant fuel cost, both fuel cost and overnight cost estimates significantly affect levelled cost. The availability of various incentives including tax credits can also impact the calculation of levelled cost. The values shown in the table 4 do not incorporate any such incentives. As with any projections, there is an uncertainty about all of these factors and their values can vary regionally and across time as technologies evolve.

However, making a long term energy policy decisions based on the current levelled cost of technologies only is completely wrong.

Firstly, levelled will costs change in the future, but once we invested large amounts of funds in a technology, we are trapped by the need to return the investment! There is no cheap and easy way out.

Secondly, levelled costs do not show aggregated economic value of investing into local renewable energy sources in the context of the national economy (even though renewables are more expensive for the time being then fossil-based sources), especially against importing fossil fuels to the value of 5-15% of GDP annually over some 30 years. With the certainty of future price increases, prospects of insecurity of supply and eventual cease of supply, set of decisions to make becomes increasingly difficult.

4. Proposed transition strategy for developing economies

The transition strategy proposed here addresses developing economies which depend on imports of fossil fuels. For these economies, annual cost of total final energy consumed is generally above 10% of GDP, and very often around 20%. The cost of imported fuels is anywhere between 30 and 70% of total energy costs, which corresponds with 5 - 15% of GDP.

Introducing systematic energy efficiency increase programmes is the first step to be taken as the transition away from imported fossil fuels because it could reduce national energy expenditures by at least 20%. These significant funds can be reinvested back into the economy and for further transition strategy implementation.

Further, over the long term perspective, as the global availability of fossil fuels shrinks, prices of fuels will increase to unsustainable levels for developing economies, because the competition for the resources will intensify. This affects security of supply, also increasing the likelihood of international conflicts about resources would be more likely to happen, and where developing countries are more likely to be put down by more powerful players.

Taking everything presented under consideration, renewable energy sources should be seen nowadays as a credible alternative to the fossil fuels. Technologies are available, and prices are falling. Every country has at least some of the renewable energy soureces in abundance. The basic development path to be taken is that local development must be based on local resources - natural and technological alike.

On the other hand, energy infrastructure development is time consuming and capital intensive. Therefore the transition away from fossil fuels should be planned right now in order to develop an energy resilient economy, able to face the more difficult situation emerging some 40 years from now.

Based on these considerations, simple transition strategy objectives can be proposed:

Eliminate gradually the need for importing fossil fuels!
Base local development on local resources!

These goals can be achieved by most developing economies by 2050, if countries seriously embark on this journey now. The goals have to be translated into sounds national policy with a perspective of supporting policy implementation on local levels, in cities, counties and regions.

Benefits from decreased energy consumption and decreased import costs for fossil fuels may not be obvious for many policy makers at the national level and even more so in the cities. Besides, eliminating imports of fossil fuels will require significant structural shifts in economies. Therefore an Energy foresight study will need to be carried out in order to charter a road map to a low carbon economy and elaborates impacts and necessary adjustments on various economic sectors in the country.

The vantage point for considerations is the current energy mix, both at the supply and demand sides (Figure 13). Total primary energy supply mix is taken into account, and shares of individual energy sources are presented. The supply mix is dominated by fossil fuels, which is still valid for all countries in the world. Increasing dependence on energy imports is also a major factor, and system losses and other inefficiencies are accounted for when determining final energy demand.

Further trends in both the supply mix and demand mix are calculated using traditional analysis.

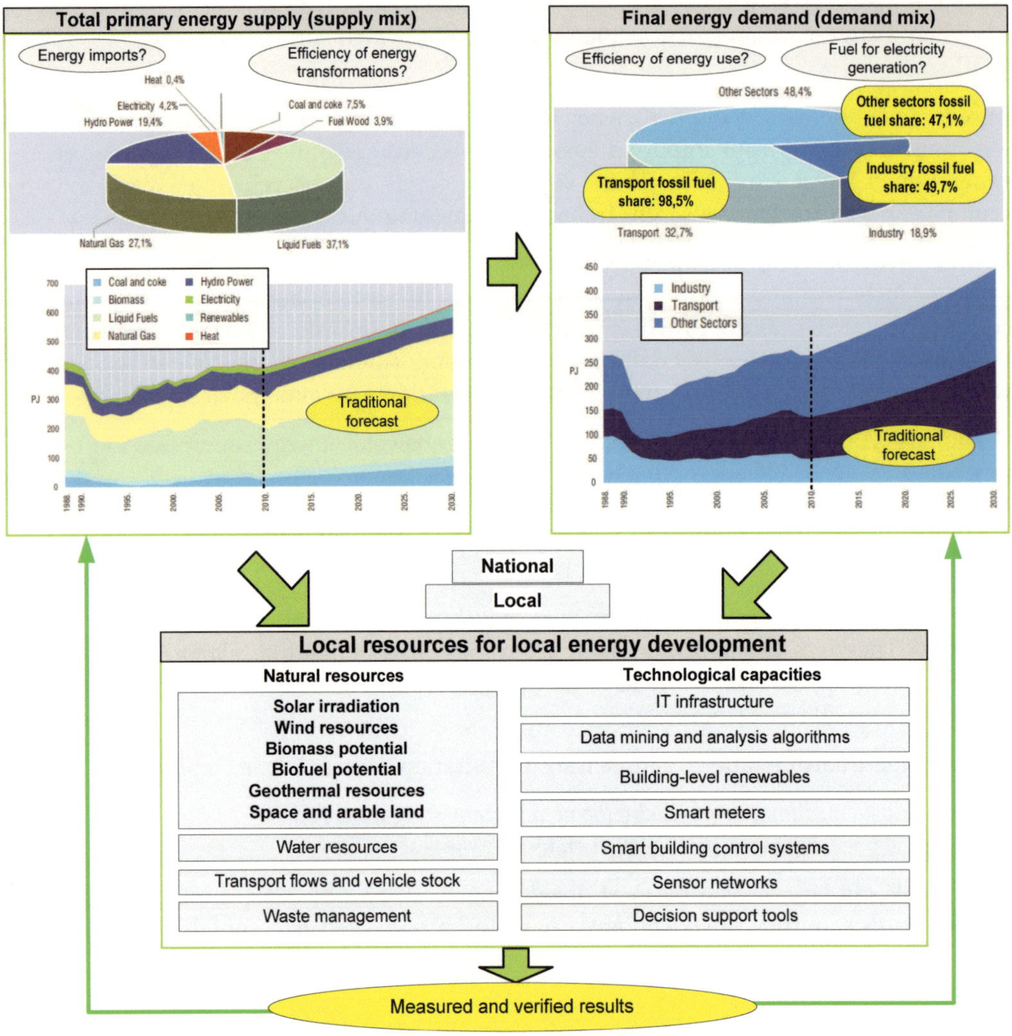

Fig. 13. Planning the utilization and development of local resources in alignment with national goals

Figure 13 represents a case that is quite common: 50% of total energy demands are fossil fuels from imports; industrial energy consumption is at 20% of the total, transport is at 30%, and buildings at 40% of total end use demand.

Targets for the transition strategy hereby are given as:

- implement a rigorous energy efficiency program aimed at improving EE in all sectors by at least 20%;
- identify local natural resources and developed related technologies so that the energy from RES can gradually eliminate all imported fossil fuels by 2050, which is 50% of current demand;
- define transition targets for particular energy end-use sectors (Figure 14).

We have underlined the word 'imported', because if a country has some fossil fuels of its own, they should be used with care and saved as a strategic reserve.

The obvious sectors to target first for the transition are buildings and transport where solutions are known and alternatives available. More difficulties are to be expected with industrial sector. But that is why we put development of an Energy foresight study as a mandatory step to get clear answers for structural changes in all sectors of the economy.

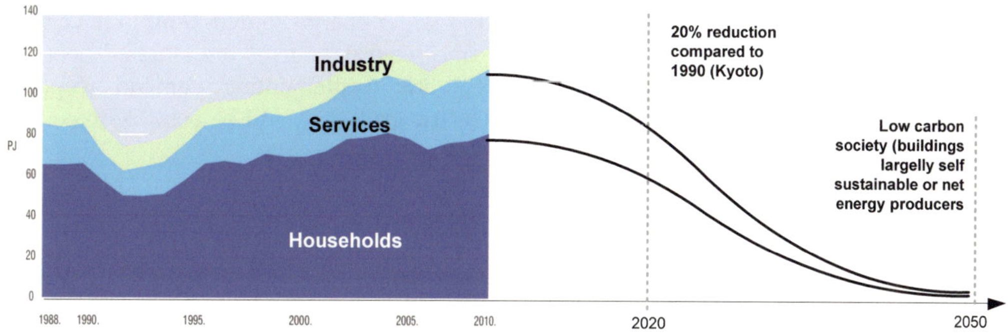

Fig. 14. Targets for transition toward a low carbon society in key consumption sectors for buildings

5. Smart energy cities - Implementation platform for transition

While concentrated action on the national level is required to develop and adopt energy policy, policy implementation has to be performed at the local levels in the cities where energy is consumed daily.

For cities which plan to apply local resource-based development approaches, the challenge will be to translate the national transition strategy into local-level projects. For this is to happen, an effective participatory local, city-level planning methodology is indispensable. Through a consultative process, involving local stakeholders from the public and private sectors, a territorial diagnosis should be carried out to assess resources, capacities and economic opportunities that can facilitate transition process - a smart energy city action plan must be produced.

The process has to optimize utilization of locally available resources and make use of the competitive advantages of a locality to stimulate productivity in selected energy value chains while promoting economic development and creating employment.

From the technology viewpoint, transition towards the smart energy city can be summed up in three basic steps, as shown in figure 15:

1. Decreasing unnecessary energy loses by implementing an energy management system and implementing measures to increase energy efficiency;
2. Managing demand to avoid consumption peaks;
3. Promoting distributed generation form renewable energy sources.

By installing smart grid technology such as home area networks, smart meters and demand side management schemes, it is possible to control and optimize energy consumption so that the maximum value of the peak demand is decreased. Smart meters along with energy management systems enable real time consumption monitoring both by consumers and utilities and enable use of smart appliances. After installing smart meters, demand response programs should be defined and implemented, which will enable an almost even consumption throughout the day.

The next step is installing renewable energy sources such as roof-mounted PV, wind turbines, biomass cogeneration plants, etc. as locally appropriate. They can be both local micro energy sources installed in the buildings and larger energy sources built in the city or nearby. This decreasew losses in transmission since energy sources are situated near the consumption area. For installing smart meters and implementing demand response programs, ICT needs to be combined with the electric grid, so it will be possible to control and use the full potential of local distributed energy sources.

Cutting losses in buildings' consumption	Optimizing consumption of energy	Installing distributed, renewable sources
• Energy efficiency • Cutting unneccessary waste of energy inputs (electricity, gas, heat, water)	• Installing smart meters • Implementing demand response program(s) • Decreasing needed capacity for peak demand	• Building-size renewable sources • Local distributed energy generation • Energy storage for power balancing

Fig. 15. Three groups of activities toward smart energy city

Monitoring the progress and verifying results are of paramount importance because this should provide feedback data on success of the transition, and enable corrective policy measures to be defined if required. Key aspect of the monitoring system is definition of performance indicators to be measured. While the list of these could be quite extensive, key performance indicators (KPI) are here simplified to the following:

- KPI1: Total Energy consumption of building surface area (kWh/m^2)
- KPI2: Thermal non-fossil energy produced locally compared to total thermal energy consumed in the city (MWh/MWh)
- KPI3: Electrical energy locally produced compared to total electricity consumed in the city (MWh/MWh)
- KPI4: Use of non-fossil fuels for transport (renewable electricity, bio fuels) compared to total energy use for transport in the city (%/%)

While other indicators are also important, these four serve to monitor two basic policy directions regarding smart energy - reducing energy consumption and increasing the share of renewable energy sources for electricity generation, heat production and transport. The presumed trend of change in accordance with the current policies in the EU of these KPI is shown in Figure 16.

Performance measurement in any process will not improve performance by itself. Performance data must be interpreted to plan and implement corrective policies and actions, and more than anything - to change the way people use energy in order to achieve lasting performance improvements.

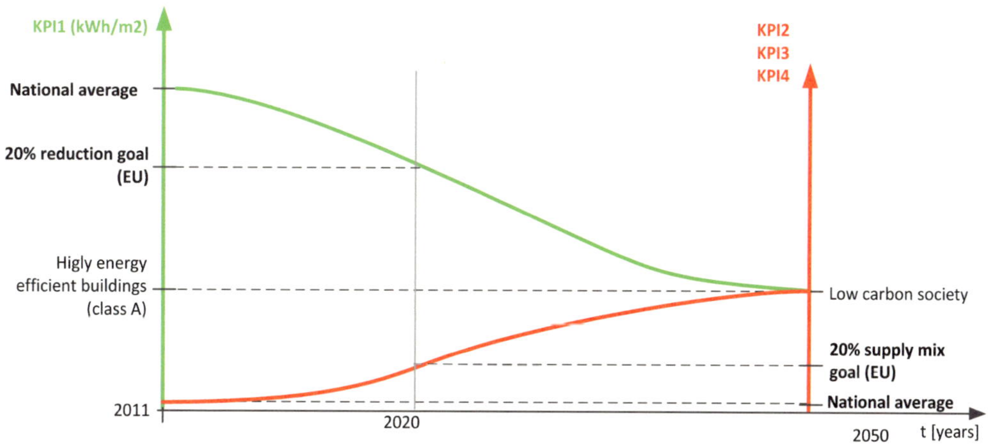

Fig. 16. Presumed trend of change in accordance with current sustainability policies

Since increasing energy efficiency at all levels is by all means the first thing to do, and since it is only natural that the public sector takes the lead, introduction of systematic energy management in all public buildings should be the driver for implementation of transition strategy. This will create necessary capacities in terms of organization, institutions, skills, competencies, awareness, knowledge, IT and energy technologies infrastructure to serve as an implementation platform for furthering the transition strategy towards achievement of its objectives (Figure 17).

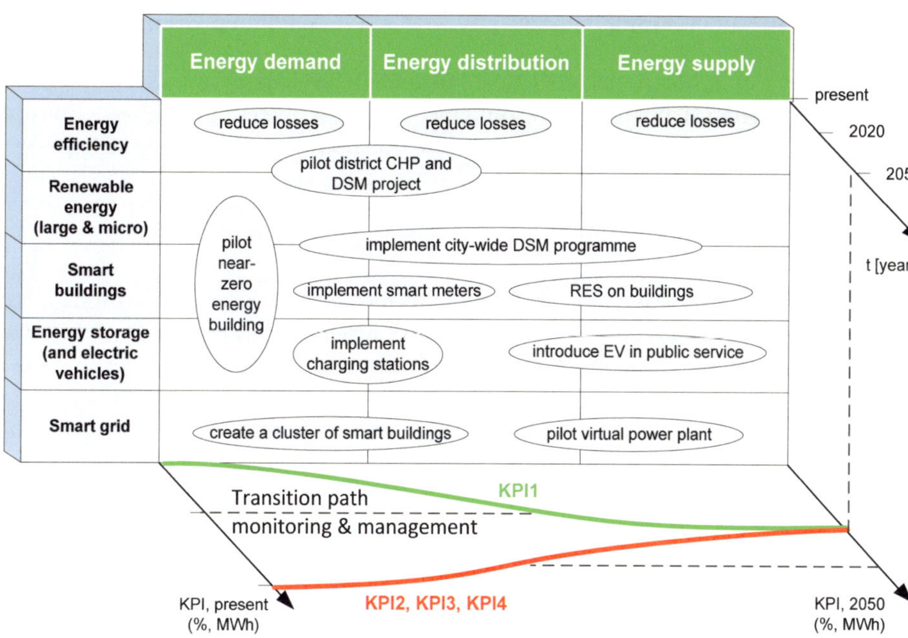

Fig. 17. Matrix structure of planning technology implementations for a smart energy city

Full implementation of energy management according to the smart building concept will gradually remove these buildings from demand for fossil fuels. This is illustrated in Figure 18.

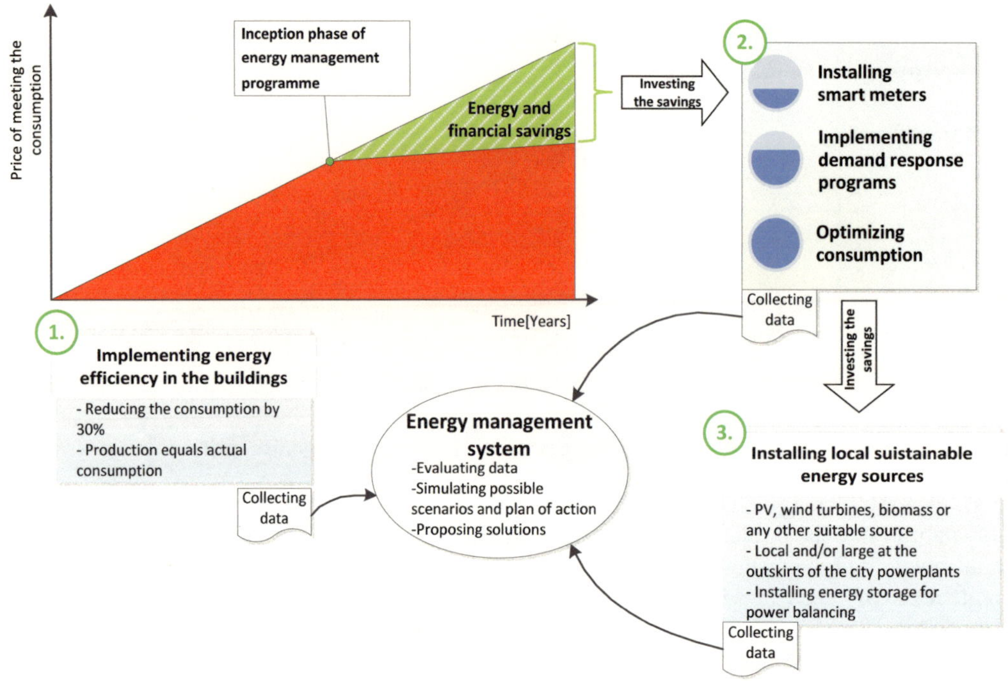

Fig. 18. Steps for transition to smart energy city with minimal initial investment

A well developed and functional energy management system in the city - inclusive of adequate organizational structures, institutional support, competent people and appropriate technology base, - presents a good foundation for transition towards the smart energy city and low carbon economic development.

6. Conclusion

We are living in an energy fossil age at the peak of its strength. There are various forecasts for how long it can still go on based on availability of fossil fuel reserves, but we believe this is extraneous. If we know that sooner or later we will run out of oil and gas, should we wait until the last drop dries out, or should we start acting earlier? Presumably, timely action is advisable, particularly in the case of the energy sector, where restructuring of energy infrastructure and changing of the energy mix is a time consuming and capital intensive process.

Nowadays there is a general awareness of the environmentally harmful side effects of using fossil fuels and geopolitical aspects of fossil fuel reserves and markets, inherent insecurity of energy supply and volatility of prices. The further we go towards scarcity of fossil fuel supplies, the greater the disturbances will be, and the higher the stakes in the struggle for securing supply.

Smaller economies and developing nations will be first to lose in this struggle if caught unprepared.

But there is also an another aspect of this situation seldom emphasized: the cost of final energy in developing economies is usually more than 15% of GDP, and often more than 20%. The money which goes for import of fossil fuels is anywhere between 30 -70% of the annual energy bill which means around 5-15% of GDP. In addition, energy efficiency in developing economies leaves a lot of potential for improvement - at least 20%. These two facts are telling us that: firstly money is being wasted due to inefficient energy consumption, and secondly there are significant capital outflows for import of fossil fuels.

With all the other concerns about the fossil fuels, these are the additional which should kick us in the action - a transition towards low carbon economies, where imported fossil fuels have to be gradually replaced by locally available renewable energy sources.

Appropriate national transition policies are required for the period of up to 2050, and cities need to lead implementation of policies by transforming themselves into smart energy cites. The first step can start now - by implementing systematic energy management in cities, aiming at eliminating energy losses, further expanded by promoting distributed energy generation from locally available renewable energy sources and finally introducing smart meters, smart homes and smart grids.

Local natural and technological resources are the basis for local low carbon development - it cannot be based on resources and technologies we don't have. Charting the transition away from imported fossil fuels and towards low carbon development, in the long run, has no alternative.

The sooner we start, the better off we will be, because there is only one thing more harmful than fossil fuels - fossilized thinking!

7. References

[1] L. Mark W. Leggett, David A. Ball, The implication for climate change and peak fossil fuel of the continuation of the current trend in wind and solar energy production, Energy Policy, Volume 41, February 2012, Pages 610-617, ISSN 0301-4215

[2] Rifkin, J. (2011) The Third Industrial Revolution: How Lateral Power Is Transforming Energy, the Economy, and the World, ISBN 978-0230115217

[3] United Nations. (2009) World Urbanization Prospects, United Nations, accessed December 2011, http://esa.un.org/unpd/wup/index.htm

[4] Girardet, Herbert. 1996. Giant footprints. Our Planet, 8(1), pp.21-23

[5] U.S. Department of Energy. (2008) Energy efficiency trends in residential and commercial buildings, Retrieved from http://alturl.com/i9afn

[6] Z. Morvaj, L. Lugaric, B. Morvaj, Smart cities, buildings and distribution networks - perspectives and significance for sustainable energy supply, presented at 2nd Croatia CIRED Conference, Umag, Croatia, 2010.

[7] Lugarić, L. et al. (2010) Smart city — Platform for emergent phenomena power system testbed simulator, Proceedings of Innovative Smart Grid Technologies Conference Europe (ISGT Europe), 2010 IEEE PES Stockholm

[8] A. Moderink et al. (2009) Simulating the effect on the energy efficiency of smart grid technologies, in Proc. 2009 Winter Simulation Conference, pp 1530-1541

[9] S. Mohangheghi, J. Stoupis, Z. Wang, Z. Li, (2010) Demand Response Architecture, IEEE
[10] Morvaj, B. (2011) Demonstrating Smart Buildings and Smart Grid features in a Smart Energy City, 3rd International Youth Conference on Energetics 2011, Leiria, Portugal
[11] Morvaj, Z.; Bukarica, V. (2010) Immediate challenge of combating climate change: effective implementation of energy efficiency policies, 21st World Energy Congress, Montreal, Canada
[12] Zoran Morvaj and Vesna Bukarica (2010). Energy Efficiency Policy, Energy Efficiency, Jenny Palm (Ed.), ISBN: 978-953-307-137-4, InTech, Available from: http://www.intechopen.com/articles/show/title/energy-efficiency-policy
[13] Z. Morvaj, D. Gvozdenac: Applied Industrial Energy and Environmental Management, a book published in September 2008 by John Wiley and Sons, UK, and IEEE press USA
[14] IEA, World Energy Outlook 2011
[15] IEA Key World Energy Statistics 2011, [Online, available 27 February 2012] www.iea.org/textbase/nppdf/free/2011/key_world_energy_stats.pdf
[16] Intergovernmental Panel on Climate Change (IPCC), Special Report on Renewable Energy Sources and Climate Change Mitigation, May 2011 [Online, available 27 February 2012] http://srren.ipcc-wg3.de/report
[17] IEA, Solar Energy Perspectives [Online, available 27 Febriary 2012] www.iea.org/Textbase/nptoc/solar2011TOC.pdf
[18] Renewable Energy Policy Network for the 21st Century, Renewables 2011 - global status report [Online, available 27 February 2012] www.ren21.net/Portals/97/documents/GSR/GSR2011_Master18.pdf
[19] Energy Information Administration (EIA), Levelized Cost of New Generation Resources in the Annual Energy Outlook 2011 [Online, available 27 February 2012] www.eia.gov/oiaf/aeo/pdf/2016levelized_costs_aeo2011.pdf

2

Urban Complexity, Efficiency and Resilience

Serge Salat and Loeiz Bourdic
Urban Morphology Lab, CSTB
France

1. Introduction

The relationships between urban forms and energy that are investigated in this chapter are an example of a more general idea: the relationships between structures and energy. This chapter aims at presenting structural laws that link urban-scale forms to their internal organization and to their energy consequences, expressing them in a simple and innovative way, and putting them in the broader context of complex systems energy. One of these complex systems is life itself. Beyond their mathematical form, the structural laws of urban energy deal with the relationship between forms and processes. If we want to create a sustainable society, then each aspect of what we do must follow living systems structural order. This structural order always results from a process. As Fritjof Capra explains in *The web of life* (1996), systems thinking requires thinking in terms of relationships and patterns.

Urban form and spatial structure constrain cities' functioning (individual spatial behaviours, land use) and cities' flows (travel, energy, water) and, retroactively, their functioning modifies both their morphology and their structure. The World Bank has recently pointed out the need for more systemic approaches, taking into account both forms and flows (World Bank, 2010). The Urban Morphology Lab works at dividing flows by a factor 2 to 4 - and thus at the same time urban footprint - just by optimizing urban forms.

What are the urban morphology parameters that influence and determine the energy flows going through cities? To answer this extraordinary difficult question, only a quantitative analysis, based on a theory of urban structures can bring clarification. There is an urgent need to address these issues. Cities are the main driver of climate change, the biggest energy consumers, and the biggest greenhouse gas emitters. Urban structures are complex artefacts that absorb energy and transform it into heat, according to thermodynamics laws.

When it comes to energy, one has to think in terms of making a more efficient use of depleting resources instead of thinking in terms of replacing energy sources one by the other. Any renewable energy cease being renewable if an intensive over-consumption is made of it. The share of renewable energy in the global figure of urban energy supply has to increase, but for renewable energy to be profitable, one should first increase energy productivity in cities. A city four times denser consumes four times less land and sixteen times less network infrastructure. And yet density variations between loose suburbs and historical cores are within a factor 16…

But this chapter does not only question the spatial aspects of urban energy. On the contrary, the approach encompasses a much broader scope. The temporal distribution of energy flows is at least as important as their spatial distribution. The temporal match between supply and demand is for example critical to develop renewable energy. The distribution of energy quantities and energy qualities is another aspect that is worth being investigated. Exergy-based approaches that put a particular focus on energy quality and degradation provide very beneficial insights to optimize energy flows within the city. Authors' main objective in this chapter is to encompass in a comprehensive way the structural parameters that make a city be sustainable.

At the crossing between thermodynamics, industrial ecology and urban morphology, this chapter summarises the lessons that can be drawn from several scientific fields and applied to urban analysis. Section 2 aims at defining what urban structure is and introduces the fundamental concept of urban complexity that is unfortunately rarely if never used - perhaps because it is hard to handle. Authors particularly focus on the hierarchy of scales within urban systems. Section 3 aims at highlighting the impact of urban structure and complexity on cities' structural efficiency and resilience. This dual approach rests upon some major scientific breakthrough of the last decades, such as fractal theory or complex systems thermodynamics. The approach though aims at a pragmatic objective, keeping in mind that urban development is in the end primarily decided by policy-makers and urban authorities. That is why authors eventually provide some examples showing the concrete and practical implications that these results have on the real urban world: bioclimatic comfort and passive urban structures, efficiency and resilience of urban transport networks.

2. About cities, urban structures and complexity

2.1 What makes a city a city?

There seems to be a great variety and complexity of cities around us. Yet approaching them with a scientific spirit means looking for what is simple behind this seeming complication. Paris and Tokyo, unlike Vienna, Barcelona or Kyoto, grew without a real general plan. But their material structure, as impermeable as it may be to all forms of topographic regularity, nonetheless evinces a very complex form of order, different for two cities, marking them with the seal of an irreducible identity. Paris remains firstly "a gigantic mosaic", closer to the structure of Pergamon than to those of Le Corbusier "Contemporary City for Three Million Inhabitants". "A sort of bit territorial weave", writes Bruno Fortier, "in which passageways established on the land of former convents, quarries turned into gardens, pagodas introduced into the civil fabric, remains of the World Map, connecting between them a few of its monuments were found intact, playing a remote score that no project really brought together." (Fortier, 1989, p. 15)

Yet what emerges from plans of Paris, as of Tokyo, is never incoherent: on different scales, the plans never cease to reveal stable structures, different for the two cities, where the course of streets as deformed as it may be by the topography or simply by history, evidences constants. The dense heart of Paris, like those of Hong Kong or Melbourne, that were conceived by Europeans, present a grid with an average distance of 120 meters between intersections, when it 50 m is more finely articulated urban settings such as Tokyo and Kyoto. In both cases, the pattern is immediately picked up, on the interior this time, by a remarkably dense interior. In every period, these cities chose in different manners, adapted to their culture, to have recourse to a limited number of schemes whose presence structure their cityscape.

If we turn our attention to these cities without preconceived plans but which, beneath the extreme variability of accidental forms, evidence astonishingly stable subjacent topological and metric structures, along with "signatures" each time that identify their natures, we can attempt to explore two questions. The first consists in asking if the invention of the city, rather than investing in isolated projects, is not firstly a matter of "defining rules of assembly and coexistence of a living, constantly open range of elements" (Fortier, 1989, p. 16). Today, the complexity of these rules of assembly has been lost in formal impoverishment of modernism that has reduced the city to isolated objects. The ideal stock of objects in the historical city had its own coherence that organized its interplay of full and empty spaces, of breaks and continuities, of sequences and views. Modernism bequeathed to us de-structured anti-forms that fail to give coherence to the city, and stand in the way of its representation. But this view of the city, this hypothesis of a pragmatics of procreation based on a coherent grammar of forms by no means excludes the considerable variations of these grammars over time and space. This then is a second level of study that opens up and that will be developed here, that of understanding the minimal threshold of complexity and of articulation that makes for rule of organization of these urban wholes that constitute an intelligible language and not a disorder of confused sounds, that produce a human environment and not a bursting where a non-qualified void distends the discordant notes of an urban harmony that seems forever lost. It is ultimately in search of these minimal rules of organization of urban areas that we must go, not to copy the past but to move toward morphologies that are at once vaster and more intimate, integrating scales never before seen, of human concentrations of tens of millions of inhabitants, in urban areas that nonetheless succeed in giving everyone the reassuring intimacy of a comprehensible neighbouring space. These rules are those of complexity

2.2 Urban complexity

"What is complexity? At first glance, complexity is a fabric (*complexus*: that which is woven together) of heterogeneous constituents that are inseparably associated: complexity poses the paradox of the one and the many. Next, complexity is in fact the fabric of events, actions, interactions, retroactions, determinations, and chance that constitute our phenomenal world." (Morin, 1990, p. 21). Two illusions, discussed by Edgar Morin, are to be avoided. The first would be to think that the complexity is such that it is impossible to draw out urban facts, clarity and distinct knowledge from the confusing and sometimes nebulous cluster.[1] The second would be to conflate complexity and completeness. We know from the start that a complete knowledge of the city is impossible: one of the axioms of complexity is the impossibility, even in theory, of omniscience.

However, the aim of a complex approach to the city is to bring together different forms of knowledge whose connections have been broken by disjunctive thinking. We are looking for a multidimensional analysis integrated by overarching universal laws that govern cities as well as the size and the distribution of clusters of galaxies, the evolutionary tree for species or the frequency and amplitude of economic cycles (Nottale et al., 2000). Complex thinking strives to establish the greatest possible number of connections between entities that must be

[1] Le Corbusier thought, by simplifying and classifying, atomizes the city into independent elements like those of a machine. Complex thought on the contrary refuses the mutilating and unidimensional conception of modernist simplification.

distinguished from one another but not isolates. In this it has the same structure as the cities that Nikos Salingaros (2006) showed to be living only if they establish a very great number of connections. In the realm of thought, Edgar Morin observed that Pascal had posited precisely that "all things being caused and causing, assisted and assisting, mediate and immediate, and all of them joined by an intangible natural bond that connects the most distant and the most variant." (Morin, 1990)

This general bond between all things brings us to pose the problem of the relationship of the whole and the parts and the links that they establish on different scales. Recent morphological theories conceive of forms not only as autonomous entities but also and especially *globally* as totalities irreducible to the sum of their parts. This is a point that Salingaros stresses: complex systems ordered by a hierarchy of coupling forces of short and long range, cannot be broken down into parts (Salingaros, 2006). This is also a point on which Ilya Prigogine insists. "One of the most interesting aspects of dissipative structures is their coherence. The system behaves as a whole, as if it were the site of long-range forces." (Prigogine & Stenger, 1984, p. 171)

Urban complexity can be understood as successive urban scales, revealing hierarchical levels of organization within a city. In these hierarchies, some sets of consecutive levels display a much better determined arrangement than others, which are much looser. The description of a "well structured" set generally introduces the notion of structure: the higher level element is broken down into lower order elements according to a well-defined scheme that can often be predicted to a great extent beforehand. The hierarchical order linking the frequency of appearance of elements to their size is, as we will see, a fractal order (see section 3.1). Generally speaking, fractal theory is a theory concerning the broken, the fractured, the scattered or yet about the granular, the porous, the tangled. But the strength of the theory is to have identified an order beneath the disorderly appearance of these irregular forms: the complex order of objects folded in multiple ways.

Urban limits, and the size and distribution of land uses and networks obey fractal laws (Frankhauser, 1994). The notion of fractal structure accounts for the economic localization of urban activities. On a still higher scale, it makes it possible to synthesize the analysis of urban density with the notion of the hierarchy of central places. Urban geography and in particular the theory of central places underscore the fact that cities exist not in isolation but rather as part of hierarchic systems that Batty and Longley (1994) demonstrate obey in rank and size a fractal distribution.

The hierarchy between urban scales, from the neighbourhood to the city, from the brick to the building, is a fundamental aspect of urban complexity.

3. Complexity, efficiency and resilience

Urban world is experiencing a never before seen growth. When put into perspective with climate change issues, fossil energy scarcity and poverty issues, this growth highlights the crucial need for more sustainable cities, be it on the energy or socio-economic side. Concerning climate change, two concepts play the major role: mitigation and adaptation to climate change. Mitigation aims at decreasing the amount of greenhouse gas emitted in the atmosphere to reduce the effects of climate change. On the other hand, adaptation is an anticipated approach to prepare to the inevitable effects of climate change.

The Urban Morphology Lab investigates these two concepts, putting them into perspective with two related ones: urban efficiency and urban resilience. Cities' efficiency is closely related to climate change mitigation. Considering cities, how can one get better services and more well-being with less resource consumption and less negative impacts? Urban resilience is related to climate change adaptation: what is the ability of a given city to resist to a series of endogenous and exogenous stresses (increase in resource prices, socio-economic instability, rise in temperature, and rise in sea level...)? Both will be crucial in the century to come in the climate change and resource scarcity compelling context.

3.1 Efficient cities

Various prisms allow investigating cities' structural energy requirements. Thermodynamics is one of them. However, using thermodynamics to assess cities energy efficiency appears to be everything but easy. Classical thermodynamics, that is widely based on the second law of thermodynamics (entropy maximization principles), fails to properly assess cities (Salat & Bourdic, 2011). Classical thermodynamics is fundamentally based on reductionist assumptions: any system can be analyzed as the sum of its elements. As authors have been explaining throughout this chapter, this reductionist approach is nothing but adequate for cities. Another reason for this failure is that classical thermodynamics has been developed to analyze closed systems. But cities are mainly driven by external flows: energy flows, material flows, information flows, etc. Cities are not closed systems. Cities exist because of their openness. As such, applying classical thermodynamics to cities is a nonsense. Fortunately, recent developments in thermodynamics provide interesting insights for open flow-driven systems such as cities.

Building on these recent developments, the Urban Morphology Lab bridges the gap between several fields of thermodynamics and shows essential results that have direct implications on urban energy efficiency issues. Salat and Bourdic (2011) base their demonstration on three main scientific areas:

- Prigogine's non-linear thermodynamics applied to open flow-driven systems (Prigogine, 1962, 1980)
- Kay's work on industrial ecology (Kay, 2002), analysing order emergence as a response from the system to make a more effective use of the available energy flows
- Bejan's "constructal theory", predicting the type of structure the most likely to emerge in a complex flow-driven system (Bejan & Lorente, 2010).

This chapter is not the place for digging into very theoretical aspects of thermodynamics and complex systems theory. That is why authors invite the reader interested in these fundamental aspects to refer to Salat and Bourdic (2011). To make a long story short, these three approaches are converging into a very same idea: open complex systems tend to be structured in the most energy-efficient way that is based on a power law distribution. In an open complex system, energy considerations impose a relationship between the different scales of the system. It imposes a mathematical relationship between the size of a given element and the number of elements of this size: few big elements, more medium-size elements, and a big number of small elements.

This power law distribution gives the number of elements (multiplicity) as a function of their size, as shown in Figure 1.

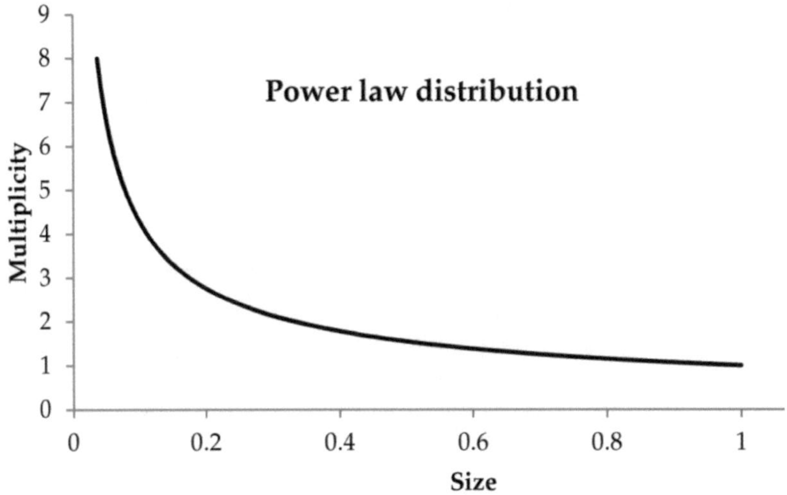

Fig. 1. Power law distribution, linking the multiplicity to the size of elements in a system.

The mathematical formula for such a power law distribution is given in Equation (1), where A is a constant and m the fractal dimension of the distribution[2].

$$multiplicity = \frac{A}{size^m} \qquad (1)$$

Power laws have a tremendous importance in many natural phenomena. They allow describing a wide range of distributions with analogue properties: many small objects and few large objects, many small events, and few large events. This structural law is omnipresent in natural phenomena involving flows: lung and river basin structures, blood system, trees… But these types of distribution are also omnipresent in man-made phenomena - under the name of Pareto distributions -, be they social, economic or cultural: size of cities, wealth within a society, or even the number of visits to internet websites. Interestingly, these types of distributions, the one structuring natural flows, the other unconsciously structuring man-made organizations, are two sides of the same coin.

After a long time of evolution, after numerous processes of construction and destruction, a whole series of systems have become more and more efficient over time, by moving toward more efficient structures. This motion has been an increase in structural complexity. To become efficient, systems have complexified at each and every scale, from the biggest elements to the smallest ones. For this purpose, authors introduce the concept of scale free complexity which speaks for itself: concerning complexity, there is no predominant scale.

Making good use of these hard-core science theories, the Urban Morphology Lab applies them to cities. There is in fact no reason why this law that applies to all complex systems

[2] It is fundamental for the reader to notice that this relationship is non-linear. If there are X elements of size 100, the adequate multiplicity for the elements of size 50 is not simply 2X. On the contrary, it is given by the equation (1). For a detailed analysis of this formula in urban context, see Salat (2011).

should not be used to improve urban efficiency. This fundamental law can indeed help make urban systems structurally more efficient, by respecting the right hierarchy of scales in urban systems: some big elements, a medium number of medium-size elements, and a very big number of small elements. As it will be shown later on in this chapter, this framework is a generic one, that applies to a wide range of parameters: transport networks, size of courtyards and buildings, socio-economic structure… To easily handle this structural law, the Urban Morphology Lab has created an innovative tool-box aiming at assessing the structural efficiency of urban structures (see section 4).

3.2 Resilient cities

Another interesting insight from the theory of complex systems deals with resilience. Understood as the ability to overcome endogenous or exogenous stresses, crisis and shocks, cities' resilience is an issue that is worth being investigated in current context. Cities will be confronted to a whole series of stresses throughout the century to come: water stress, increase in urban population, socio-economic crisis, natural resources scarcity, climate change, etc. This section aims at presenting the influence of urban structures on cities' resilience.

For this purpose, let us briefly open a parenthesis to introduce and explain the difference between a tree and a leaf. Mathematically speaking, a tree and a leaf have extremely different structures. Let us consider a small branch in a tree. It belongs to one, and only one bigger branch (see Figure 2). If you cut the bigger branch, the small branch falls and dies. The leaf structure on the contrary gives rise to a much bigger complexity. A small vein does not only "belong" to one bigger vein in the leaf, but to several (see Figure 3). A leaf is entirely structured by interconnected loops at every scale: there is a scale-free feedback looping. This point constitutes the fundamental difference between leaves and trees. If you cut a vein in leaf, the sap flow will be entirely compensated through the upper and lower levels of veins: the leaf survives.

Fig. 2. A tree structure (Portoghesi, 1999).

Fig. 3. A leaf structure (Portoghesi, 1999).

Surprisingly, this point has direct implications on cities and urban networks. Since Edison, electricity and energy networks have classically been structured like trees: a big remote power plant unit pouring the electricity flow into overhead high voltage power transmission lines, eventually reaching the consumer after having been cascaded into a series of lower voltage power lines. Tree-like structures are not resilient: cutting a branch in the tree leads to the loss of all the small branches belonging to this branch. Damage in the big remote power plant or in the high voltage line impacts a whole part of the network. On the contrary, leaf-like structures are resilient: damage in a vein of the leaf is immediately compensated by flows in parallel circuits, causing less if not no damage in the rest of the leaf. Analyzing leaves and trees' structures, Corson (2010) shows that redundancy[3] within leaves' venations improves the tolerance to damages.

This result has direct implication on transport networks (Dodds, 2010; Katifori et al., 2010) and can also be transposed to all sorts of urban networks: electricity, energy, water, waste, etc. The tolerance to damages and shocks can be interpreted as the adaptation and resilience ability of urban systems. Multi-scale interconnected loops, redundancy and connectivity could thus lead to an improvement of the resilience and adaptability of urban networks. In the climate change and resource scarcity context, instabilities and shocks will become more and more frequent, and adaptability and resilience become all the more crucial. Theoretically speaking, creating interconnected loops at every scale of urban networks correspond to a move toward a leaf-like structure at the urban scale, and therefore a move toward more adaptive and resilient structures.

Urban tissues resilience is an indicator for cities' stability and has therefore a strong influence on long term economic value. Resilience of urban systems is heavily dependent on its level of redundancy. In a highly dense and connected city with high levels of complexity, functional mix allows sparing significant amounts of inputs (materials, energy…). Furthermore, high levels of complexity and density make it easier to manage residual needs in a circular economy structured by feedback loops at every scale.

[3] Redundancy stands for the multiplication of elements or functions of a system to improve its stability and its reliability. Since each element rarely fails, and is supposed to fail independently from the others, the probability of all redundant elements failing is extremely small.

4. Concrete implications for the urban world

The investigations presented earlier in this chapter may appear dry and theoretical to the reader. However, they have very sound, concrete and direct implications on cities.

4.1 Bioclimatic comfort, heating and cooling energy efficiency

Heating and cooling requirements represent a very significant part of urban energy consumption, respectively in cold and hot climates. The current trend is to foster energy efficiency of systems (heating and cooling systems) and buildings (insulation and glazing). The approach of the Urban Morphology Lab rests on a wider understanding of urban efficiency. This approach, inspired from von Weiszäcker et al. (1997), is based on 4 leverages to improve urban efficiency, as shown in Figure 4.

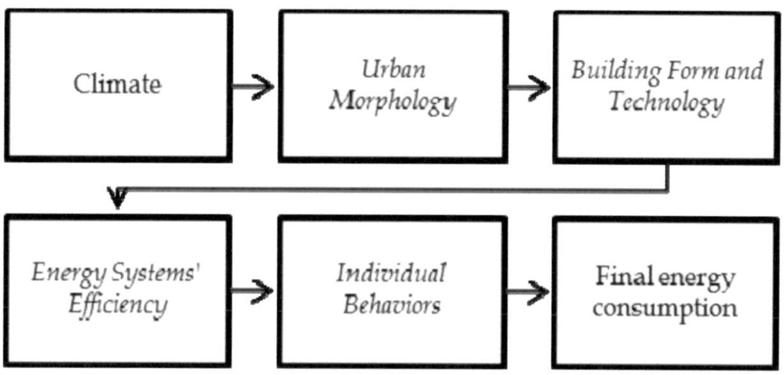

Fig. 4. Four leverages to improve urban energy efficiency (displayed in italics).

Whereas most of the current efforts aim at improving buildings' technology and energy systems' efficiency, very significant reductions in final energy consumption can be achieved by tackling the two other leverages that are urban morphology and individual behaviours. The Urban Morphology Lab mainly focuses on the first leverage that is responsible for a factor 2 to 2.5 in the final energy consumption. In other words, everything else being equal, a city with an appropriate urban morphology has a structural energy consumption that is 50 to 60% smaller than another city with a "bad" urban morphology. This section shows how complex urban structures can be structurally more efficient than simple ones.

Taken as a whole, a city is nothing else but a membrane exchanging a wide range of flows with the outside: air, heat, solar radiations, etc... The following analysis aims at showing how fractal theory can help optimize the interface between the building and the outside, with the example of passive zones. The concept of passive zone is described in the LT-method (Baker & Steemers, 1996) as being the area in the building within a distance from a perimeter wall, usually between 6 and 8 meters, depending on the floor to ceiling height (see Figure 5). These passive zones benefit from natural lighting and natural ventilation, but also from useful solar gains in winter. The energy consumption associated with lighting and ventilation is thus expected to be lower in these zones, an important part of lighting and ventilation being 'free'. On the contrary, these zones suffer from heat loss through the envelope and from unwanted solar gains in summer.

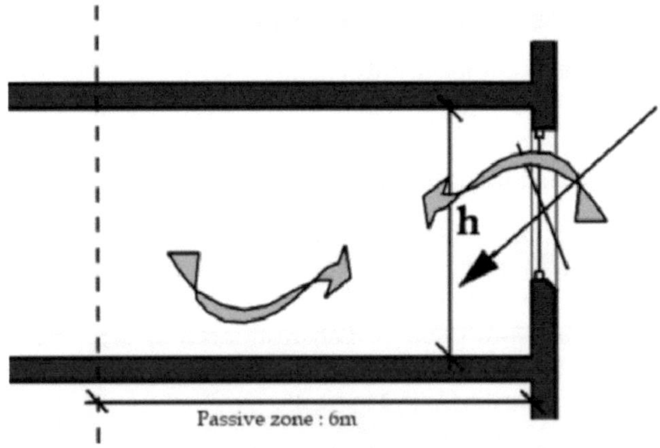

Fig. 5. The passive zone is located less than 6 meters from the façade (Ratti et al., 2005).

But as building technologies improve significantly at the present time, notably concerning glazing and insulation, the share of this unwanted phenomenon in the overall energy consumption figure will tend to diminish significantly in the future. In the office buildings, energy consumption is mostly associated with lighting, ventilating and cooling, even though the outside temperature is low. Concerning residential buildings, improved glazing and insulation will diminish the share of heating in the overall energy consumption figure in a close future. As it is already the case in office buildings, the share of ventilation, lighting and cooling will increase.

Strategically speaking, the role of passive zones will become more and more significant in the coming years and decades, as the benefits from improvements of insulation and glazing will become marginal. The more passive zones in the building, the better. Unfortunately, it is much harder to improve the passive volume ratio[4] of a building than its insulation. This ratio entirely depends on the original form of the building. If the passive volume ratio of a building is low, it is almost impossible to change it, but to destroy and rebuilt. Whereas improving insulation or glazing is a matter of months or years, improving passive volume ratios is a matter of several decades, *i.e.* the lifespan of the building.

The approach defended by the Urban Morphology Lab though rests upon an ability to scale up urban issues. Passive volume ratios are a characteristic on the building scale. But considering this issue from the neighbourhood or the district scale provides interesting insights. The following analysis is based on the neighbourhood scale. It aims at showing how passive volume ratio may increase as urban fabric becomes more complex. In the six situations, the zone under consideration is a 200x200m square, in which the building occupies 70% of the available floor area. The first three examples display simple urban organizations on which most of modernist cities have been based.

Figure 6 displays a mono-block structure, typically a tower. Passive zones are in green whereas non-passive zones are in black. The passive volume ratio (PVR) is only 17%, which is extremely low and leads to high energy consumptions notably for lighting, ventilation

[4] The passive volume ratio corresponds to the ratio of the volume of passive zones within a building over the total volume of the building.

and cooling (even in cold climates). The reader will certainly notice that unfortunately most of the office buildings - where energy consumption is mainly associated with lighting, ventilation and cooling - are towers...

Fig. 6. One block, PVR=17%.

Figure 7 and Figure 8 display two other structures, with the exact same floor area ratio. The passive volume ratio remains below 60% in both cases.

Fig. 7. 9 blocks, PVR=46%.

Fig. 8. Linear buildings, PVR= 58%.

Figure 9, Figure 10 and Figure 11 show three structures based on square courtyards, with a growing complexity. The construction is directly inspired from fractal theory, and more precisely from a Sierpinski carpet[5]. Figure 9 displays a massive building with only one

[5] For further information on fractal theories applied to urban structure, we invite the reader to refer to Batty and Longley (1994) and Salat (2011).

block, with one big courtyard: the passive volume ratio is low. In Figure 10, a second level of smaller courtyards has been added in the building. This leads to an almost doubling of the passive volume ratio. Finally, another level of courtyards is added in the building (Figure 11), leading to a passive volume ratio of 100%.

Fig. 9. One courtyard, PVR=33%.

Fig. 10. Two levels of courtyards, PVR=60%.

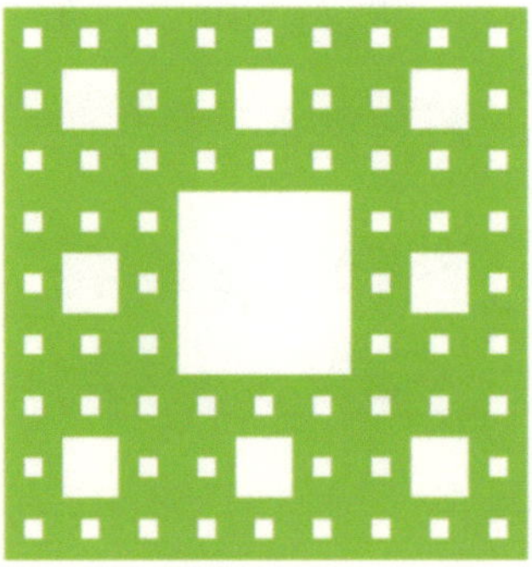

Fig. 11. Three levels of courtyards: , PVR=100%.

This simple geometric analysis shows that complex urban fabrics, based here on fractal theory, display a much higher passive volume ratio than simple ones. Fractal theory is a way to optimize the "urban membrane" - the interface between the inside and the outside. In Figure 11, the pattern is distributed over three scales, instead of one in Figure 9. The careful reader will then certainly notice that the multiplicity-size distribution of courtyards in Figure 11 follows a power law. Further research is currently carried out to understand how size and scale hierarchy of courtyards impact on energy consumption patterns according to the different climates.

Pushing further this geometric analysis, authors have investigated numerous urban tissues, historical and modernist ones, in cold and hot climates. When analyzing real cities, the same kind of results emerge: the more complex the urban tissue, the higher the passive volume ratio. The four following figures display two modernist districts (800x800 m squares) and two historical ones. Passive zones are in dark grey, whereas non passive zones are in light grey. The two first districts are made of simple blocks, without any courtyard. In Shanghai Lujiazui Central Business District (Figure 12), elements are so massive that the passive volume ratio is smaller than 50%. In Thianhe district (Figure 13), there are two predominant scales of buildings. The small ones have an acceptable passive volume ratio, but the big ones have a dramatically low one, leading to an average passive volume ratio of 66%.

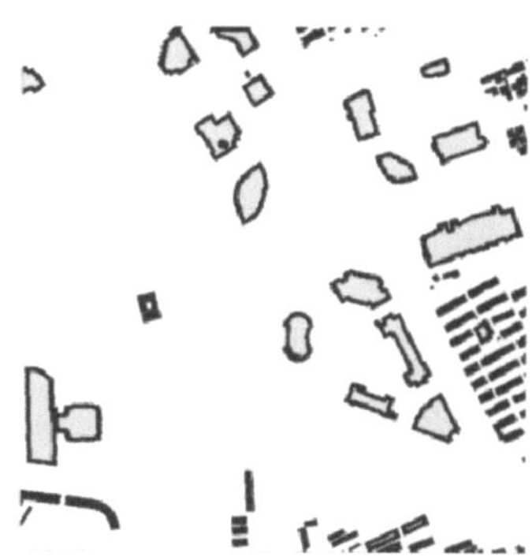

Fig. 12. Lujiazui (CBD), Shanghai. Passive Volume Ratio = 43%.

Fig. 13. Tianhe district, Guangzhou. Passive Volume Ratio = 66%.

In the two historical urban tissues, Shanghai's Lilongs on Figure 14 and a Parisian district on Figure 15, there are still some big elements. But they are organised around numerous courtyards of all scales that allow a much better interface with the outside, and a passive volume ratio higher than 80%. The analysis of the building size distribution and of the courtyards size distribution shows that the two historical urban fabrics display a high scale hierarchy, close to an optimal power law distribution (see section 3.1).

Urban scale-free complexity is a way to optimise passive volumes in the urban fabric. Urban complexity is not about scattering numerous small elements, but on the contrary about respecting an adequate scale hierarchy: a small number of big buildings and courtyards, a medium number of medium size elements, and a big number of little elements. Modernist urban fabrics based on one scale (see Figure 12) are structurally speaking unsustainable. On the contrary, urban fabrics based on several scales (up to three or four fundamental scales in Figures 14 and 15) allow optimizing crucial parameters for sustainability, such as the passive volume ratio.

Fig. 14. Lilongs, Shanghai. Passive Volume Ratio > 80%.

Fig. 15. Paris district, 19th century. Passive Volume Ratio > 80%.

A proper urban complexity is a way to improve the passive volume ratio, and thus to optimize the interface between the city and the outside. Pushing the thought further, this approach aiming at optimize the urban envelope can have implications on the renewable energy potential of urban structures. An optimized and complex interface on the district and city scale is a way to increase, with the same land footprint, the available envelope area, and thus the available area for solar energy. Complexification of urban structures may thus also reveal to be a partial answer to the higher land footprint of renewable energy compared to fossil fuels.

4.2 Tools to asses urban networks' structural efficiency

The complexity analysis is transposable to other aspects of urban sustainability such as urban networks. Efficiency is crucial for designing urban transport networks. An efficient urban transport network aims at providing a service - make every location in the city easily accessible from any other location - with the least energy consumption. Based on the theoretical analysis presented in section 3, the Urban Morphology Laboratory has developed a tool-box to assess urban transport networks' structural efficiency, notably with a tool assessing the scale hierarchy of the network. This tool measures the distance (or deviation) between the network and the associated optimal one. A low value insures that no scale in the network is underrepresented: the highways, the large scale transport infrastructures, the medium streets and the bicycle and pedestrian networks are then in the right proportions. On the contrary, a high value shows that one scale of the network is either over or under represented in the network: the network is then structurally inefficient.

Equation (2) shows how to calculate this indicator for a system with N scales, each scale gathering n_i elements of size x_i: [6]

$$S = \frac{1}{N}\sum_{i=1}^{N}\left[1 - \frac{n_i x_i^m}{A}\right]^2 \qquad (2)$$

For each scale i (i going from 1 to N) the relative distance between the number of elements of scale i and the optimal one is calculated. The indicator is then the sum of the squares of these relative distances. The closer this indicator to zero, the closer the actual network to the structurally optimal one. But if some scales are over-represented, under-represented or missing, the value of this indicator increases. The Urban Morphology Lab has analyzed two city-scale road networks to compare a historical city (Paris, see Figure 17) with the archetype of many urban cities (Contemporary City for 3 Million Inhabitants, Le Corbusier, see Figure 18).

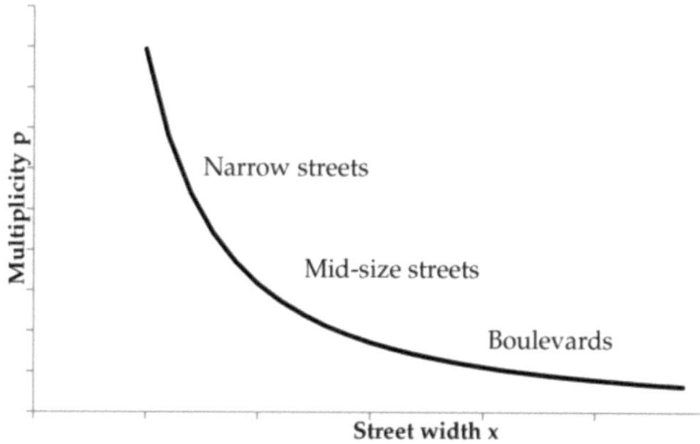

Fig. 16. Paris road size distribution.

[6] For further details on the calculation of the constants A and m, we invite the reader to refer to Salat et al. (2010) and Salat (2011).

The Parisian network displays numerous scales of streets, and each scale is present in a proportion that is relatively close to the optimal one: there are some large boulevards, more mid-size streets, and a very big number of narrow streets (see Figure 16). The Parisian road network has evolved from a vernacular structure to a highly scale hierarchic one. Baron Haussmann, who designed Paris wide avenues in the late 19th century, gave Paris his current structure by superimposing a larger scale on the historical network. Far from destroying historical urban complexity, Baron Haussmann added one more scale to the Parisian network by cutting wide boulevards through the old urban fabric. Unconsciously increasing scale hierarchy, he gave a new coherence to the city, opening at the same time a new era for motorized transports (see Figure 17). The indicator presented here above highlights the good scale hierarchy of the network: it is equal to 0.17, which means the structure of the network is very close to the optimal power law distribution; The Parisian network respects a fundamental scale hierarchy, from the pedestrian pathways to the wide Haussmannian boulevards.

Fig. 17. Transformation of the street system, Paris, by Pierre Ladevan (Salat, 2011).

On the other hand, Le Corbusier designed in the early 20th century an abstract of what would become an archetype of modern cities: a regulated and geometric urban scheme, with little ground coverage but great height, and a mix of cruciform towers, setbacks, cellular units and extensive empty spaces (see Figure 18). Le Corbusier sought to decongest city centres, augment their density, increase means of circulation, and increase open spaces. Purism informs his architectural choices. The Athens charter adopted by the fourth International Congress of Modern Architecture in 1933 artificially separated four urban functions - living, recreation, working, and circulation - in opposition to the existing urbanism that was characterized by mixed-use and tightly interwoven functions. It is to this mixed-use model that sustainable urbanism today is seeking to return.

Le Corbusier explicitly proclaimed his desire to destroy the street. The Athens charter recommended replacing house-lined streets in the living areas with tall buildings set at some distance from each other to free ground space for big landscaped areas. The freed spaces were also meant to be utilized for playgrounds, promenades, and sports. In actual fact, when these modernist principles were applied, enormous motorways and parking lots occupied most of the freed ground space. People were driven off the streets of the city by cars. Indeed, the need to speed up traffic was behind the idea of destroying existing urban fabrics, too cramped to meet the new needs of automobile traffic. The fact that functional zoning would lengthen distances between living and working areas and increase travel time was already noted in the charter. Instead of reconsidering functional zoning, its authors advocated the massive development of automobile transportation, replacing the many traditional streets of different sizes and the many crossroads with straight wide arteries.

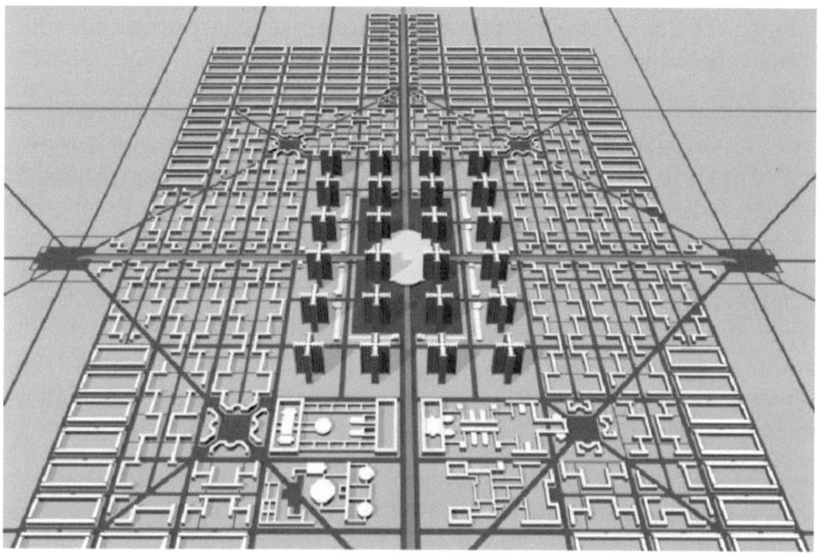

Fig. 18. Le Corbusier's Contemporary City for Three Million Inhabitants (Salat, 2011).

This destruction of the road network scale hierarchy is made obvious by the indicator assessing the distance between the network and the optimal power-law distributed network: it is equal to 509. It is 3000 times bigger than for the Parisian network. The Corbusean fabric is only made of three street scales: 240 km of 10 m wide streets, 220 km of 30 m wide streets, and 1,640 km of 50 m wide streets. All other scales are missing. The extremely high value of the deviation indicator shows that the scale hierarchy is reversed. Whereas the Parisian network distribution is very close to the theoretical optimum, Le Corbusier's is selective, discontinuous and reversed, allowing neither complexity nor coherence for the pedestrian, and offering no complexification potential over time.

Alongside with this major tool assessing networks' structural efficiency, the Urban Morphology Lab has used a wide range of complementary tools to assess both efficiency (proximity, scale hierarchy, ...) and resilience, with several tools that assess the level of connectivity (intersection intensity) and the redundancy (cyclomatic number, feedback loop intensity, etc...). Numerous urban tissues in Europe, Asia and America have been analyzed using these methods. Among other results, the analysis shows very significant differences between historical cities and "modernist" cities (see Salat, 2011). All the indicators show that most of the "modernist" cities eliminate essential scales in the network hierarchy, structurally banishing the low (or zero) energy transportation means (pedestrian and cycling) from the city. At the same time, the structural redundancy of urban networks tends to decrease, inducing a loss of resilience.

4.3 Economic insights

4.3.1 Path redundancy instead of element redundancy

Analyzing urban efficiency and urban resilience is not without raising a series of economic and financial issues. The first issue at hand concerns the concept of multi-scale redundancy that we showed earlier to be crucial for networks' resilience and stability. The role of redundancy is already well known and recognized as a key point for electricity networks. The main brake though on redundancy is of course the economic affordability. A twice redundant network induces twice higher costs.

Prima facie, this assumption seems to be true. But the approach developed in this chapter is a little bit more subtle. This approach requires going further than the reductionist paradigms and being able to grasp several scales at the same time. Authors have explained throughout this chapter the influence of complexity on efficiency and resilience, notably through the prism of scale hierarchic networks. Scale hierarchy does not induce higher costs. On the contrary, it allows sparing money. Let us take the example of a road network. If this road network displays only one scale (one street width) - as in many square grid based cities – the required amount of asphalt to achieve the same access to any location in the city is much higher than in a scale hierarchic grid – inspired from a lung for instance. The networks structured with only one fundamental scale are dysfunctional because most of the streets are either over or undersized.

Coming back to redundancy, let us just consider once again figures 2 and 3 that display a tree and a leaf structure. In a mono-scale network, increasing path redundancy means increasing the number of elements. But on the contrary, in a multi-scale and scale hierarchic network, a path redundancy does not increase the number of elements. It only curls the elements. Let us consider a tree structure. There is only one path to go from point A to point B. Now consider the same tree and curl the branches to connect the elements within each scale. The number of elements is the same, but the redundancy has extraordinarily increased. There are now dozens of paths to go from point A to point B. Increasing resilience in an urban network is thus not about an element redundancy, but a path redundancy instead. It is about connecting elements on all the scales instead of sprawling them apart.

4.3.2 The challenge of stability and resilience in the future

Stability and resilience will get a bigger and bigger importance in a close future. They have not been such a big deal so far, as most of the energy network depend on some big remote power plants, which energy production can roughly follow the demand. But the takeoff of renewable energy will induce a fundamental questioning of this paradigm. A crucial question of the next decades is how to increase the share of renewable sources in the energy portfolio. The main brake though on the renewable energy takeoff lies in their inherent instability and unpredictability. Over a 20% share of renewable energy in the portfolio, energy supply becomes fundamentally unstable. And this instability is expected to induce extremely high marginal costs, as very few consumers will accept random blackouts. The renewable energy takeoff is impossible and incompatible with the current energy supply (and demand) structure.

The current fashion around smart grids provides interesting insights on this issue. The concept of smart grids aims at making all the individual objects in the city (electric cars, boilers, heaters, washing machines...) active participants of electricity supply, by making them communicate with each other. The implicit objective of these 'smart' approaches is to improve the overall resilience of urban systems. Smart grids are presented as the best solution so far, but remain *theory*. In fact, this concept relies on a series of assumptions that will be nothing but obvious in the coming years, notably a rapid and high market penetration of hybrid and electric vehicles. That is why there is an urgent need for solutions to improve the structural resilience and stability of energy networks in a close future. We hope the approaches we propose in this chapter will feed the thought.

5. Conclusions

Historical cities, from Sienna to San Gimignano, from Suzhou to Beijing, from Tunis to Jerusalem, are a vast laboratory for examining the relations between people, the climate and the urban environment. Faced the forces of nature - the soil, the sun, and the wind - these fractal cities were the outcome of generations of patient efforts. Conversely, the planners of the modernist city set out to raze real cities in the name of abstract principles and unreal theories about the primacy of right angles and simplicity, when all historical cities are multi-scale systems complexified by their irregular topography and hydrography, and by the curving paths marked out by human beings focalized on such centres of attraction as marketplaces or mosques. Whereas urban development used to be a movement towards complexification, modernism has lead to a violent break leading to an extreme simplification.

The major problem of the contemporary city is the disconnection between scales. The 20th century technicist urban planners who ignored the fractal structure of historical cities divided the city into two spatial scales dedicated to two types of relations and behaviours: the greater metropolitan region traversed and structured by large transit infrastructures dedicated to speed and summarily zoned; and the neighbourhood, celebrated as the building block of the sustainable city, when its concept, boundaries and limits remained blurry and ill-defined. Two stances were adopted as a result. The first involved razing the old fabric and inordinately enlarging the urban grid to bring it in line with the major regional throughways. This was the position taken by Le Corbusier (Le Corbusier, 1942), modernism and the new towns in France. We know today that this approach is a failure, that it engenders inhuman cities, entirely given over to speed and to the ever-growing intensification of transports and energy consumption. This floating city, drifting in a territory that is too big for it, loses all urbanity, all identity, and all definition. It stops being a city. In this sense, the 20th century will have been the century of the demise of cities. The innovative insights coming from recent scientific breakthrough presented in this chapter allow extending this thought in a more quantitative way. The reductionist approach associated with modernism has not only leaded to a dehumanization of cities. It has also leaded to structurally inefficient urban tissues. Modernist planning has been unable so far to grasp the complexity of historical urban structures that make them be climaxes of efficiency, of interaction between people and of value creation.

The capacity to survive disasters and even to rise out of its ashes, like Lisbon after the 1755 earthquake, London after the Great Fire in 1666, Kyoto after the fires in the Middle Ages, Tokyo after the 1923 earthquake, is what authors call urban resilience - a complex concept related to the permanence of a memory at once social, symbolic and material. Partly because of their leaf-like structure and of their extremely high level of redundancy, the vast majority of historical cities is resilient and has managed to survive the centuries, often outlasting the civilizations that built them. Cities worldwide will be confronted to various types of perturbations and chocks in the century to come. Will modernist cities manage to survive the century and hold out against the growing risks linked to climate change? How will their structure evolve and behave if confronted to a rise in prices due to natural resources scarcity? This adaptation ability, or resilience, that is rarely -if not never - taken into account in urban policy processes, should be given the attention it deserves.

6. References

Baker, N., & Steemers, K. (1996). LT Method 3.0 - a strategic energy-design tool for Southern Europe. *Energy and Buildings 23* , pp. 251-256.

Batty, M., & Longley, P. (1994). *Fractal cities: a geometry of form and function.* London Academic Press.

Bejan, A., & Lorente, S. (2010). The constructal law of design and evolution in nature. *Philosophical transactions of the Royal Society B , 365*, pp. 1335-1347.

Capra, F. (1996). *The Web of Life: A New Scientific Understanding of Living Systems.* New York: Anchor Books, Doubleday.

Corson, F. (2010, 29 January). Fluctuations and redundancy in optimal transport networks. *Physical Review Letters , 104.*

Dodds, P. (2010, 29 January). Optimal form of branching supply and collection networks. *Physical Review Letters , 104.*

Fortier, B. (1989). *La Métropole Imaginaire. Un Atlas de Paris.*

Frankhauser, P. (1994). *La fractalité des structures urbaines.* Paris: Anthropos.

Katifori, E., Szöllösi, G., & Magnasco, M. (2010, 29 January). Damage and fluctiuations induce loops in optimal transport networks. *Physical Review Letters , 104.*

Kay, J. (2002). *On complexity theory, exergy and industrial ecology: some implications for construction ecology*". in C. Kibert, J. Sendzimir & B. Guy, Construction Ecology: Nature as the basis for green buildings, pp. 72-107: Spon Press.

Le Corbusier. (1942). Charte d'Athènes. *CIAM IV 1933.* Paris.

Morin, E. (1990). *Introduction à la pensée complexe. In english: On Complexity, 2008.*

Nottale, L., Chaline, J., & Grou, P. (2000). *Les arbres de l'évolution.* Paris: Hachette Littératures.

Portoghesi, P. (1999). *Natura e architettura.* Milano: Skira.

Prigogine, I. (1962). *Introduction to non-equilibrium thermodynamics.* New York: Wiley-Interscience.

Prigogine, I. (1980). *Self-Organization in Non-Equilibrium Systems.* Freeman.

Prigogine, I., & Stenger, E. (1984). *Order out of Chaos.* Bantam Books.

Ratti, C., Baker, N., & Steemers, K. (2005). Energy Consumption and Urban Texture. *Energy and Buildings , 37* (7), pp. 762-776.

Salat, S. (2011). *Cities and Forms.* Hermann.

Salat, S., & Bourdic, L. (2011). Scale Hierarchy, Exergy Maximisation and Urban Efficiency. *2nd International Exergy, Life Cycle Assessment, and Sustainability Workshop & Symposium.* Nisyros.

Salat, S., Bourdic, L., & Nowacki, C. (2010, december). Assessing urban complexity. *SUSB journal , 1* (2).

Salingaros, N. (2006). *A Theory of Architecture.* Umbau-Verlag, Solingen, Germany.

Von Weizsäcker, E. U., Lovins, A. B., & Lovins, L. H. (1997). *Factor Four, Doubling Wealth, Halving Resource Use - A Report to the Club of Rome.* Earthscan.

World Bank. (2010). *Eco2 Cities: Ecological Cities as Economic Cities.*

3

Evaluation of Energy Efficiency Strategies in the Context of the European Energy Service Directive: A Case Study for Austria

Andrea Kollmann and Johannes Reichl
Energy Institute at the Johannes Kepler University Linz, Austria

1. Introduction

We evaluate possible strategies to improve end-use efficiency under Directive 2006/32/EC of the European Parliament and Council on energy efficiency and energy services (ESD) in Austria. The main objective of the ESD is to invite the Member States of the European Union to cut their final energy consumption by 9% compared to the business-as-usual development within a period of 9 years ending in 2016. The emphasis is on 'invite' because the European Parliament and the Council could not agree on mandatory energy savings to be fulfilled by the Member States. The aim of the ESD is therefore indicative and Member States are basically only asked to undertake sufficient measures to increase the end-use energy efficiency and submit interim reports to the European Commission every three years. Currently, a new energy efficiency directive (COM(2011)370) proposed by the European Commission in which an even higher saving target auf 20% to be reached in 2020 is demanded, is under review. This new directive is not yet in force.

Basically, the purpose of the ESD is to increase the efficiency of energy end-use cost-effectively. For that reason the ESD sets up a framework for creating the conditions for the development and promotion of a market for energy services. The directive applies to (see Article 2, ESD)

- provider of energy efficiency improvement measures, energy distributors, distribution system operators and retail energy sales businesses;
- final customers, with the exception of companies that are involved in the European Emissions Trading Scheme;
- the armed forces, if its application does not conflict with the nature and primary purpose of the activities of the armed forces and with the exception of material that is used exclusively for military purposes.

In determining the manner in which the ESD is implemented, two options from which the individual Member State can choose are given. The first possible option is to select one or more of the following measures, to be performed by energy distributors, distribution system operators and / or energy trading companies:

- Promotion and offer of competitive priced energy services;

- Promotion and offer of competitive priced energy audits conducted in an independent manner and / or energy efficiency improvement measures;
- Participation in funds and funding procedures.

The second option is to adopt and implement voluntary agreements and / or other market-based instruments such as white certificates. Austria chose the second option. In January 2010, voluntary agreements with several trade associations and interest groups of the energy industry were signed (see Rezessy & Bertoldi (2011) for a discussion of the benefits and draw-backs of voluntary agreements and an overview of such agreements made in other European countries). In late 2009, Austria's Federal Ministry of Economy completed voluntary agreements with the association of the gas and heat supply companies, the association of the Austrian electricity companies, the trade association of the petroleum industry and the trade association of energy trading. The saving target defined in the agreements sum up to 3,020 GWh (14% of the total saving target, see below). This amount must be reached through energy efficiency measures that result in energy savings for end users and the achievements are subject to on-going monitoring. The Austrian calculation methods for energy efficiency measures are described in Adensam et al., (2010). Reichl & Kollmann (2010) give a general discussion of different monitoring paradigms in the context of the ESD.

Interesting (and crucial) about the ESD is the calculation of the overall savings target, which is described in Annex I to the Directive: 9% of the average of the total final energy consumption of a country in the five years prior to the Directive (or the last 5 years for which data are available, in the case of Austria 2001 to 2005) are to be used as the savings target. The total final energy consumption must be corrected of the energy consumption in the segments air traffic, heavy oils, which are used in maritime navigation (not relevant for Austria), and of the final energy consumption of all companies participating in the European emission trading scheme.

For Austria, an indicative savings target of 22,330 GWh to be reached by 2016 was calculated (see Bundesministerium für Wirtschaft und Arbeit (2007)). Thus, the Austrian final energy consumption in order to successfully comply with the directive in 2016 must be 22,330 GWh below the value that would have been the Austrian end-energy consumption without implementing specific energy efficiency measures in the period 2008 to 2016. It also means that the Directive does not require an absolute reduction of final energy consumption under the 2008 level, but simply an attenuation of energy consumption compared to the business-as-usual consumption. Obviously, the assumptions under which the business-as-usual scenario is calculated have significant impact on the size of the savings target. We will not go into detail on this subject, but refer the reader to Reichl & Kollmann (2010) for a more detailed discussion about defining this baseline. Boonekamp (2006) gives a comprehensive overview of the issues involved in quantifying the effects of energy savings and Thomas et al., (2009) discuss the importance of accuratly measuring the effectiveness of the ESD.

A central aspect in this context is that each member state may include improvements in energy efficiency, which had been initiated before 2008, when reporting the energy efficiency improvements achieved. In the case of Austria significant early actions or so-called early savings were achieved through measures such as the Austrian housing subsidies scheme in the past two decades. The responsible Federal Ministry for Economy reports that early savings of 9,200 GWh were achieved and are accounted for in the National Energy Efficiency Action Plan.

Our article is divided into three parts: firstly, we define and analyse potentially appropriate energy efficiency measures and develop policy strategies to support the implementation of these measures. These policy strategies encompass the development of a framework regarding funding and eligibility criteria or the setting of energy efficiency minimum standards. And finally, we evaluate these energy-efficiency strategies according to how they can support the achievement of the ESD target and their impact on key variables of the Austrian economy.

The analysis of energy efficiency measures and strategies extends both on the household sector and industrial and service sector. Thus, an almost complete coverage of the policy-relevant energy consumption is given. However, it must be noted that measures analysed in this study are those that the authors identified as the most effective options for a rapid reduction of final energy consumption, but by no means claim to have a complete list of all potential measures. Especially in the corporate sector large energy savings are often possible by changes in the processes or by restructuring production facilities (see Tanaka (2011) for a meta-study of energy efficiency policies in the industries of IEA countries). A study like this can only analyse energy efficiency measures with a relatively uniform effect on the energy consumption of final customers. Furthermore, the aim of this article is not on showing how the saving effect of individual energy efficiency measures was calculated (see Reichl et al., (2010) for a detailed description of the individual energy efficiency measures underlying our strategies) but on discussing the challenges of a sound analysis of the overall effects of a national energy efficiency strategy and the necessary assumptions that need to be made in doing so.

2. Developing the energy efficiency strategies

We define and analyse five different energy efficiency strategies for Austria. The basis for the development of these strategies is the current subsidy environment in Austria in the year prior to the directive's introduction (2007) as well as the current state of discussion concerning energy efficiency polices in Austria.

Currently the following mechanisms to promote energy efficiency are in use in Austria:

- Ecological tax reform
- Energy Efficiency Information campaigns
- Tax relief for setting energy efficiency measures (especially in the building sector)
- Monetary support for energy audits for companies (especially small and medium sized companies)
- Monetary support for energy efficiency measures of households (for example when inefficient white goods are replaced).

The currently existing measures and mechanisms to promote energy efficiency are not sufficient under the given conditions and requirements to fulfil the ESD target which is strongly undermined when looking at the pre-ESD period: between 1990 and 2007, Austria`s overall final energy intensity has decreased by only 4% (compare Jellinek (2009) for a description of sectoral energy intensities in Austria). We therefore assume that technical progress is accelerated by the funding measures in our strategies. Also, additional strategies are proposed to promote energy saving measures in the industrial sector. Firstly, to increase funding rates, energy efficiency funds are launched, which support the implementation of energy efficiency technologies in domestic companies. Energy efficiency funds have already been introduced in Denmark, Norway, Great Britain and some U.S. States (Irrek et al., (2004) give a comprehensive discussion of the potential effectiveness of energy efficiency funds and an overview of international experiences.) They not only offer funding but also consultancy, present energy-efficiency-related information and have a wide range of motivational incentives.

Secondly, minimum standards are a tool to increase energy efficiency and are based on the market's best technologies. In developing the scenarios it is assumed that all eligible firms meet the requirements for the use of funds. We assume that only cost-effective energy efficient technologies are promoted.

All scenarios are based on existing technologies only. Calculations for the development of energy saving in the industrial and service sector are based on literature reviews and a survey of opinion leaders from the energy sector. In our framework - starting in the base year 2008 - for the two strategies *'Intensive monetary promotion of energy efficiency enhancing technologies'* and *'Intensive monetary subsidies of energy efficiency enhancing technologies with simultaneous introduction of minimum standards'*, two scenarios for possible energy savings are designed. For the strategy *'Introduction of minimum standards for energy efficiency-enhancing technologies'*, only one scenario was developed. These scenarios are supposed to show different developments, depending on the strategy chosen.

The five evaluated energy efficiency strategies are:

- Introduction of energetic minimum standards (*Min Stand*)
- Intensive monetary subsidies of energy efficiency enhancing technologies, moderate intensity = funding quote is increased to 40% (*IMF mid*)
- Intensive monetary subsidy with simultaneous introduction of minimum standards, moderate intensity = funding quote is increased to 40% (*IMF-US mid*)
- Intensive monetary subsidies of energy efficiency enhancing technologies, high intensity = funding quote is increased to 50% (*IMF high*)
- Intensive monetary subsidies of energy efficiency enhancing technologies with simultaneous introduction of minimum standards, high intensity = funding quote is increased to 50% (IMF-US high)

The scenarios *IMF mid* and *IMF high* focus on the optimization of the promotion policy by increasing the rate of funding. The scenario *Min Stand* focuses on the impact of the introduction of minimum standards for energy-efficient technologies and the potential savings achievable. The scenarios *IMF-US mid* and *IMF-US high* combine the use of intensive monetary support with the introduction of minimum standards.

We assume that all companies that apply for funding are eligible for funding and thus get the full financial support. Furthermore, we assume that new investments have to result in at least 15% energy savings compared to the industry average and for replacements over 20% of energy savings have to be achieved in order to be eligible for funding (compare Emnid, 2005). According to this, by optimizing support policies the implementation rate of energy efficiency measures in companies is expected to increase. We analyse these effects in the two scenarios *IMF mid* and *IMF high*. In the first case, the predicted funding rate is 40% and in the second case companies receive a funding of 50% of their investment costs. The following table gives an overview of the five strategies as well as of their underlying assumptions.

Measures		*Min Stand*	*IMF mid*	*IMF-US mid*	*IMF high*	*IMF-US high*
Renovation of buildings/heat dissipation	From 2011	No change to 2007	FQ: 60% of investment	FQ: 50% of investment	FQ: 60% of investment	FQ: 50% of investment
	From 2012	Buildings built before 1980 have to be renovated within 15 years	No change to business as usual	Buildings built before 1980 have to be renovated within 15 years	No change to business as usual	Buildings built before 1980 have to be renovated within 15 years
Heat supply	From 2011	No change to 2007	FQ: 30% of investment (45% for combined solar thermal systems)	FQ: 15% of investment (30% for combined solar thermal systems)	FQ: 30% of investment (45% for combined solar thermal systems)	FQ: 15% of investment (30% for combined solar thermal systems)
	From 2012	Obligatory use of condensing heating technology; in case of building renovation obligatory installation of solar heat systems	No change to business as usual	Obligatory use of condensing heating technology; in case of building renovation obligatory installation of solar heat systems	No change to business as usual	Obligatory use of condensing heating technology; in case of building renovation obligatory installation of solar heat systems
Lighting & Stand By	From 2011	No change to business as usual				
	From 2012	Directive Nr. 244/2009 EG Directive Nr. 1275/2009	No change to business as usual	Directive Nr. 244/2009 EG Directive Nr. 1275/2008	No change to business as usual	Directive Nr. 244/2009 EG Directive Nr. 1275/2008
White goods	From 2011	No change to 2007	FQ: 30% of investment with linear reduction to 0% in 2014	FQ: 30% of investment with linear reduction to 0% in 2014	FQ: 30% of investment	FQ: 30% of investment with linear reduction to 0% in 2014
	From 2012	Dynamic standards from 2011 on, A++ within 4 yours	No change to business as usual	Dynamic standards from 2011 on, A++ within 4 yours	No change to business as usual	Dynamic standards from 2011 on, A++ within 4 yours
individual transport	From 2011	No change to 2007	750 € for each full hybrid and natural gas-dedicated vehicle	750 € for each full hybrid and natural gas-dedicated vehicle	750 € for each full hybrid and natural gas-dedicated vehicle	750 € for each full hybrid and natural gas-dedicated vehicle
Public transport	From 2011		Funding of 40% of costs	Funding of 40% of costs	Funding of 40% of costs	Funding of 40% of costs

Table 1. Main characteristics of the strategies; FQ: funding quota

The strategy *MinStand* is the only one that does not set new incentives by imposing new subsidies, while the other strategies all use new subsidies as incentives for energy efficiency actions. Another aspect of the strategy *MinStand* is that all costs of the energy efficiency measures are carried by end-customers. Especially for households, this may lead to social imbalances but does not need to. Schiellerup (2002) for example shows in a case study for Great Britain, that the introduction of minimum standards for cold appliances let to considerable savings on the consumers' side. In contrast, Galvin (2010) calculate that raising the minimum standard that needs to be reached in the thermal insulation of an existing house to be eligible for subsidies may prevent renovation activities among low-income households significantly. Concerning the firms that are subject to European and international competition, the cost shifting that is possibly associated with the proposed energy efficiency measures can negatively influence their cost structure. Any introduction of minimum standards without financial support for end-customers thus needs to be carefully analysed.

For each strategy under consideration, the evaluation consists of three steps:

1. a comparative-static analysis of the total potential energy savings achieved by a promotion of end-use energy efficiency,
2. a comparative-static analysis of the investment costs and the induced changes in energy demand in the respective strategy ,
3. a dynamic analysis of the economic effects induced by the respective end-energy-efficiency strategy in Austria.

The considered strategies and packages of measures are compared to a baseline scenario in each case. This baseline scenario anticipates the development of energy consumption and economic development of Austria without any measures to promote energy efficiency.

2.1 Definition and Analysis of appropriate energy efficiency measures

The aim of the analysis is to determine whether the implementation of the energy efficiency strategies can be realized cost-effectively or not. For that, the cost-efficiency of a single energy efficiency measure has to be assessed. Basically, an energy efficiency measure is cost-effective when:

$$C_K \leq V_K = \sum_{t=1}^{T} = \left(\left[P_t \left(F_C - F_N \right) \right] / \left(1 + r \right)^t \right) \tag{1}$$

with C_K being the total cost of the energy efficiency measure, V_K representing the present value of energy savings that can be achieved by the measure, P_t the energy price at time t, F_C the amount of the energy source that is used before the implementation of the measure, F_N the amount of the energy source, which is used after implementation. T is the lifetime of the measure and r is the discount rate. Equation (1) was used for the calculation of all analysed energy efficiency measures. When interpreting the cost-efficiency of an energy efficiency measure, the following aspects need to be considered:

- The more the energy price P_t increases over the lifetime of the energy efficiency measure, the more likely is the energy efficiency measure cost-effective.
- The higher the interest rate, the lower the present value of expected energy savings.

- The calculation assumes that the annual savings remain constant and equal to the initial value, while in reality it may be more fluctuant and depending on external influences (for example on climatic influences).

In the calculations, any annual maintenance or repair costs are neglected. The disposal of old appliances is free of charge (except only problematic substances like asbestos). The calculations assume that the installer of the new appliance ensures the proper disposal of the replaced appliance.

All of the potentials presented throughout this article represent realistic, socio-economic potentials. That is, for every period considered, those potentials are considered that are realizable by ambitious policies. The behaviour of a particular stock of energy consuming objects over time is shaped by several factors. Probably the most important parameter is the lifetime of an object. If an object reaches the end of its life, is must be replaced. Assuming a uniform lifetime, excluding any exchanges before the end of the lifetime, any given stock is completely replaced within its lifetime. Therefore, the parameter '*life time*' is especially important for the analysis of the savings potentials.

In addition to the technical lifetime of an object, also the rate of early replacement has a major influence on how long it takes to replace the stock of objects completely. It is not trivial to determine this parameter, because data of product lifetimes already include these early exchanges; i.e. the replacement of the product before it no longer meets its technical function. Therefore, within this project any early replacement is considered as a rise above naturally defined replacement (i.e. by historical experience) rates. In the field of renovation of buildings and heat supply systems, the potential shown in this article represents such a rise in early replacements. At times, this rate fluctuates between the different measures. The reasons for this are not only the different cost-efficiencies of the measures but also the fact that the rates of early exchanges are higher, especially when the user experiences a noticeable increase in comfort of setting a measure. If such a comfort factor is not given - the insulation of hot water pipes may serve as an example - actions are perceived as less urgent as they are less visible.

Another distinguishing criterion is the required investment. With increasing investment, risks also increase. Furthermore, the market penetration of technologies and the extent of disturbance during implementation of the measures have an impact. For the topics *electrical load* and *traffic*, we assume that there is no premature replacement of products.

2.2 Energetic and Economic Effects of relevant energy efficiency measures

For this study, a variety of energy efficiency measures was analysed. For each of these measures, the expected energy savings per implementation (households) or per unit of current energy consumption (services and goods production) was calculated. Based on these individual savings, subsequently, the potential of these measures for the reduction of annual energy consumption in the target year 2016 was calculated and comprehensive strategies were designed. As mentioned before, we do not go into detail about individual measures, but present the overall saving potentials of a whole group of measures. For example, the replacement of inefficient white goods as presented below is the sum of five individual energy efficiency measures: the replacement of inefficient refrigerators, of inefficient washing machines, of inefficient laundry dryers, of inefficient dish washers as well as of inefficient electric-cookers.

The potentials presented here are to be interpreted as additive. This means that for example, the potential savings realized by a partial restoration of the building envelope are calculated in a way so that further savings can be credited for an additional installation of a highly efficient heating system. If, for example by replacing all windows and doors in a house from the building period 1945 to 1970 savings of 3,600 kWh per year are achieved and at the same time a solar thermal plant with 20 m^2 is installed, the creditable savings are 3,600 kWh + 6,000 kWh = 9,600 kWh.

Again it should be noted that these savings each represent average values and are therefore suitable for the consideration of energy efficiency measures that are implemented in high numbers and yield a comparably uniform saving effect, but cannot be applied to individual objects for business considerations.

2.2.1 Saving potential of energy efficiency measures in the household sector

The highest potential in the household sector comes from space heating. Under the headline *space heating* thermal preparation, thermal rehabilitation of buildings, an efficient building envelope for new buildings and heat distribution are comprised. With full implementation of the realistically feasible measures in this area the final energy consumption in 2016 would be around 20,176 GWh lower than in the business-as-usual scenario.

Concerning *electric lighting* the replacement of bulbs (without lamp replacement) is analysed for the household sector. We calculate a potential for reducing the annual energy consumption for lighting in 2016 by 1,126 GWh compared to the business-as-usual scenario for 2016.

Under the topic *electrical applications* (except electric lighting) in households a potential for reducing the annual energy consumption in 2016 by 1,885 GWh is calculated compared to the business-as-usual scenario for 2016. These potentials are composed of a reduction of the stand-by consumption of household appliances including consumer electronics and a rapidly increasing penetration of highly efficient white goods.

By maximizing the activities in the area of *individual transport*, savings in energy consumption of 1,173 GWh can be achieved in 2016 compared to the business-as-usual scenario. The following table 2 gives an overview of the possible savings realized by measures in household measures.

2.2.2 Saving potential of energy efficiency measures in the production of goods and services

We now turn to the realisable potentials in the industrial and service sector which are shown in Table 3. Again, energy efficiency measures in *space heating* yield the highest savings (space heating preparation, distribution and thermal remediation). These savings amount to 5,355 GWh. Concerning *electrical applications* (lighting, compressed air optimization, ventilation and air conditioning systems, optimization of pump systems and process refrigeration) annual savings of 4,960 GWh are possible in 2016.

Measure	*Potential Savings*
Domestic water system	1,806 GWh
Heating systems (of which 75% are combination of domestic water and heating systems)	6,504 GWh
Building envelop in new buildings (without savings from efficient in-house systems	1,661 GWh
Thermal building renovation	6,622 GWh
Heat distribution system	3,583 GWh
Electric Lighting	1,126 GWh
White goods	812 GWh
Stand-by consumption	1,073 GWh
Individual Transport	1.173 GWh
Total	***24,360 GWh***

Table 2. Measures related to the total potential reduction of annual energy consumption in the household sector in 2016.

High energy savings are also possible in *freight transportation*. But these savings are only achievable through radical structural measures with long-term nature. The presented potential to reduce annual energy consumption in 2016 is thus only 720 GWh compared to the year 2008 therefore relatively low compared to the total consumption in this sector of about 26,000 GWh (without kerosene and LPG).

Measure	*Potential Savings*
Lighting	1,150 GWh
Compressed air optimization	460 GWh
Air conditioning and ventilation	1,600 GWh
Optimization of pumping systems	650 GWh
process cooling	1,102 GWh
Thermal building Renovation	2,403 GWh
Preparation of space heating and its distribution	2,952 GWh
Freight transport	720 GWh
Total	11,037 GWh

Table 3. Measures related to the total potential reduction of annual energy consumption in the services and goods production in 2016, own calculations

Of all the measures presented in Table 3, freight transport points out as it is a highly important sector for long-term energy efficiency improvement as well as for significant reductions in greenhouse gases. Obstacles to rapid structural changes are manifold: Liimatainen & Pöllänen (2010) argue that on the one hand, policy makers are reluctant to touch upon these sensitive issues as they need significant behavioural changes of companies and customers. On the other hand, they show in an analysis of the freight transport sector in Finland that this sector is highly dependent on the overall economic circumstances and was therefore severely hit by the recent economic crisis during which indicators like average payload and vehicle utilization rate significantly dropped.

2.3 Underlying assumptions

From a scientific perspective, it is indispensable to make a number of assumptions when analysing the economic effects of the studied energy-efficiency strategies for Austria. In the following these assumptions - both for the comparative-static as well as for the dynamic economic analysis are - summarized.

Assumption 1: Reaction of the households to changes in their consumption patterns

The reaction of households to measures initiated by changes in the consumption structure (e.g. increased non-energy consumption by purchasing new circulating pumps) requires assumptions about how those expenditure are financed by each household. Here, a central factor is the choice between substitution within the non-energy consumption and financing out of the total savings of households. Similarly, an assumption is needed on how the monetary savings are used. Depending on the strategy, we vary the proportion of financing from savings from 33% to 100%.

Assumption 2: Reactions of companies to changes in their investment behaviour

The need for assumptions on the response to measures that initiate changes in their investment behaviour is as important for companies as it is for households. Companies decide whether to respond to changes in their cost structure with a substitution within the investment or a change in their reserves. Our assumptions are based on a survey undertaken by Brüggemann (2005) in 2005. She interviewed 4,100 companies in Germany and showed that 79% of those companies financed energy efficiency measures with own resources. 42% of the companies also took bank loans. In our analysis we vary the proportion of funding from reserves, depending on the strategy, between 33% and 100%.

Assumption 3: Borrowing of the public sector

The additional expenditure of the public sector (direct subsidies, infrastructure investments, etc.) during the evaluation of the strategies does not lead to cutting of other spending in the public budget: There is no reaction or adaptation of other public expenditure and revenues in the analysis; they are funded by additional debt. However, in the individual evaluations of interventions also the induced changes in public revenues from energy taxes and taxes of VAT revenue from private consumption and wage tax revenue are calculated.

Assumption 4: Simulation horizon to the year 2020

All analysed energy efficiency measures start on January 1, 2008 and end on December 31, 2016. After this date, no further additional strategies are realized, the effect of the strategies installed prior to that date will be displayed in the dynamic economic analysis, however, until 2020.

Assumption 5: Time costs and external effects (like abatement costs of greenhouse gas emissions)

As part of the dynamic economic analysis neither time costs, which particularly exist in the transport segment, nor any external costs such as damage costs of greenhouse gas, air pollutant emissions or noise costs are included. The analysis of these parameters cannot be part of the present study for reasons of capacity. The fact that these potential costs are not included does not mean that they do not exist.

Assumption 6: Technological Progress

For the overall observation period, technology that was state of the art in 2008 is used. The technological developments to be expected by 2016 are not predictable, but due to the comparably short time horizon are considered of low importance for the sake of this study. New – improved – technologies would only be of major importance in our analysis if a high penetration rate was to be expected.

Assumption 7: Funding rates and tax rate of the public authorities in the measures

The analysis of the individual measures uses - over the entire observation period – the support schemes valid in Austria in the year 2008. Due to the partly inhomogeneous systems in Austria's nine federal states weighted averages were used. Similarly, taxes and levies are given by the values used in 2008 and remain constant for the entire observation period.

Assumption 8: Development of energy prices until 2016

By assumption, the price of electricity is 18.26 €cent/kWh; the price of heat energy is 6.5 €cent/kWh, which reflects the Austrian price level in the year 2008. For an analysis of the influence of energy prices, discount rates and inflation on the cost-efficiency of the measures underlying the energy efficiency strategies, see (Reichl, et al., 2010).

Assumption 9: Customer adoption of different measures and additional overhead costs

In the course of this study, no analysis of the customer acceptance regarding the specific actions is conducted. This would need to be done by the use of elaborate methods (interviews ...). Such surveys could be conducted during the time in which the individual strategies are carried out to evaluate customers' acceptance. Furthermore, it should be noted that additional overhead costs such as consulting, training; PR, etc. were not included in the study.

Assumption 11: Economic cycles, and current international economic crisis

A study with a medium-to long-term time horizon in the observation period, as represented by this study, cannot comment specifically on exact future or current business cycles. As a consequence, the present study - also because of the lack of available data - does not consider the current financial and economic crisis. In an economic crisis, especially if the financing of certain activities or actions (including households and businesses) are significantly negatively concerned, it is likely that there may be delays in the implementation of individual measures and strategies. Marino et al., (2011) who give an overview of the European energy service market mention that the economic crisis made the access to finance more difficult and has negatively influenced the initiation and development of new energy efficiency projects.

Assumption 12: Energy savings through renovation of old buildings

These calculations are based on energy performance certificates (heating demand calculation) and do not consider the individual user behaviour (expansion of living space, ventilation behaviour, heating characteristics, etc.). Studies show that the user's behaviour can significantly reduce the potential savings due to the heating requirements of building renovations (compare Sanders & Phillipson (2006) for a review of related papers). For this study, due to the lack of a publicly available comprehensive and statistically usable data material, no adjustment of the shifts of the heat consumption through a change in the individual user behaviour is considered.

3. Costs and realized energy savings of the strategies

In the following, the effects of the analysed energy efficiency strategies and the associated capital costs are presented.

3.1 Energetic effects of the strategies

Firstly, in Table 4, the expected savings under each strategy for the household sector, the service sector, and the production of physical goods are shown. The sum of savings in these sectors allows meeting the savings targets of 22,330 GWh, but the target is not achieved under all strategies. However, it should also be pointed out explicitly that Austria has already undertaken efforts to improve energy efficiency before 2008, and these early actions can be credited to the ESD target. The extent to which Austria can succeed in being allowed to count its pre-2008 energy efficiency improvements on to the target, contributes significantly to how ambitious the savings target has to be interpreted. Should the savings target solely have to be achieved by the strategies analysed here, and have to be realized only in the period 2008 to 2016, then only the strategies *IMF US mid*-and *IMF-US high* deliver overall savings that are high enough to meet the target.

Year	*MinStand*		*IMF mid*		*IMF-US mid*		*IMF high*		*IMF-US high*	
	HH	*Service/Goods*	*HH*	*Service/Goods*	*HH*	*Service/Goods*	*HH*	*Service/ Goods*	*HH*	*Service/Goods*
	in GWh									
2008	462	316	638	506	638	633	638	696	638	886
2009	925	633	1,315	1,013	1,315	1,266	1,315	1,392	1,315	1,772
2010	1,597	949	2,677	1,519	2,635	1,899	2,677	2,089	2,635	2,659
2011	2,288	1,266	4,213	2,026	4,048	2,533	4,213	2,786	4,048	3,546
2012	3,280	1,582	5,897	2,531	7,794	3,164	5,897	3,480	7,794	4,430
2013	4,350	1,898	7,777	3,037	10,329	3,796	7,777	4,176	10,329	5,315
2014	5,489	2,215	9,829	3,543	13,023	4,429	9,829	4,872	13,023	6,201
2015	6,589	2,531	12,102	4,049	15,741	5,062	12,102	5,568	15,741	7,086
2016	7,746	2,848	14,602	4,557	18,656	5,696	14,602	6,266	18,656	7,974
Total in 2016	**10,594**		**19,159**		**24,352**		**20,868**		**26,630**	

Table 4. Reductions in energy demand in the energy efficiency strategies, without early-actions.

3.2 Necessary expenditures

Table 5 shows the expected expenditure for the implementation of the respective energy efficiency strategies in the period 2008 to 2016. These expenditures are the sum of consumers' as well as corporate spending and all possible subsidies. Not included in these expenses are the expenses for the administration of the strategy.

For all strategies in the household sector, a strong rise in spending on energy efficiency measures is expected. Acceptance and implementation readiness is assumed to rise with the duration of the strategies. In the industrial and service sector, given assumptions about energy prices and a constant funding quota, no dynamic development is expected in the investment strategies. The impacts of the programs are therefore considered to yield a uniform effect over the whole observation period. The anticipated expenditure (sum of all expenditures between 2008 and 2016) lies between € 13.7 billion and € 34.8 billion.

How do these numbers compare to current Austrian activities? Between 1993 and 2008 3,781 projects were realized with the support of public funding that targeted the efficient use of energy. The overall investment amounted to 467 Mio. € and was support with 94 Mio. €. In the year 2008, funding of 22.1 Mio. € was given to 881 projects (see Bundesministerium für Land- und Forstwirtschaft, 2008).

	MinStand		*IMF mid*		*IMF-US mid*		*IMF high*		*IMF-US high*	
Year	*HH*	*Service/Goods*	*HH*	*Service/ Goods*	*HH*	*Service/Goods*	*HH*	*Service/Goods*	*HH*	*Service/ Goods*
					in Mio. €					
2008	648	325	992	519	992	649	992	714	992	909
2009	648	325	1,074	519	1,074	649	1,074	714	1,074	909
2010	648	325	2,010	519	1,654	649	2,010	714	1,654	909
2011	736	325	2,288	519	1,883	649	2,288	714	1,883	909
2012	1,370	325	2,552	519	3,707	649	2,552	714	3,707	909
2013	1,539	325	2,889	519	3,932	649	2,889	714	3,932	909
2014	1,666	325	3,221	519	4,148	649	3,221	714	4,148	909
2015	1,744	325	3,652	519	4,420	649	3,652	714	4,420	909
2016	1,779	325	4,125	519	4,768	649	4,125	714	4,768	909
Total	***10,778***	***2,925***	***22,803***	***4,671***	***26,578***	***5,841***	***22,803***	***6,426***	***26,578***	***8,181***

Table 5. Expenditures for the energy efficiency strategies; sum of private expenditures and public funding; HH: Households; Services/Goods: production of goods and services.

3.3 Costs of the strategies and expected reduction of energy costs

Table 6 shows the expected expenditure or investment in energy efficiency measures compared to the consequential reduction in expenditure for energy in the respective years. The expected expenditures exceed the expected reduction in energy bills in every year in the period under consideration. This mainly shows that the short horizon of the directive of 9 years, is suitable to force rapid action from the Member States, but is too short in connection with assessing the economic viability of energy efficiency programs.

Even though many energy efficiency measures are paid off from a business point of view after a few years, in the context of the ESD, efforts must be intensified until the last year of the Directive to achieve the targets. This necessarily results in an investment that exceeds the expected reduction of energy expenditure in the term of validity of the ESD.

	Min Stand		*IMF mid*		*IMF-US mid*		*IMF high*		*IMF-US high*	
Year	*Exp.*	*Cost Red.*	*Exp.*	*Cost Red.*	*Exp.*	*Cost Red.*	*Exp.*	*Cost Red.*	*Exp.*	*Cost Red.*
	in Mio. €									
2008	973	55	1,511	92	1,641	103	1,706	109	1,901	125
2009	973	110	1,593	188	1,723	211	1,788	222	1,983	255
2010	973	211	2,529	339	2,303	404	2,724	389	2,563	471
2011	1,061	298	2,807	500	2,532	587	3,002	567	2,791	676
2012	1,695	403	3,071	671	4,356	920	3,266	754	4,615	1,031
2013	1,863	516	3,408	858	4,581	1,176	3,603	958	4,841	1,309
2014	1,990	635	3,740	1,057	4,797	1,444	3,935	1,173	5,057	1,599
2015	2,069	753	4,171	1,270	5,069	1,708	4,366	1,403	5,329	1,889
2016	**2,104**	**875**	**4,644**	**1,497**	**5,417**	**1,992**	**4,839**	**1,647**	**5,677**	**2,191**

Table 6. Expenditures for energy efficiency strategies and induced reductions in expenditures for energy; Exp.: Expenditures for energy efficiency strategy; Cost Red.: Reduction of costs.

Table 7 presents the cost parameters of the analysed energy efficiency strategies in relation to each unit of energy. The average cost of energy-efficient technologies varies, depending on the strategy, from 129 to 143 cents per $kWh_{creditable}$ in 2016. This means that on average, from 2008 to 2016 135 cents must be invested for 1 $kWh_{creditable}$ to the savings target in 2016. The installed energy efficiency measures are in effect before, during and after 2016 so that these costs are distributed on the life of the energy efficiency measure in a business perspective (and not with regard to the policy). Especially with long-life measures often high rates of return are noted.

Strategy	*Total credible energy savings in 2016*	*Total expenditure for energy efficiency measures*	*of which public funding*	*Total expenditures per kWh credible*	*of which public funding per kWh credible*
Minimum Standards	10,593 GWh	13,699 Mio. €	876 Mio. €	129.3 Cent	6.4 Cent
IMF mid	19,156 GWh	27,476 Mio. €	13,442 Mio. €	143.4 Cent	48.9 Cent
IMF-US mid	24,349 GWh	32,420 Mio. €	12,708 Mio. €	133.1 Cent	39.2 Cent
IMF high	20,865 GWh	29,229 Mio. €	14,785 Mio. €	140.1 Cent	50.6 Cent
IMF-US high	26,627 GWh	34,756 Mio. €	14,460 Mio. €	130.5 Cent	41.6 Cent

Table 7. Creditable reductions in the target year of 2016, Policy and monetary parameters of the respective strategy, own calculations.

Table 7 shows the expected eligible savings of each strategy in the target year 2016 and the cost parameters of their implementation. The total expenditure per $kWh_{creditable}$ describes the necessary investments to reduce energy consumption compared to the business-as-usual scenario by exactly 1 kWh in 2016. Since this unit of energy is, however, saved not only in 2016 (even if it is credited to the policy goal only in 2016), these measures save energy costs depending on the nature and durability of action for up to 25 years.

It is obvious, that the scenario *Min Stand* is the most cost-effective option regarding the total necessary funding. The total expenditure required for the implementation is almost entirely covered by the consumer and companies - and not the funding bodies. Acceptance and financial viability and social equity, however need to be considered critically. Furthermore, this strategy even though it ambitiously adopts minimum standards can only achieve savings in the target year of the directive of 10,593 GWh. Even with full credit for early actions claimed by Austria and early savings amounting to approximately 9,600 GWh, the savings target of 22,330 GWh is only achievable by an unexpected positive development of end-use efficiency in the period 2008 to 2016 with this strategy.

3.4 Energy savings by energy source

Table 8 shows the expected savings of the analysed strategies separated by energy sources. Of particular interest are the percentage shares of the savings generated by the strategies of the (policy relevant) consumption in 2005. While in the most ambitious strategy, *IMF-US high*, a reduction of energy consumption for all energy sources by more than 10% is expected compared to the business-as-usual scenario, it is not possible to reduce vehicle fuel consumption in sufficient dimensions. As described before, structural measures are essential for the reduction of fuel consumption in the transport sector, so that a real turnaround can be achieved. Especially in the field of freight logistics, but also in the field of public transport, the horizon until 2016 appears to be insufficiently short.

Furthermore, it is obvious that electricity consumption which has an average yearly growth rate of about 1.5% (10-year average in Austria), cannot be reduced below the level of 2008 even in the ambitious *IMF-US high*. For energy sources like coal and coal products, a significant decrease in the absolute consumption at 2008 levels is to be expected.

Strategy	*Electricity*	*Natural Gas*	*Vehicle Fuels*	*Fuel oil*	*Renewables*	*Coal and Coal Products*	*Others*
				in GWh			
Min Stan	3,160	2,385	108	1,820	2,658	246	216
IMF mid	4,100	2,860	2,892	3,512	4,994	434	364
IMF-US mid	6,210	3,769	2,935	4,251	6,189	540	454
IMF high	4,677	3,351	2,957	3,589	5,315	511	463
IMF-US high	6,980	4,424	3,022	4,354	6,618	642	586
	in % of the end-energy consumption 2005 which is relevant for the ESD savings target						
Min Stan	7.0%	6.0%	0.1%	6.8%	6.4%	12.0%	6.2%
IMF mid	9.1%	7.2%	2.9%	13.1%	12.0%	21.2%	10.4%
IMF-US mid	13.7%	9.4%	3.0%	15.8%	14.9%	26.3%	12.9%
IMF high	10.3%	8.4%	3.0%	13.4%	12.8%	24.9%	13.2%
IMF-US high	15.4%	11.1%	3.1%	16.2%	15.9%	31.3%	16.7%

Table 8. expected creditable energy savings in the target year 2016 realized by the energy efficiency strategies.

Our results, especially concerning shifts in the energy source composition of end-energy demand, are solemnly based on the analysis of the impacts of the ESD. Apart from the improvement of energy efficiency, the European Union has set two other major energy policy goals: 1) a reduction in EU greenhouse gas emissions of at least 20% below 1990 levels as well as a 20% share of EU energy consumption to come from renewable resources. With a look at the CO_2- and renewable energy targets, complementary end-energy efficiency improvement policies would support target achievements, especially if the energy efficiency measures - as presented in our approach - are realized cost-effectively. As the interaction of the various European Union energy policies is not the subject of our analysis, we will not go into detail on this matter but point the reader to Philibert (2011).

4. Economic effects of the strategies

In the following, the effects of the energy efficiency strategies on the three variables *gross domestic product*, *employment* and *energy-related public revenue* are presented.

When interpreting the results it must be stressed again that they are to be understood in comparison to the business-as-usual scenario. The reported values thus give the difference to the development of the observed variables in each year, in a situation without the implementation of the analysed energy efficiency strategies.

4.1 MOVE

For the assessment of the economic effects of the energy efficiency measures, the economic model MOVE is used. MOVE is a macroeconomic time series model and was developed for modelling the effects of economic and energetic measures in Austria. MOVE provides forecasts for macro-economic parameters like GDP, budget, inflation, employment, investment, energetic and non-energetic consumption, and increase of transport costs for different economic sectors. MOVE includes an economy module accounting for 12 economic sectors, an energy module, and an ecological module. The energy part is not limited to energy consumption; it also accounts for energy flows to generate secondary energy sources, the production of primary energy as well as energy imports and exports. A further feature of the MOVE model is an emission tool showing the changes in all relevant type of emissions, since energy use implies an impact on the environment in most cases (compare Tichler (2008) for a detailed presentation of MOVE).

4.2 Economic effects of the energy efficiency strategies

The analysis shows that the two key economic variables, *gross domestic product* and *employment* are positively influenced by increased investment in energy-efficient technologies. An investment in this area is effective in the domestic economy in an above-average level, as these measures are often provided by labour-intensive industries. The energy-efficiency strategies analysed here can therefore be understood not only as instruments to achieve the objectives of the ESD, but rather as an integral part of a future-oriented and ecologically sustainable economic policy.

The effects of the analysed strategies on the gross domestic product are shown in Table 9. They are each calculated as the difference to the business-as-usual scenario, that is the development of the gross domestic product without the investments made to improve energy efficiency that were initiated by the strategies under consideration. To interpret these results in each case the assumptions about the financing of the measures are of great importance. With the exception of the strategy *MinStand* - which transmits almost all the investment burden on the consumer - common to all strategies is that both the end user and the state are acting as financing partners. The actual amount of investment on the part of consumers and the state's participation in the form of grants is given in Table 7.

Effects on the gross domestic product (in% of nominal GDP at an assumed 2.5% annual growth)										
	MinStand		*IMF mid*		*IMF-US mid*		*IMF high*		*IMF-US high*	
	in %									
2008	0.29	0.47	0.53	0.77	0.62	0.89	0.74	0.98	0.88	1.17
2009	0.39	0.61	0.72	1.02	0.82	1.16	0.97	1.29	1.15	1.51
2010	0.44	0.67	1.07	1.39	1.03	1.46	1.35	1.69	1.39	1.84
2011	0.47	0.72	1.26	1.6	1.15	1.63	1.55	1.9	1.52	2.02
2012	0.55	0.92	1.38	1.75	1.61	2.2	1.67	2.05	1.99	2.6
2013	0.6	1.04	1.5	1.89	1.82	2.48	1.79	2.19	2.2	2.87
2014	0.64	1.11	1.61	2.02	1.93	2.62	1.91	2.33	2.31	3.02
2015	0.66	1.16	1.73	2.17	2.01	2.74	2.03	2.48	2.4	3.14
2016	0.68	1.19	1.86	2.33	2.1	2.86	2.16	2.65	2.49	3.26
*Effects on the labour market (change in% points to the unemployment rate *)*										
2008	-0.08	-0.15	-0.15	-0.25	-0.18	-0.29	-0.22	-0.32	-0.26	-0.39
2009	-0.11	-0.2	-0.21	-0.32	-0.24	-0.38	-0.29	-0.42	-0.34	-0.49
2010	-0.11	-0.2	-0.26	-0.38	-0.26	-0.41	-0.34	-0.47	-0.37	-0.53
2011	-0.12	-0.2	-0.31	-0.44	-0.28	-0.45	-0.39	-0.53	-0.39	-0.57
2012	-0.13	-0.25	-0.33	-0.47	-0.36	-0.57	-0.41	-0.56	-0.48	-0.7
2013	-0.15	-0.3	-0.36	-0.51	-0.44	-0.68	-0.44	-0.61	-0.56	-0.81
2014	-0.16	-0.31	-0.39	-0.56	-0.45	-0.7	-0.48	-0.65	-0.58	-0.84
2015	-0.17	-0.33	-0.43	-0.61	-0.48	-0.74	-0.51	-0.71	-0.61	-0.89
2016	-0.17	-0.35	-0.47	-0.67	-0.51	-0.79	-0.56	-0.77	-0.64	-0.94
*Effects on energy-related taxes and charges (in% of revenue value 2008 *)*										
2008	-0.07	-0.05	-0.12	-0.08	-0.13	-0.08	-0.11	-0.07	-0.11	-0.07
2009	-0.2	-0.18	-0.34	-0.31	-0.38	-0.34	-0.38	-0.34	-0.43	-0.39
2010	-0.41	-0.39	-0.77	-0.73	-1.01	-0.97	-0.84	-0.8	-1	-0.96
2011	-0.59	-0.57	-1.21	-1.18	-1.62	-1.57	-1.33	-1.29	-1.57	-1.51
2012	-0.82	-0.78	-1.69	-1.65	-2.5	-2.43	-1.84	-1.8	-2.41	-2.34
2013	-1.06	-1.02	-2.22	-2.17	-3.28	-3.21	-2.41	-2.36	-3.13	-3.06
2014	-1.31	-1.26	-2.78	-2.73	-4.12	-4.04	-3.01	-2.96	-3.9	-3.82
2015	-1.57	-1.52	-3.37	-3.32	-4.94	-4.86	-3.64	-3.58	-4.66	-4.58
2016	-1.83	-1.78	-3.99	-3.94	-5.8	-5.72	-4.3	-4.24	-5.45	-5.36

Table 9. Economic effects of the energy efficiency strategies.

The financing of government subsidies can always be done in two ways: 1) by a shift in expenditures and 2) by a corresponding (new) debt. The calculations presented here are based on alternative 2), ie there is no response or adaptation of other public expenditures and revenues.

The financing of measures on the side of end customers (households and businesses) can occur in three ways: 1) by access to capital reserves and savings, and 2) through debt financing, and 3) a reduction of expenditure for other goods / investment. As part of the calculations performed here, two different rates of funding from reserves / savings and loans are presented: one scenario with a 33% ratio, and a scenario with a 100% ratio.

A comprehensive overview of ways of financing energy efficiency measures and overcoming barriers is given in Rezessy & Bertoldi (2010).

With a 33% funding ratio, 67% of the investments needed on behave of the end customers (ie after deduction of subsidies) are funded by a reduction in the expenditures for other goods. At a 100% rate, there is no such shift in expenditures and the investment in energy efficiency is fully mobilized additionally. Table 9 shows that all strategies have a significantly positive impact on gross domestic product, and thus on economic growth.

From the figures it is apparent that the funding of measures on the part of end users through recourse to reserves / savings or by borrowing has a higher positive influence on GDP, than a high proportion of financing through expenditures shifting. It is also obvious that the progressive alignment of strategies with a constant intensification of investment incentives accelerates economic growth annually until 2016. Again it should be noted, that the GDP effects in Table 9 are only fully effective if the subsidies are to 100% financed by (re-) debt. The simultaneous implementation of the strategies described here and political measures for counter financing, such as the greening of the tax system - keyword: CO_2 tax - requires separate consideration.

Table 9 also shows that all strategies have a significantly positive impact on employment. Investments in energy efficiency are effective in labour-intensive industries. Furthermore, by the savings in energy costs, additional funds are released which supports a further shift of consumption and investment from the less labour-intensive energy production and energy import to more labour-intensive market areas.

Overall, the strategies analysed here show that achieving a reduction in the unemployment rate in the order of 0.5 to 1.0%-points below the level of the business-as-usual scenario is possible by 2016. The presented energy efficiency strategies, especially those with intensification of funding, can thus be understood as a contribution to long-term recovery of the labour market.

Furthermore, Table 9 illustrates the effects of the analysed strategies on energy-related public revenues between 2008 and 2016. These effects are again each calculated as the difference to the business-as-usual scenario. The state receives significant public revenues from the taxation of energy sources. The reduction of energy consumption therefore inevitably leads to a reduction of public revenues. The strategies analysed here furthermore aim at substantially reducing fossil fuels consumption that is subject to higher taxation than the consumption of renewable energy sources.

The shortfall in public revenue resulting from this adds up to about 5% of the public revenues from energy-related taxes and fees in the business-as-usual scenario.

5. Conclusions

Achieving the objectives of the European Directive 2006/32/EC on energy efficiency is possible for Austria, but requires an intensification of existing activities. The potential for energy efficiency measures analysed in this report exceed Austria's savings target by almost 50%, but these potentials need to be exploited by 2/3 to reach the target. A potential utilization of 2/3 is only possible by maximizing the efforts and by intensive funding policy instruments.

The analysis shows that funding policy instruments are most effective when combined with the introduction of minimum standards. Such systems, minimum standards for energy efficiency of appliances and energy consumption, both in households, in the business sector and the Public sector, displace inefficient consumption units and fully support the development of new energy-efficient technologies. Special attention should be placed on a European solution for the introduction of minimum standards to broaden the pressure on manufacturers of energy-consuming equipment and systems. Besides the presentation of potential savings and the effectiveness of different intensities of monetary support of energy efficiency measures, this study also examines the economic effects of these measures. The analysed energy-efficiency strategies are not only effective tools for reducing energy consumption and emissions of greenhouse gases, but also as programs to boost the economy and the labour market. This is mainly due to the fact that effective investments in energy efficiency have to be made in labour-intensive sectors. Furthermore, these energy savings also have a positive effect on disposable income, which supports a further shift of consumption and investment in the less labour-intensive energy production and energy imports into more labour-intensive sectors.

From the results of this study, the authors derive the following priority to-do's for Austria's energy policy to achieve the objectives of the directive on energy efficiency and to sustainably reduce energy consumption in Austria.

Rapid decision-making for an energy efficiency strategy: The results presented in this study clearly show that the objectives of the ESD are accomplishable only if the actions are carried out rapidly and progressive. In this sense, a strategy has to be defined as soon as possible so that its entry into force is not further delayed.

To-Do 2: Rapid decision for an energy efficiency strategy. Only with undelayed action, the success of the strategies described can be achieved. Especially the comparably short duration of the ESD needs to be kept in mind.

Precise preparation and definition of individual instruments: In this study we describe possible strategies and the nature of the necessary minimum standards. But a detailed elaboration of the funding instruments and limits for energy-consuming appliances cannot be made in a study like this. Especially when it comes to minimum standards there barely are adequate structures (outside the construction area) to establish such a system in Austria. To achieve the savings targets of the ESD in due time, a rapid implementation of such structures is necessary. Above all, a broad involvement of stakeholders is mandatory.

To-Do 3.a: Creation of an orderly process for the preparation of specific educational tools and limits for energy consuming appliances and their applications with the involvement of stakeholders.

To-Do 3.b: Rapid creation of structures for the on-going establishment and administration of minimum standards for energy consuming appliances and systems. The Japanese Top Runner program can serve as a model, if minimum standards are introduced at European level.

Social compatibility of the measures examined: The introduction of minimum standards forces households to purchase equipment that is probably more expensive than one with lower energy efficiency. On the one hand this may lead to a situation in which inefficient units remain longer in the household; on the other hand, it may also lead to additional financial burdens for economically less well-off households. To avoid jeopardizing the introduction of minimum standards, keeping the measures socially balanced, these aspects must be taken into account in developing the instruments and measures.

To-Do 4: Consideration of social aspects in the design of minimum standards and funding instruments.

Financing of Grants: depending on the energy efficiency strategy chosen, state funding is needed on different levels. In this study, it is not expected that this funding is cost-neutral - that is through redeployment of the budget. As a way to finance the subsidies, the introduction of a tax on greenhouse gases is proposed. In this context it is essential to determine both the economic effects and the achieved additional savings in advance to ascertain that the economically most efficient strategy is chosen.

To-Do 5: Development of instruments to finance the subsidies and accurate analysis of the economic effects of these instruments to optimize their impact and compatibility.

6. Acknowledgment

The authors work on this study was funded by the Austrian Climate and Energy fund under the grant no. 815587.

7. References

Adensam, H., Bogner, T., Geissler, S., Groß, M., Hofmann, M., Krawinkler, R., et al. (2010). *Methoden zur richtlinienkonformen Bewertung der Zielerreichung gemäß Energieeffizienz- und Energiedienstleistungsrichtlinie* 2006/32/EG.

Boonekamp, P. (2006). *Evaluation of methods used to determine realized energy savings.* Energy Policy(34), 3977-3992.

Brüggemann, A. (2005). *KfW-Befragung zu den Hemmnissen und Erfolgsfaktoren von Energieeffizienz in Unternehmen.* Publikation der Volkswirtschaftlichen Abteilung, KfW-Bankengruppe.

Bundesministerium für Land- und Forstwirtschaft (2008). *Umweltförderungen des Bundes.*

Bundesministerium für Wirtschaft und Arbeit (Federal Ministry of Economy (2007). *1. Nationaler Energieeffizienz-Aktionsplan (1st National Energy Efficiency Action Plan).* Abgerufen am 5. August 2011 von http://www.monitoringstelle.at.

Emnid, T. (2005). *dena Unternehmensbefragung zum Thema Energieeffizienz.* DENA.

Galvin, R. (2010). *Thermal upgrades of existing homes in Germany: The building code, subsidies, and economic efficiency.* Energy and Buildings(42), S. 834–844.

Irrek, W., Thomas, S., Barthel, C., Kirchner, L., Spitzner, M., Wagner, O., et al. (2004). *Energieeffizienz-Fonds.* Wuppertal Institut für Klima, Umwelt, Energie GmbH, Wuppertal.

Jellinek, R. (2009). *Energy Efficiency Policies and Measures in Austria in 2007.* Wien: Austrian Energy Agency.

Liimatainen, H., & Pöllänen, M. (2010). *Trends of energy efficiency in Finnish road freight transport 1995-2009 and forecast to 2016.* Energy Policy(38), 7676-7686.

Marino, A., Bertoldi, P., Rezessy, S., & Boza-Kiss, B. (2011). *A snapshot of the European energy service market in 2010 and policy recommendations to foster a further market development.* Energy Policy(39), 6190-6198.

Martin Jakob, M., Primas, A., & Jochem, E. (2001). *Erneuerungsverhalten im Bereich Wohngebäude.* Working paper series 01-09, Center for Energy Policy and Economics, Zürich.

Philibert, C. (2011). *The Interaction of Policies for Renewable Energy and Climate.* Paris: IEA.

Reichl, J., & Kollmann, A. (2010). *Strategic homogenisation of energy efficiency measures: an approach to improve the efficiency and reduce the costs of the quantification of energy savings.* Energy Efficiency(3), S. 189-201.

Reichl, J., & Kollmann, A. (2010). *The baseline in bottom-up energy efficiency and saving calculations – A concept for its formalisation and a discussion of relevant options. Applied* Energy(88), S. 422-431.

Reichl, J., Kollmann, A., Tichler, R., Pakhomova, N., Goers, S., Moser, S., et al. (2010). *Analyse der Wirkungsmechanismen von Endenergieeffizienz-Maßnahmen und Entwicklung geeigneter Strategien für die Selektion ökonomisch-effizienter Maßnahmenpakete.*

Rezessy, S., & Bertoldi, P. (2010). *Financing Energy Efficiency: Forging the Link between Financing and Project Implementation.* ISPRA: Joint Research Centre of the European Commission.

Rezessy, S., & Bertoldi, P. (2011). *Voluntary agreements in the field of energy efficiency and emission reduction: Review and analysis of experiences in the European Union.* Energy Policy(39), 7121-7129.

Sanders, C., & Phillipson, M. (2006). *Review of Differences between Measured and Theoretical Energy Savings for Insulation Measures.* Glasgow Caledonian University, Centre for Research on Indoor Climate and Health.

Schiellerup, P. (2002). *An examination of the effectiveness of the EU minimum standard.* Energy Policy(30), S. 327–332.

Tanaka, K. (2011). *Review of policies and measures for energy efficiency in industry sector.* Energy Policy(39), 6532-6550.

Thomas, S., Boonekamp, P., Vreuls, H., Broc, J.-S., Bosseboeuf, d., Lapillonne, B., et al. (2009). *How much energy saving is 1% per year? We still don't know, but we know better how to find out.* ECEEE Summer Study 2009. Act! Innovate! Deliver!

Reducing Energy Demand Sustainably, *ISBN 978-91-633-4454-1.* La Colle sur Loup, France.

Tichler, R. (2008). *Optimale Energiepreise und Effekte von Energiepreisveränderungen in der oberösterreichischen Volksiwrtschaft.* Linz: Energieinstitut an der Johannes Kepler Universität.

4

Promoting Increased Energy Efficiency in Smart Grids by Empowerment of Customers

Rune Gustavsson
KTH School of Electrical Engineering
Sweden

1. Introduction

The *EU Climate and Energy package* is setting the 20-20-20 targets of future energy systems by 2020 and will change the landscape of future energy system in Europe and worldwide. The package sets the following objectives;

- reduce greenhouse gas emissions by 20%,
- increase the share of renewable energy to 20, and
- to make a 20% improvement in Energy Efficiency.

The European Parliament has continuously supported these goals. It is a common understanding that this change will require a transition from *monopolised hierarchically controlled* Power networks to *customer oriented Smart Grids* operating in *deregulated energy markets.* However, this poses several regulatory, organizational and technical challenges to be identified and addressed. To that end several international Smart Grid projects have been launched worldwide in EU, the US, and in China.

In a follow-up Proposal by EC is the *Energy Efficiency Plan 2011.* That is a Directive of the European Parliament and for the Council on Energy Efficiency[1] based on assessing results and findings so far versus the stated 20-20-20 targets. The assessments shows that the *major concerns* related to the expected fulfilment of the Energy packet target is related to meeting the Energy Efficiency (EE) goals, Figure 1.

Increased Energy Efficiency is expected be enabled by the following actions and their combinations:

- *Active intelligent* Distribution grids incorporating vast amounts of RES
- *Active and empowered* customers
- *Active operations* of markets

The term "active" in this setting refers to "smart" or "intelligent". We suggest that careful selection and implementation of Multi-Agent technologies is crucial for Services supporting active customers, intelligent distribution grids and the two other recommended actions [2] (Section 2).

---

[1] Home page: http://ec.europa.eu/energy/efficiency/index_en.htm

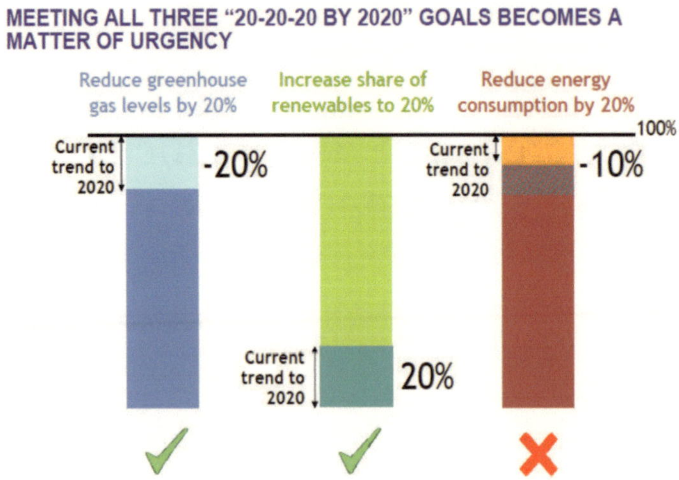

Fig. 1. Assessments of progress towards 20-20-20 objectives.

The Energy Efficiency Plan document makes the following explicit observations and recommendations concerning *Energy Service Companies* (ESCOs), *Customer empowerment, Smart grids and Smart meters* as a backbone for *smart appliances* thus enabling increased EE.

"Smart grids and smart meters will serve as a backbone for smart appliances, adding to the energy savings obtained by buying more energy efficient appliances. New services will emerge around the development of smart grids, permitting ESCOs and ICT providers to offer services to consumers for tracking their energy consumption at frequent intervals (through channels like the internet or mobile phones) and making it possible for energy bills to indicate consumption for individual appliances. Beyond the benefits for household consumers, the availability of exact consumption data through smart meters will stimulate the demand for energy services by companies and public authorities, allowing ESCOs to offer credible energy performance contracts to deliver reduced energy consumption. Smart grids, meters and appliances will allow consumers to choose to permit their appliances to be activated at moments when off peak cheaper energy supply or abundant wind and solar power are available - in exchange for financial incentives. Finally, they will offer consumers the convenience and energy saving potential of turning appliances on and off remotely.

Delivering on this potential requires appropriate standards for meters and appliances, and obligations for suppliers to provide consumers with appropriate information (e.g. clear billing) about their energy consumption including access to advice on how to make their consumption less energy intensive and thus reduce their costs. To this end, the Commission will propose adequate measures to ensure that technological innovation, including the roll-out of smart grids and smart meters fulfils this function. These measures will include minimum requirements on the content and format of information provision and services.

Further, the Commission needs to ensure that energy labels (energy performance certificates) and standards for buildings and appliances reflect, where appropriate, the incorporation of technology that makes appliances and buildings "smart grid ready" and capable of being seamlessly integrated into the smart grid and smart meter infrastructure. Appliances such as fridges, freezers and heat pumps could be the first to be tackled.

Improvements to the energy performance of devices used by consumers - such as appliance and smart meters - should play a greater role in monitoring or optimizing their energy consumption, allowing for possible cost savings. To this end the Commission will ensure that consumer interests are properly taken into account in technical work on labelling, energy saving information, metering and the use of ICT.

The Commission will therefore research consumer behaviour and purchasing attitudes and pre-test alternative policy solutions on consumers to identify those which are likely to bring about desired behavioural change. It will also consult consumer organisations at the early stage of the process. Consumers need clear, precise and up to date information on their energy consumption - something that is rarely available today. For example, only 47% of consumers are currently aware of how much energy they consume. They also need trustworthy advice on the costs and benefits of energy efficiency investments. The Commission will address all of this in revising the legislative framework for energy efficiency policy."

A summary of benefits for consumers through provision of proposed tailored Energy Services and Information is given in Figure 2,

The remaining part of the chapter is organized as follows.

- Section 2 Smart grids - Architecture, Stakeholders, Challenges, Barriers and Solutions. We follow up some of the issues addressed in previous Section 1 Background.
- Section 3 Requirements engineering and Validation. In this section we are addressing some aspects of system Interoperability and customer acceptance based on found requirements. We also outline some engineering principles of trustworthy Smart grids.
- Section 4 Customer empowerment. In this section we give a high-level architectural view of customers in a Smart grid. In particular we differentiate between different types of customers (consumers), that is, consumers of home - centric energy-based services and consumers of business-based energy services. This section also outlines challenges related to implementations of the Energy Efficiency Plan 2011 refereed to above.
- Section 5 Information processing systems and sharing and protection of information. We address issues related to information sharing (interoperability) and information protection (security and integrity).
- Section 6 Cyber security and privacy. Assuring information security and privacy in Smart grids are major concerns for acceptance by all stakeholders, mot the least, customers.
- Section 7 Use cases, This section illustrate our approach and findings by two use cases related to empowerment of customers. That is, use cases related to Smart homes and Green energy. Use cases are the drivers of successful business cases of Smart grids. The use cases are basis for requirement engineering and validations of interoperability. Based on the use cases we identify corresponding Service Level Agreements (SLAs) supporting coordination of stakeholders providing the intended services respecting agreed upon Quality of Service (QoS). To enable reconfiguration of use cases supporting different types of Self healing and/or new business opportunities we also discuss the role of meta modeling in Smart grids.
- Section 8 Tools and Environments, We briefly describe some tools and environments supporting structured requirement engineering, design, development, monitoring,

maintenance and assessments of Smart grid pilots. Proper tools and environments are crucial for successful approaches of Smart grid solutions.

- Section 9 Conclusions and future work.

The chapter ends with a list of *References*.

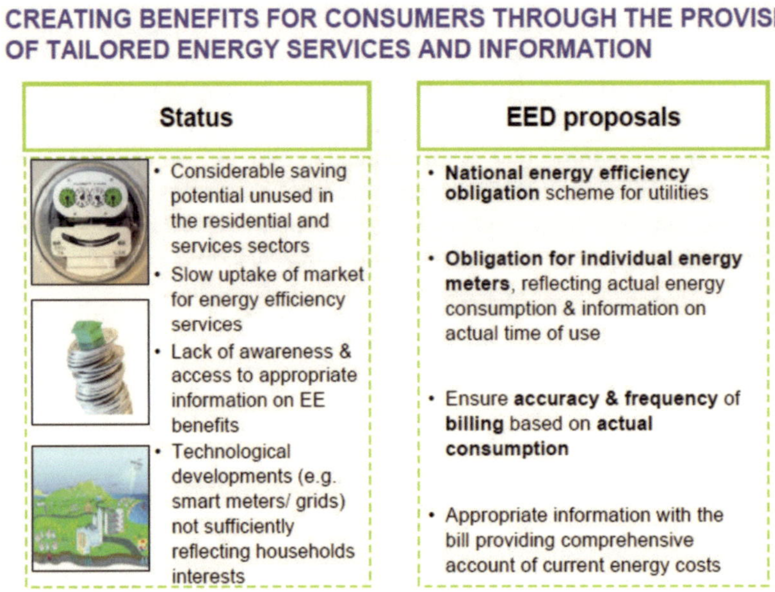

Fig. 2. Promoting Active Consumers participating in EE efforts.

2. Smart grids – architectures, stakeholders, challenges, barriers and solutions

Modeling and Optimization of Sustainable Energy Systems poses several challenges. A starting point for our investigation is the *NIST Framework and Roadmap for future Smart Grids* [30]. The document identifies *seven domains* within the Smart Grid—*Transmission, Distribution, Operations, Bulk Generation, Markets, Customer*, and *Service Provider*. A Smart Grid domain is a high-level grouping of organizations, buildings, individuals, systems, devices, or other *actors* with similar objectives and relying on—or participating in—similar types of applications. Across the seven domains, numerous actors will capture, transmit, store, edit, and process the information necessary for Smart Grid applications.

The NIST Framework shows that future Smart grids support the following two flows:

- *Power (electrical) flows* (generation, transmission and distribution)
- *Information processing flows* (collecting, processing and distribution). The information flow has the following *two* objectives:
  - Monitoring and control of the energy flows (c.f., classical SCADA)
  - Monitoring and control of future and new energy based services in Smart grids

Both flows require protection and ancillary systems to support interoperability and Quality of Service.

The NIST Framework has also been adopted by IEEE in IEEE Guide for Smart Grid Interoperability for Energy Technology and Information Technology Operation with Information with the Electric Power System (EPS), End-Use Applications, and Loads (2011)[2]. The guide gives architectures for the different Domains as well as identified interfaces between them related to the information flows. Based on the ANSI/ISA-99 (now IEC 62443.02.01) protocols we can introduce levels of segmentation and traffic control inside control systems creating multiple separated Zones and conduits supporting "defense in the depth strategies" providing Cyber security (Section 6).

In general, actors in the same domain have similar objectives. To enable Smart Grid functionality, the actors in a particular domain often interact with actors in other domains, as shown in Figure 3. However, communications within the same domain may not necessarily have similar characteristics and requirements. For example, for communications or information within the Customer domain, simple meter readings have simple characteristics and requirements such as a meter communicates with a specific utility head-end system, while a customer portals need to have multiple users accessing it at the same time and to different accounts (*role based access*).

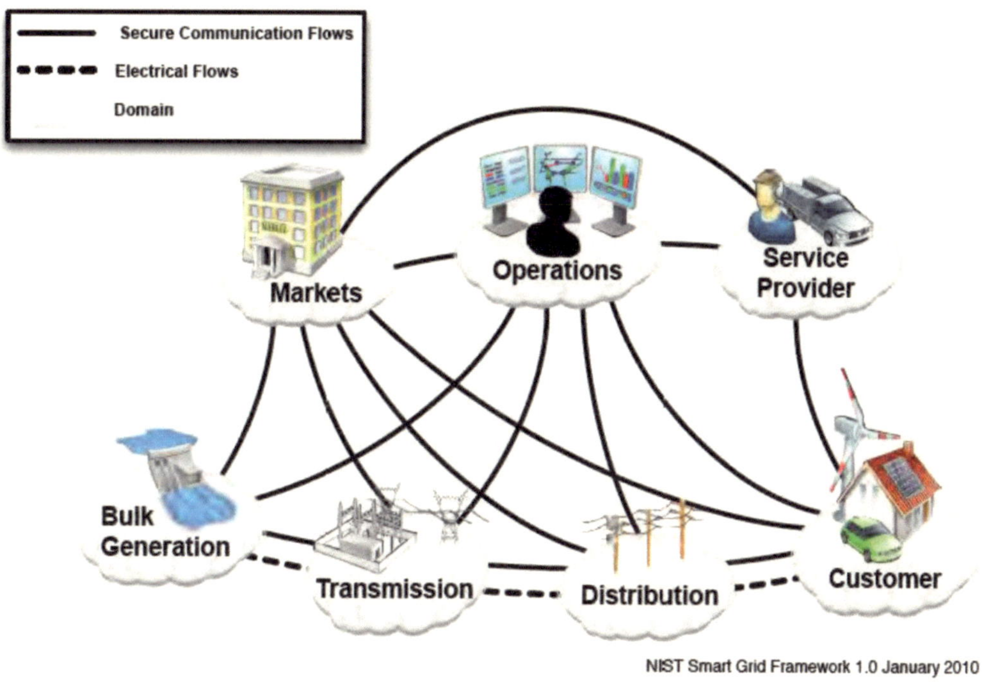

Fig. 3. NIST Framework of Smart grids as composed of seven domains.

Moreover, particular domains may contain components of other domains. For instance, the ten Independent System Operators and Regional Transmission Organizations (ISOs/RTOs) in North America have actors in both the Markets and Operations domains. Similarly, a distribution utility is not entirely contained within the Distribution domain—it is likely to

[2] Home page: http://standards.ieee.org/findstds/standard/2030-2011.html

contain actors in the Operations domain, such as a distribution management system, and in the Customer domain, such as meters.

The core of present day power systems is the *EMS – Energy Management System* monitoring and controlling the performance of production, transport and distribution of power. The EMS is supported by *SCADA systems* for monitoring and control as well as *support systems* for protection, optimization and billing. The stakeholders are *TSOs (Transmission System Operator)* responsible for the generation and transport grid, *DSOs (Distribution System Operators)* responsible for the distribution grid and service to *customers*.

We note the following inherent increased complexities of Smart grids compared to present day grids:

- *Increased number of Stakeholders* with new capabilities, roles and responsibilities
- *Increased complexity* of the electric grid that now has to incorporate vast amounts of *Renewable Energy Resources (RES)* and *Distributed Generation (DER)*
- A need of a *complementary ICT* information management system to support information exchange and sharing between stakeholders providing energy-based services in a secure and trusted way

More efforts are required by all stakeholders to enable improved future energy production, distribution and usage [1]. Furthermore, novel business models are required to support the transition from today's situation to Smart grid based on markets of energy-based services [7, 12, 13, 14]. Providing novel services based on setting of customer comfort is one identified area by the EU projects FENIX[3] and SEESGEN-ICT[4]. Of particular interest to us is the *Customer Domain* given in Figure 4. From this figure it follows that we can have two kinds of Customers, *Home-based Consumers* and *Business-based Consumers* that can interact with actors of other domains in their business processes.

Proper Empowerment of those two kinds of customers can largely contribute to increased Energy Efficiency. However, we will in this chapter concentrate on the Home-based Consumer (Customer).

Given the central role of Customers and supporting infrastructures in the Energy Efficiency Plan 2011 (above) it is also clear that Customers could be important partners with other Stakeholders and ESCOs in providing substantial increases in EE in the future. However, clearly this take-up will to a large degree depend on the *trustworthiness of the supporting infrastructures*.

To support further investigations in this chapter along those lines we introduce different coordination views of the Smart grid Socio-technical system, Figure 5. One of the identified barriers of Smart grids is the inherent inflexibility to add more and flexible business cases in today's power systems, due to tight coupling of the Grid hardware and SCADA systems [25, 26].

Future energy systems will become robust and efficient with a careful *supplement* of the SCADA systems with specifically designed and implemented ICT systems *ensuring Smart*

[3] http://www.fenix-project.org/

[4] http://seesgen-ict.erse-web.it/

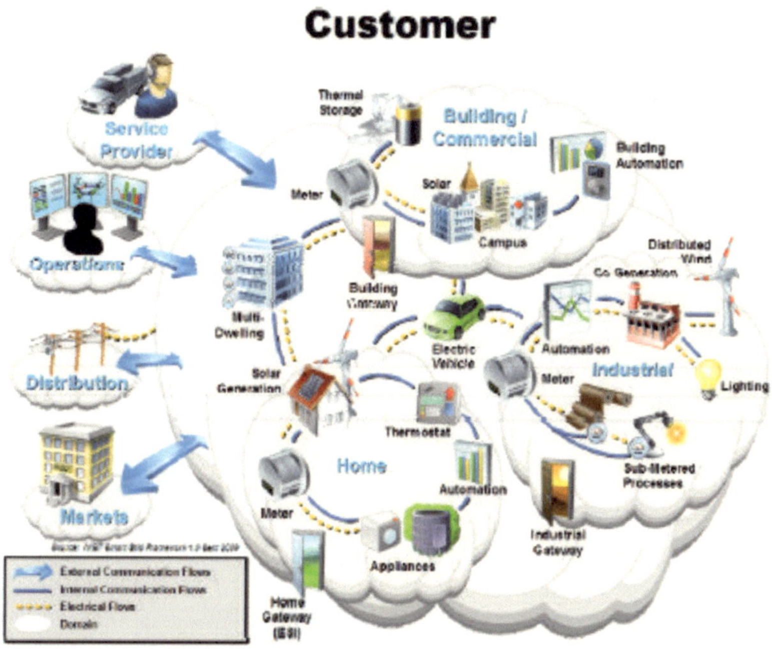

Fig. 4. The Customer Domain including different types of Consumers.

grid Interoperability (Figure 6). In this Chapter, we expand some novel ideas introduced in SEESGEN-ICT [1], deliverables D3-2, D3-3 and to assess identified barriers and implement relevant ICT solutions for future pilots of Smart Grids.

Figure 5 captures some of the challenges identified in [1] To cope with coordination of different sets of stakeholders we propose introduction of *Service Level Agreements* (SLAs). SLAs are identified and set up supporting business cases with identified stakeholders. *Key Performance Indicators* (KPIs) are identified and monitored during deliverance of the agreed upon energy based services. In Figure 5 some challenges are identified in transforming Business models, related Stakeholders and relevant Infrastructures into SLAs. Challenges related to supporting ICT infrastructures are also indicated, e.g., *real time dependencies and data management*. In particular the *cross-point* between high-level and low-level system views is indicated. That is, high-level business oriented infrastructures such as web-services and low-level infrastructures supporting distributed systems such as OPC have to be suitable *interoperable* (Section 7 Use cases), It should be noted that for classical SCADA systems, we have a *bottom up integration* of signals from the EMS system to the top operator level for management of monitoring and control with no need to address this cross-point.

We will come back to Figure 4 and Figure 5 in Section 3 Requirements engineering and Validation as well as in Section 4 Customer empowerment and Section 5 Information processing systems and sharing and protection of information. Issues related to the cross-point will be addressed in Section 7 Uses Cases. Basically, we will introduce Service Level Agreements (SLAs) to coordinate stakeholders providing energy-based services by selected clusters. Furthermore monitoring of Key Performance Indicators (KPIs) of SLAs will enable validating selected Interoperability criteria and Quality of Service (QoS).

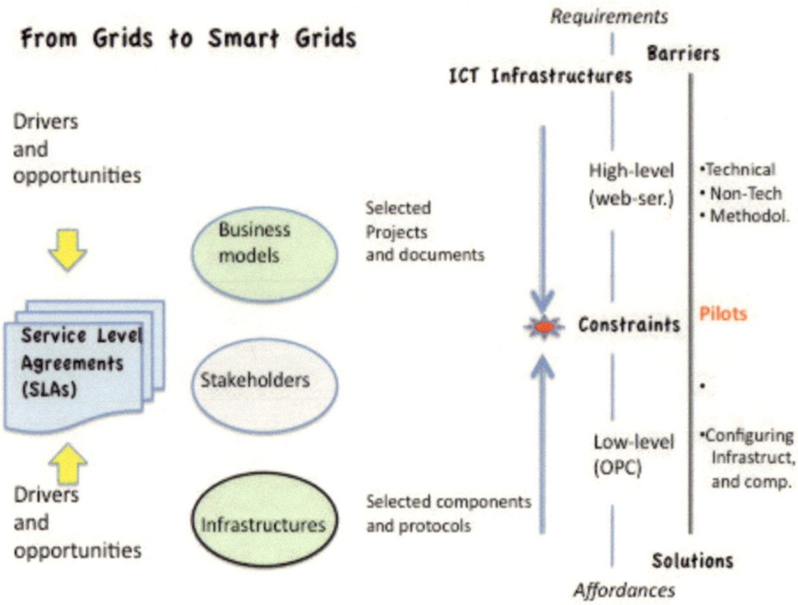

Fig. 5. Coordination aspects of future Smart grids.

One of the identified *barriers* in this transition of systems into Smart grids, is the inflexibility to add more and flexible business processes into today's power systems, due to the tight vertical (in voltage levels) coupling of the Grid hardware and operator stations by SCADA (Sensory control and Data Acquisition) presented in [4, 5]. Future energy systems will hopefully become robust and efficient with careful integration of ICT (Information Communication and Technology). However, due to the difficult to predict nature of changes in the infrastructures and business models as well as regulatory uncertainties, the pace of uptake and implementation of Smart Grid is hard to predict.

The scope and purpose of monitoring has lately, however, changed towards *ensuring interoperability* of systems due to *increased complexity* of the systems at hand. As a matter of fact, analysis of larger blackouts, such as the August 14, 2003 blackout in northeast United Stated and Ontario, has shown that this kind of event can be attributed to sequences of *interoperability failures*[5] of related systems. The systemic property of *Interoperability* has been proposed by organisations such as NIST[6] and GridWise[7] in the US and is also adopted by EU.

NIST has the following definition of Interoperability:

"The capability of two or more networks, systems, devices, applications or components to exchange and readily use information, securely, effectively and with little or no

[5] GridWise Architecture Council Report: *Reliability Benefits of Interoperability*, 2009, pp. 7 – 9.

[6] Home page: http://www.nist.gov/smartgrid/

[7] Home page: http://www.gridwiseac.org/

inconvenience to the user. The system will share a common meaning of the exchanged information and this information will elicit agreed-upon types of response." [NIST[8]]

The following additional requirements are put forward by GridWise Architecture Council (GWAC[9]):

- "an agreed expectation for the response of the information exchange"
- "requisite quality of service in information exchange: reliability, fidelity, security"
- "the results of such interactions enables a larger system capability that transcends the local perspective of each participating subsystem"

GWAC has proposed the following *Interoperability Framework* consisting of three *Interoperability Categories* (Technical, Informational and Organizational) and *Crosscutting Issues* related to *non-functional* requirements, such as *Energy Efficiency* (EE). The Technical interoperability is enabled by *proper open protocols and network technologies.* In order to *verify or validate* interoperability of Smart grid systems we have to identify *suitable views* of those systems. We argue that such views can be provided and monitored by suitable *Service Level Agreements* (SLAs) [3, 14, 27, 28, 29, 30]. The SLAs will take into accounts business cases and involved stakeholders to assure relevant Quality of Service (QoS). That is, *also* take into account the *Informational and Organizational* categories of the GWAC Framework (Section 3).

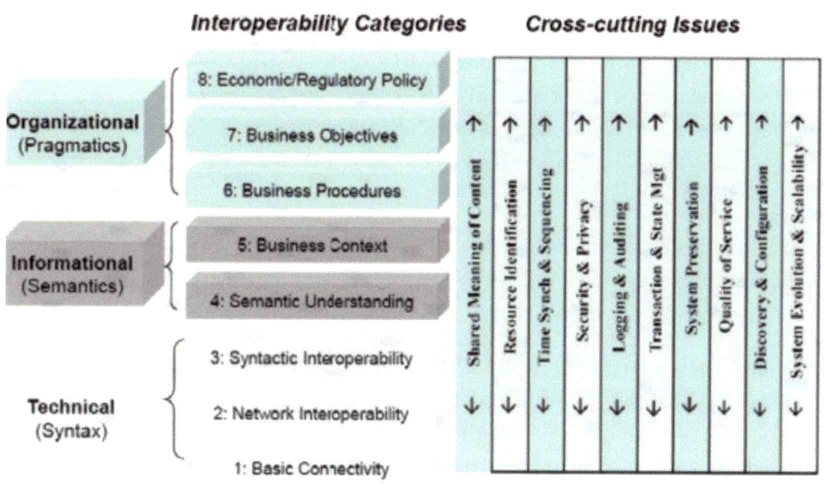

Fig. 6. GWAC Interoperability Framework with a layered set of Categories and non-functional Cross-cutting Issues.

The concept of "smart" in Smart Grids refers mostly to "Smart distribution grids" utilizing smart energy system components, empowered and active customers and flexible and resilient systems. This is enabled by a transition of today's hierarchical and mostly proprietary systems to open, loosely coupled and flexible service oriented systems. Obviously, flexible pattern oriented interaction models are here key enabler. In short, a transition to sustainable Smart Grids will benefit from utilization and use of agent

[8] Home page: http://www.nist.gov/smartgrid/

[9] Home page: http://www.gridwiseac.org/

technologies. We have during the last decade addressed different views on technologies underpinning design and implementation of Interoperability according to Figure 6. That is, *agent systems* [3, 5, 7] , *service oriented systems* [4, 6] , *critical infrastructures* [8, 9, 11, 16, 24], *experimental environments* [8, 9, 15, 24, 25, e, g], and SLAs [12, 26, 27, i]. In short, in order to design and validate interoperability of views of smart grids, we configure and monitor SLAs based on agent-oriented services. Design, implementation and validation are utilizing aggregation tools and experimental environments.

Introducing agent-based services or implementing *Service Oriented Multi-agent Systems (MAS)* facilitates taking into account intelligence or smartness of future Smart Grids. Agent technologies allow modelling systems as configurations of smart flexible components, e.g., *Active Network Management* (ANM) of Distribution Grids [18]. *The IEEE Power and Energy Society Multiagent Systems Working Group*[10] aims to promote the openness of agent architectures within the power domain. In this paper we argue that *Service Level Agreements (SLAs)* provides a *control and monitoring structure of MAS* assuring inter-operability and QoS.

3. Requirements engineering and validation

Proper Requirements Engineering is the underpinning of design, implementation, validation and maintenance of systems. This is especially important for future Smart grids due to the inherent complexities of *coordinating different sets of stakeholders* in *changing market environments* with *internal and external threats.* In short, among desired system requirements complementing selected functional and non-functional requirements, we also have to address different aspects of *self-healing in cases of breakdowns* of SLAs and methods to support *adjustments and reconfigurations of business cases.* In short we need to address *meta-models* of SLAs to allow *resilience and flexibility.*

In Section 5 Information processing systems and sharing and protection of information is addressed as enabling technologies of Interoperability. The important issues of security, privacy and vulnerabilities of Information processing systems (ICT and SCADA) will be further addressed in Section 6 *Cyber security and privacy.*

Figure 7 expands upon Figure 3 and depicts a composite high-level view of the actors within each of the Smart Grid domains. This high-level diagram is provided as a reference diagram. Actors are devices, systems, or programs that make decisions and exchange information necessary for executing applications within the Smart Grid. The analysis and discussions later in this chapter expand upon this high-level diagram and include logical interfaces between actors and domains.

From Figure 3 and Figure 7 we have a high-level conceptual architecture of Smart grids domains and the power and information flows between domains and actors. Furthermore, in Figure 7 some high level components and their interfaces are depicted.

Requirements engineering concerns meeting *functional requirements and constraints* of the flows as well as means of *instrumentation, monitoring and controlling* sensors and actuators regarding *Key Performance Indicators* (KPIs).

---

[10] Home page: http://ewh.ieee.org/mu/pes-mas/

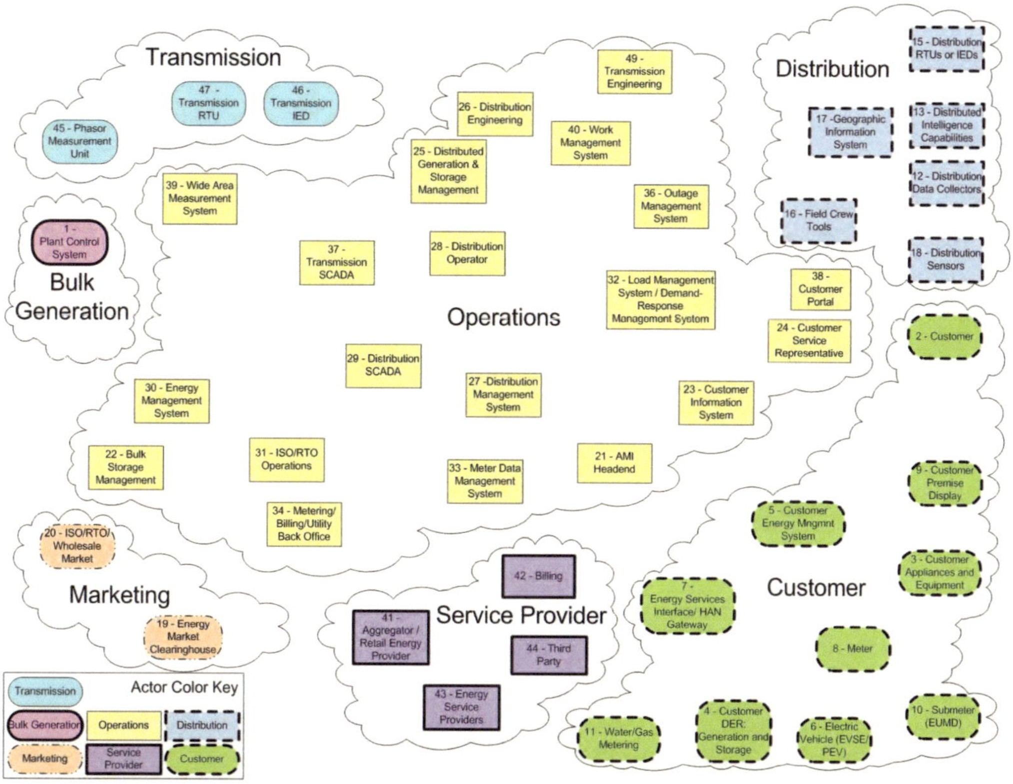

Fig. 7. Composite High-level View of the Actors within Each of the Smart Grid Domains.

In Figure 5 of Section 2 we illustrate some views and challenges to be addressed in requirements engineering for Smart grids. It captures the *constraints that Interoperability criteria* (Figure 6) have to *comply with concerns* from the different Interoperability Categories *with affordance*s from the supporting infrastructures, for instance at Customers premises (Section 4 and Section 7). The high-level demands have to meet the low-level affordances and constraints. Compare with Figure 6, where the Organizational and Information Categories have to meet the Technical Category as well as relevant Cross cutting Issues while meeting interoperability goals. The Service Level Agreements involves concerned stakeholders as well as Key Performance Indicators (KPI) to be monitored to ensure interoperability and Quality of Service. In [1] a set of *Barriers* and *Solutions* related to Smart Grids have been identified as well as suitable ICT systems.

It should be noted that in the classical grid the information processing system is by and large proprietary SCADA systems. The SCADA system *integrates* information *bottom-up* from the grid to the system operators and allows sending control signals *top-down* to the grid components: In short, a *stove-pipe* system! In Smart Grids we also need ICT systems providing horizontal as well as vertical interoperable information exchange between stakeholders (Section 5). In Figure 5 the *interactio*n between low-level and high-level SLAs is indicated. We will address this challenge later in the case study *Customer empowerment* (Section 4 and Section 7).

Identified challenges include coordination of sets of stakeholders and monitoring of processes related to new energy-based business processes. To that end we have advocated introduction and use of mechanisms based on Service Level Agreements (SLAs). Introduction of SLAs also enables a principled structuring of Smart Grid systems and related data flows [26, 27, i]. We will also address some of those topics in this Chapter.

Our approach towards modelling and implementing Smart Grid is utilizing carefully chosen infrastructures in flexible couplings and integrations (configurations) of system components [4, 6, a]. The configurations should support monitoring of processes by *clusters of SLAs implementing selected scenarios* of Smart Grids.

The *power flows* of power systems must fulfil the *electric constraints* balancing *active* and *reactive* power in real time. This balance in terms of KPIs includes voltage, current and frequency. Introduction of DERs will potentially affect voltage and frequency and has to be controlled. This issue will further be discussed in *Section 7 Use cases.*

Furthermore power could be of *two types*: *DC* or *AC*. *Transformation* between *power types* and *voltage levels* (low, medium, high) will assure proper functioning of the grid infrastructures.

The trustworthy information processing and information sharing system of Figure 3 has also to take into account *similar constraints* as the power system. Firstly the information can be modelled in the following three different *types* [9]:

- I_1, *Information payload*. This is the information type shared among stakeholders.
- I_2, *Communication related information*. This type of information that manages the networking tasks, e.g., by middleware.
- I_3, *Processing information*. Running code of executable tasks for stakeholders. Cyber attacks are targeting at manipulating I_3 by exploiting vulnerabilities (Section 6).

Usually, information security focuses on protection of type Information payload since it primary concerns *Confidentiality, Integrity, and Availability* (CIA) of stakeholder centric information. However, external cyber attacks usually try to take control of the processes by *manipulating* the run-time stack containing code (I_3) (e.g., Buffer overflow). We can also have directed attacks on the SCADA system itself (e.g., the worm Stuxnet[11]) (Section 6).

Obviously, the different information types I_2 and I_3 can be *compared with Reactive and Active power of AC grids*. They divide the processing and electric power into two parts that must interact to enable working systems. In the case of Information systems, given a fixed amount of computational power, there is a *trade-off* between communication processing and task solving processing allowing a trade-off between communication costs and local task solving capacity in distributed systems. We also only have only one chargeable part in each case (active power and task processing). The reactive power is used to maintain energy balance during changing loads, but is not directly chargeable to customers. However, the customers have to pay transmission and distribution cost beside the energy consumption costs, For information systems there is in the same way separate costs for *communication (networking)* and *customer services*.

[11] http://en.wikipedia.org/wiki/Stuxnet

Finally, as with power systems, information can be transformed into *levels*. The GWAC *Interoperability Framework* (Figure 6) identifies the following eight *Interoperability Categories (levels)* that fall into the following categories (bottom up).

- *Technical*: Basic Connectivity, Network Interoperability, Syntactic Interoperability
- *Informational*: Semantic Understanding, Business Context
- *Organizational*: Business Procedures, Business Objectives, Economic/Regulatory Policy

The organizational categories emphasize the *pragmatic* aspects of interoperation. They represent the policy and *business drivers* of interoperation. The information categories emphasize the *semantic* aspects of interoperation. They focus on what information is being exchanged and its *meaning*. The technical categories emphasize the *syntax* or format of the information. They focus on how the information is *represented* within a *message exchange* and on the *communication medium*.

Information types consequently have *data formats* and *exchange protocols* for the technical categories. Interoperability on higher semantic categories has to be supported by *dialogue-based* protocols and *semantic annotations* (Section 4, Section 5, and Section 6).

Figure 5 complements Figure 3 and Figure 6 in providing a *structured approach* towards tool-based Design and Implementation and Validation of Smart grid *Pilots* (Section 8).

We can now *rephrase* the proposal Energy Efficiency Plan 2011 and recommendations of Section 1 (Figure 2) as follows. Future Smart grids should support:

- Empowerment of Customer to *automatically manage and control clusters* of customer-centric *smart appliances* and to dynamically change user profiles to *meet user preferences* and market models.
- *Trustworthy* and transparent business and use-case *information management* and *information exchange* with customers and other stakeholders.

Requirements engineering will *match* the relevant business case with stakeholders' capabilities and concerns with affordances from the infrastructures (Figure 5). A selection of relevant components of the Interoperability Categories together with a selection of relevant Cross- cutting Issues give the input to negotiating and setting up a suitable SLA. To support this activity we have developed a suitable tool (Section 8). I should be noted that adding the constraints (e.g., crosscutting issues) might result in that we do not have *any* initial solution. Addressing interoperability can add or delete constraints towards an acceptable solution. *Pilots and validations*, based on requirements, can now be addressed (Section 8).

The following Sections will address challenges related to the recommendations above in more detail.

4. Customer empowerment

The concepts and ideas of *Customer empowerment* are arguably the most revolutionary in the transition towards Smart grids (Section 1). The traditional roles of customers have been as *passive loads* of the grid. Sometimes even refereed to as "two holes in the wall". The billings by the DSOs have been based on *fixed tariffs* and *measured consumption at customer sites* done *at pre-defined* intervals using *Automatic Meters Reading* systems (AMR). During the 90ties, some efforts where made to allow DSOs to *control* the consumption of customers by introducing different schemas of *Demand Side Management (DSM).* By different time-based fixed tariffs the DSOs could to some extent *alter* the amount of energy delivered to customers during agreed upon conditions and intervals. During the beginning of this century a new generation of meters allowing *remote reading* were introduced and deployed in several countries. However, most of these meters only allowed *unidirectional* information flows from meter to DSOs. *Bidirectional* enabled metering systems allowing true communication between customers and DSOs is the focus of future *Smart Metering Systems* (SMS). In fact those mew kinds of Smart Meters could be seen as *Intelligent access tools* between *prosumers* (a mixture of producer and consumer) and other stakeholders in a local energy market (Figure 7).

The changing views of customers from passive "two holes in the wall" to active empowered stakeholders of future Smart grids requires changes of mindsets and implementation of supporting technologies and legal frameworks. These changes of mindsets are pre-conditions for acceptance and uptake of Smart grids. The underpinnings here are *trustworthiness, usefulness* and *added value.*

NIST provides the following definition of the Customer.

"Customer is an entity that pays for electrical goods or services. A customer of the utility, including customers who provide more power then they consume (prosumers)".

Customer empowerment aim at:

1. Identifying and eliminate or circumvent identified shortcomings by customers in pursuing selected tasks.
2. Enable trustworthy information exchange with other stakeholders and with smart appliances

The solutions are development of context sensitive tools and environments (item 1) and validation of appropriate views of Interoperability (item 2). We will address some of those aspects in Section 5 Information processing systems, Section 7 Use cases, and Section 8 Tools and Environments. As Use cases we address the following aspects of customer empowerment:

- Support for customer to *include/change* energy provided by *Renewable Energy Sources (RES/DER)* in the customer profile.
- Support for customer to take part in agreements in trustworthy curtailment of energy,

The first use case address increasing user involvement in selecting energy sources. The second use case illustrates trusted behaviour in service break-downs, Both are handled by setting up and monitoring suitable SLAs.

In both those use cases we have to take into the *dependency* between business cases and its impact by the physical status of the energy system (voltage control) (Figure 5).

From Figure 4 and Figure 7 we can identify some of the *Actors (tasks) related to the Customer Domain* of Smart grids. From the NISTIR 7628 document we have the following high- level listing of Actors, tasks, and infrastructure components based on Figure 7:

- *Customer Appliances and Equipment.* A device or instrument designed to perform a specific function, especially an electrical device, such as a toaster, for household use. An electric appliance or machinery that may have the ability to be monitored controlled and/or displayed.
- *Customer Distributed Energy Resources (DER) Generation and Storage.* Energy generation resources, such as solar or wind, used to generate and store energy (located on customer site) to interface to the controller (HAN - Home Area Network / BAN) to perform an energy related activity.
- *Customer Energy Management System (EMS).* An application service or device that communicates with devices in the home, This application service or device mat have interfaces to the meter to read usage data or the operations domain to get pricing or other information to make automated or manual decisions to control energy consumption more efficiently. The EMS may be a utility subscription service, a third party, offered service, a consumer-specified policy, a consumer-oriented device. Or a manual control by the utility or consumer
- *Customer Premises Display.* This device will enable customers to view their usage and cost data within their home or business
- *Sub-Meter - Energy Usage Metering Device (EUMD).* A meter connected after the main billing meter. It may not be a billing meter and is typically used for information monitoring purposes.
- *Electric Vehicle Service Element/Plug in Electric Vehicle (EVSE/PEV).* A vehicle primary driven by a rechargeable battery that may be recharged by plugging into the grid by recharging from a gasoline-driven generator.
- *Home Area Network Gateway (HAN Gateway).* An interface between the distribution, operations, service provider, and customer domain and the services within the customer domain.
- *Meter.* Point of sale device used for transfer of product and measuring usage from one domain/system to another.
- *Customer Premise Display.* This device will enable customers to view their usage and cost data within their home or business.
- *Sub-Meter – Energy Usage Metering Device (EUMD).* A meter connected to the main billing meter. It may or may not be a billing meter and is typically used for information-monitoring purposes.
- *Water/Gas Metering.* Point of sale device used for the transfer of product (water and gas) and measuring usage from one domain to another.

The most important Customer related infrastructures potentially supporting increased EE are selections of:

- Customer EMS
- HAN Gateway
- Customer DER Generation and Storage
- Smart devices and appliances

Configurations of those components of Figure 7 correspond to *Slices* of Figure 6 that can be recast in terms of Cloud Computing [28]. Cloud Computing comes basically in three types:

- Software as a Service (SaaS)

- Platform as a Service (PaaS)
- Infrastructure as a Service (IaaS)

A Current example of PaaS is *Windows Azure Platform*[12] a supplementary SaaS is Microsoft Online Services[13]. An other example is *Amazon Elastic Compute Cloud* EC2[14]. Cloud computing allows *resource sharing and outsourcing*. Well-known examples include sharing of large data centres that implies large IT-based increases in EE.

So far, it has been very few investigations of the impact on Cloud computing technologies on future Smart grids. Obviously Customer based infrastructures and services could benefit from *different types* of SaaS, PaaS or IaaS solutions with *selected combinations of stakeholders*. The selected customer actors could then play different roles in the corresponding Service Level Agreements.

The different Cloud Computing types correspond to *different slices* of the GWAC Interoperability Framework Figure 6. A SaaS offers an environment (device) with a *fixed* integration of Interoperability categories, allowing easy plug in of software services. An interesting trend is here Smart phones (e.g., iPhone) with Apps[15]. Paas and IaaS allows not only plug in of Apps but also customized configuration of higher-level Informational and/or Organizational Categories. Devices accessing a PaaS or IaaS thus support the user with a richer environment than a device accessing a PaaS (single service support).

It should be noted from Figure 5 that a *fixed physical infrastructure* could support *several separated virtual overlay infrastructures* allowing structures reuse of physical resources in virtual settings. Of course, there is a *price to pay* (*performance and security*) for this flexibility.

For a Customer accessing a set of Smart home actors he/she can choose to access each of them individually as smart services (Apps). However, due to complexities of management or lack of overview, this solution can create cognitive overloads and become contra-productive. We should aim at flexible grouping of actors with common interfaces to customers including support tools.

In Section 6 we take a selection of selected actors in addressing these challenges. But as a preparation, we need to take a deeper view on challenges and solutions related to customized information processing systems supporting information sharing and learning. The material in those sections takes into account referenced produced thesis work and papers related to several international and national R&D project from the R&D Group Societies of Computation (SoC) at Blekinge Institute of Technology (BTH) [22, 32, a, b, c, d, f, h].

5. Information processing systems and sharing and protection of information

The following basic relation, Figure 8, between *Information, Representation* and *Interpretation* is captured from [19, 20] on *meaning* of *Situations and attitudes*. It captures the relation between Information (data) and its Representation (text, video, graphics) that has to be

---

[12] Home page: http://www.microsoft.com/windowsazure/
[13] Home page: http://www.microsoft.com/online/
[14] Home page: http://aws.amazon.com/ec2/
[15] Home page: http://www.apple.com/iphone/apps-for-iphone/

Interpreted by an agent with certain *capabilities*. The meaning by the receiving agent of the Representation should reflect the intended meaning by the sending agent. To assist the receiving agent some times *empowering* tools are provided to support interpretation.

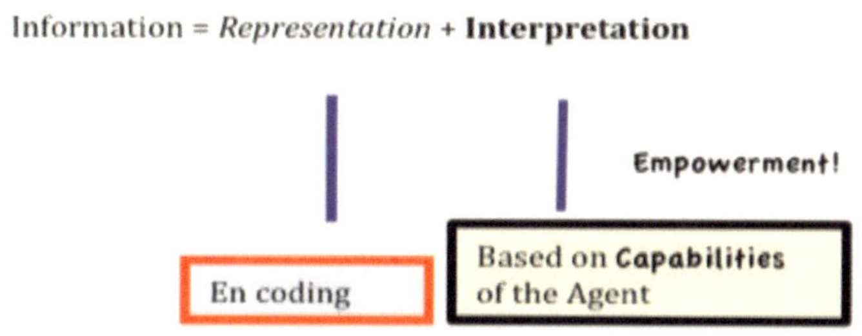

Fig. 8. Relations between Information (data), its representation and Interpreted meaning by an agent.

Basic challenges to be addressed in *designing and validating Interoperability* (Figure 6) are:

- Information sharing in teams requires *common understanding* (Interoperability)
- Information *hiding* is based on cryptographic representation
- Information *management needs tools* (*role and credential based access control* and data flow assurance). Enabling technologies are here *information models and structures.*

Addressing and validating Interoperability based on the frameworks by NIST, GWAC and IEEE requires standards related to connectivity (protocols), Information sharing and contextual issues (Figure 6). Protocols establish *horizontal connectivity* between stakeholders at different Interoperability Categories.

Standards supporting different aspects of Interoperability include *IEC protocols* IEC 61850[16] (connectivity) and IEC 61970-302 & 61968-11 (Common Information Model[17]), *MultiSpeak protocols* [18]and *OPC UA*[19] *protocols.*

The most important feature of those standards is the introduction of structured Information Models. In the IEC case the modeling techniques from UML (including XML Schema and RDF Resource Description Framework) are used with implementation techniques such as web services. MultiSpeak and OPC UA also use similar approaches. Common abstract data modeling supports translations between the different implementations. However, the different implementations have different performance profiles.

The following Figure 9 shows a CIM model of a Transformer consisting of four objects with attributes and attribute values, The UML models uses Class Hierarchies and UML Class diagrams together with *Inheritance, Associations, Aggregations* and *Compositions.*

From Figure 9 we can search for and find, or set, attributes belonging to a given object. This allows us to compose attributes belonging to a given object (stakeholder) from other objects.

---

[16] User Group: http://iec61850.ucaiug.org/default.aspx
[17] CIM User Group: http://cimug.ucaiug.org/default.aspx
[18] Home Page: http://www.multispeak.org/Pages/default.aspx
[19] OPC UA Home Page:
http://www.opcfoundation.org/Default.aspx/01_about/UA.asp?MID=AboutOPC

Fig. 9. CIM model of the object Transformer with four sub objects and attributes.

In fact we can give a semantic mapping of object attributes onto other attribute. Hence a *semantic mapping of attributes both horizontally and vertically* across the Interoperability Framework of Figure 6.

Modeling the relevant CIM components (Figure 9) gives us basic *Representations* of Information to be *interpreted and validated* by the different stakeholders to ensure Interoperability according to Figure 6 and Figure 8. The CIM models have to be *interpreted* with the correct semantics, given by the *competencies* and supporting *tools* of the stakeholders in case.

To address these challenge we have developed a methodology supporting *trustworthy engineering* of complex systems. The main components of our model for *engineering trustworthy systems* are as follows (Figure 10).

In setting up a Service Level Agreement (SLA) among stakeholders in delivering a energy based service we will use a negotiation tool for that purpose (Section 8 Tools and Environments). To allow for Interoperability we use the GWAC Framework of Figure 6. The different stakeholders bring upfront their *concerns* regarding selected Interoperability Categories and restricting Cross-cutting Issues.

During the negotiating phase agreements are found among stakeholders on *Trust aspects* to be monitored during deliverable of services. An important finding here is to agree upon the

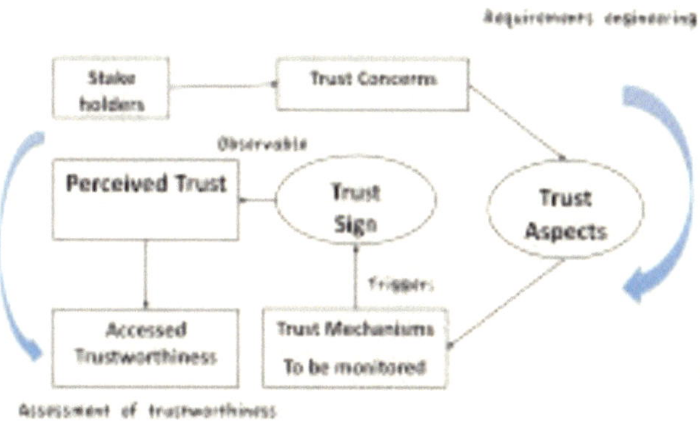

Fig. 10. Main components of Engineering trustworthiness.

shared meaning of related Key Performance Indicators (Figure 8!), Selected *Trust aspects* are translated into *Trust Mechanisms* to be implemented and monitored, The status of the trust mechanisms are implemented as *Trust signs* to be observed by related stakeholders. The trust signs are implemented using the *CIM modeling tools. Observing and interpreting* those signs can enable a shared awareness of the state of the system by the stakeholders. The design can then be *validated* against the chosen business case. The common understanding of the signs ensures that we have *interoperability and trustworthiness.* The following Figure 11 summarizes the achieved solution to *interoperability and shared semantics* illustrated by the red vertical line between two signs.

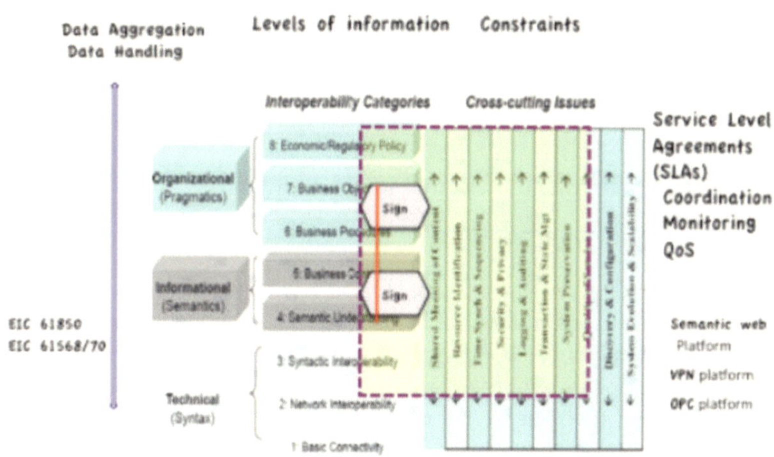

Fig. 11. The Interoperability Framework revisited. The vertical red line between signs illustrate verified interoperability.

In the process we have *identified a common semantics* for stakeholders in a *selected context.* Furthermore we can design and implement *empowerment tools* and *validation procedures* based on those findings (Figure 8).

To summarize: The output of the SLA negotiation process is a specification of a business process with *identified effect* (mission statement) delivered by identified stakeholders (roles, capabilities, concerns) and with indentified KPIs assuring trustworthiness and QoS to be monitored. The requirements also constitute a *meta-model* of the business process. This allows for identification and implementation of *self-healing mechanisms* maintaining a selected set of KPIs as invariants. Meta models also allows adjustments of SLAs, again maintaining some invariant properties, reflecting *controlled flexibility* of business cases.

6. Cyber security and privacy

Assuring Cyber security and privacy are arguably the most challenging and demanding tasks underpinning trustworthiness and thus acceptance and uptake of Smart grids. Both concepts are examples of cross-cutting issues of the GWAC Framework (Figure 10). Of particular importance are those issues in cases of Customer empowerment. Addressing aspects of cyber security and identifying countermeasures we can use the model supporting engineering of trustworthiness given i Figure 10.

The following Figure 12 sets the scene for Cyber attacks and other vulnerability related threats to Smart grids and its stakeholders.

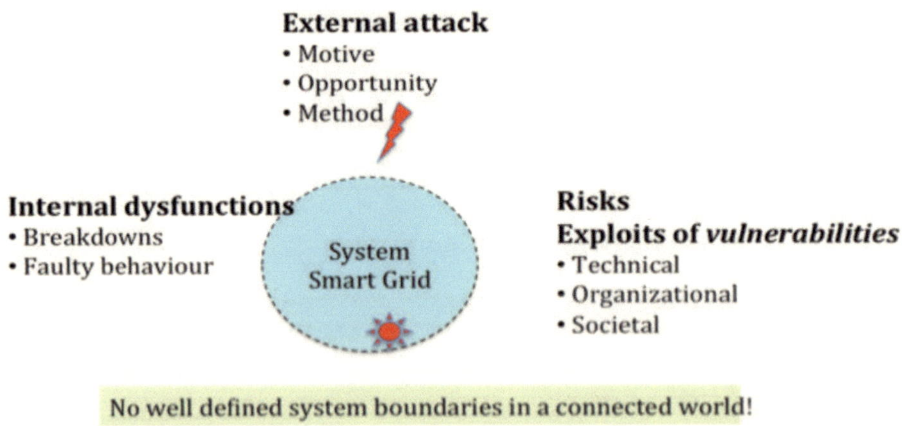

Fig. 12. Threats and exploitable weaknesses of Smart grids.

Threats could be realized given the Motive, Opportunity and Methods to exploit system vulnerabilities or by system dysfunctional behaviors. To cope with those threats we can do high-level attack three analysis or bottom up system hardening or a combination. In either case the core problem appears when a component, software, agent get improper access to data storage or exchange (Section 5). To remedy this several access control policies have been proposed.

To cope with these and related cyber threats The White House has issued a *National Strategy for Trusted Identities in Cyberspace (INSTIC)* in April 2011.The Strategy vision is:

Individuals and organizations utilize secure, efficient, easy-to-use, and interoperable identity solutions to access online services in a manner that promotes confidence, privacy, choices, and innovation.

The US Federal Government is initiating two short term actions to implement the Strategy. These are to:

- Develop an implementation Roadmap
- Establish a National Program Office (NPO)

The main purpose is to provide means and policies supporting trusted *Role based Access Control* with mechanisms supporting *revocation of identities and credentials.*

Issues related to security threats involve new kinds of recent attacks (since 2010). That is *Advanced Persistent Attacks* (APT), or advanced and targeted cyber attacks on infrastructures (*sabotage, business intelligence, thefts*).

Examples include:

- *Stuxnet* - industrial sabotage of Siemens (Distributed Control Systems) DCS in Iran
- *Ghostnet* - theft of diplomatic information
- *Aurora* - theft of source code and IPR at Google
- *Night Dragon* - industrial and commercial intelligence of large oil companies
- *PS3/PSN attack* - business sabotage on Sony Play Station Networks

Also under attack:

- RSA
- Intellicorp

These kinds of targeted attacks, of course, also pose cyber threats via systems aiming at empowering the customer, such as AMI systems.

With thousands of workstations and servers under management, most enterprises have little to no way to effectively make sure they are free of malware and Advanced Persistent Threats (APTs). APTs are broadly defined as sophisticated, targeted attacks (as opposed to botnets, banking Trojans and other broad-based threats) that rely heavily on unknown (zero-day) vulnerabilities and delivery via social engineering.

The reminding part of the section will give a short summary and lessons learned from the Stuxnet attack followed by highlights from a report from McAffe (August 2011) on *Operation Shade RAT* (Remote Access Tools). We also give a short overview of challenges related to privacy and security. The Section ends with some recent technologies to successfully address APT and RAT threats.

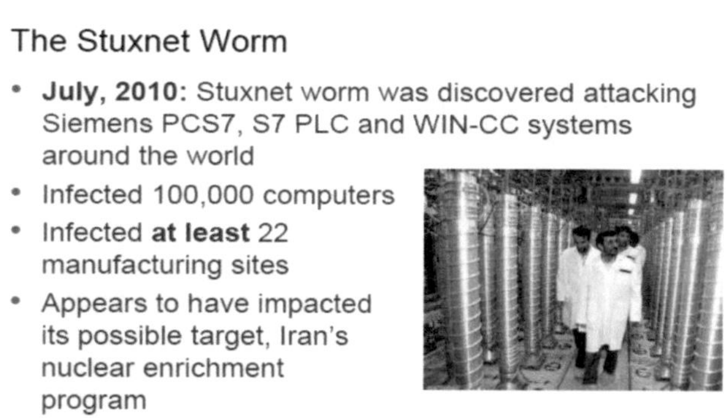

Fig. 13. The attacks by the Stuxnet worm.

The Stuxnet attack has been analyzed, for instance, by its detector Symantecs in the *W32. Stuxnet Dossier*[20]. The following Figure 14 gives the different propagation methods used by the worm. The starting point was an infected USB flash drive, followed by attacks on Local area Networks including SQL connections. The final steps of the attack were on Siemens WnCC and STWP 7 files.

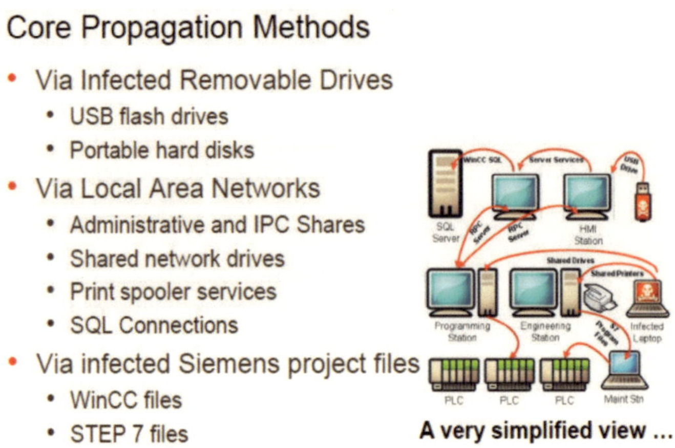

Fig. 14. Propagation methods of the Stuxnet worm.

Some lessons learned are:

- A modern Industry Control System (ICS) is highly complex and interconnected
- Multiple potential pwthways exists from the outside world to the process controllers
- Assuming an air-gap between ICS and corporate networks is unrealistic
- Focusing security efforts on a few obvious pathways (such as USB storage drives or the Enterprise/ICS firewall) is a flawed defense
- A perimeter defense is not enough (firewalls)
- We must harden the entire system
- We need Defense in Depth

Siemens gives an illustration of Defense in Depth in a recent (post Stuxnet) report, Figure 15. In the Figure are architectures for the functions Data Exchange, Real time data, Real time controlling, Maintenance, and Support given.

Threats to *privacy* are based on *misuse* of information related to individuals. This information can either be *generated* as footprints by individuals or *gathered* by tracing the behavior of the individual in cyber space using different kinds of *spyware*. Theft of identities or credentials is often a staring point in attacks on privacy.

A background to NSTIC Proposal is hat *identity theft* is costly, inconvenient and all-too common:

- In 2010, 8.1 million U.S. adults were the victims of identity theft or fraud, with total costs of $37 billion.

---

[20] Home page: http://www.symantec.com/connect/blogs/w32stuxnet-dossier

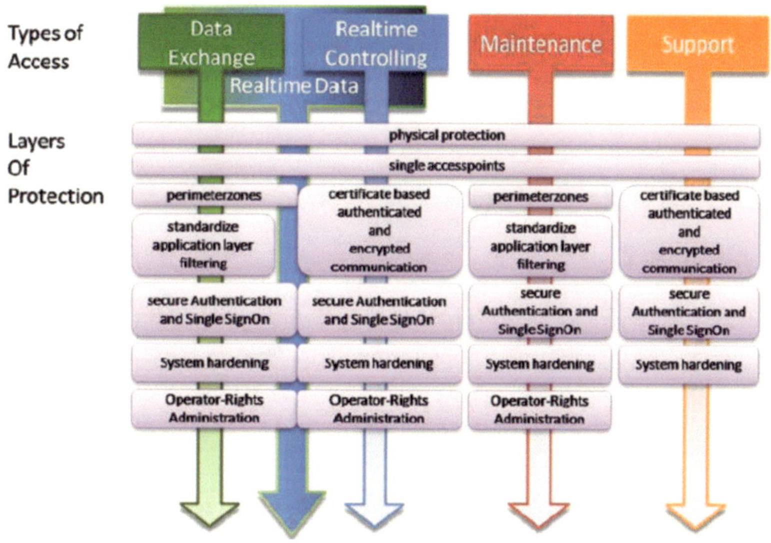

Fig. 15. Defense in depth architectures for different functionalities of Smart grids.

- The average out-of-pocket loss of identity theft in 2008 was $631 per incident
- Consumers reported spending an average of 59 hours recovering from a "new account" instance of ID theft.

Threats on information (information security) is usually described by the CIA model as attacks on *protection* of:

- Confidentiality
- Integrity
- Availability

Privacy is then an aspect of confidentiality. Information integrity assures that information is not tampered. Availability assures access by stakeholders to information. Obviously the CIA highlights three potentially conflicting concerns. The inherent conflict between information sharing and information protection is also illustrated in Figure 8, In short, there is *no generic "best" solution to proper information sharing and information security*, We can only aim at trustworthy system interoperability (Figure 11).

Due to this *inherent conflicting complexity* there are a wealth of R&D efforts on these subjects. A good source is the IEEE journal *Security & Privacy*. From our perspective, trustworthiness of SLAs based on context of business cases (use cases), is a *promising strategy* to gain more context dependant views on cyber security and privacy (Figure 11).

Empowered stakeholders, for example customers, have access to tools that could *potentially* be used for remote access (*Remote Access Tools – RAT*). Stuxnet was exploiting vulnerabilities to configure a *remote access and attack* tool. The following facts are condensed from the recent McAffee[21] White paper (August 2011) *Revealed: Operation Shady RAT* based on collecting and analyzing logs from a Command & Control (CC) server *used by intruders* since

[21] Home page: http://www.mcaffe.com/

2006. Figure 16 gives an overview of attacked organizations, companies and agencies by RAT. The attacks had duration of months to years and were not detected.

Fig. 16. Log based RAT attacks on selected organizations, companies or agencies.

- Vast amounts of data (petabytes) has been *lost* to (unknown) users
- The loss represent a *massive economic threat* to individual companies and industries and even countries that face the prospect of decreased economic growth in a suddenly more competitive landscape and the loss of jobs in industries that lose out to unscrupulous competitors in other part of the world

The McAffe report illustrates a large amount of undetected cyber attacks. Methods to identify attacks rely on *detection of known attack signatures* by firewalls or other techniques. However, the numbers of identified signatures are growing almost exponentially as is illustrated by the following Figure 17.

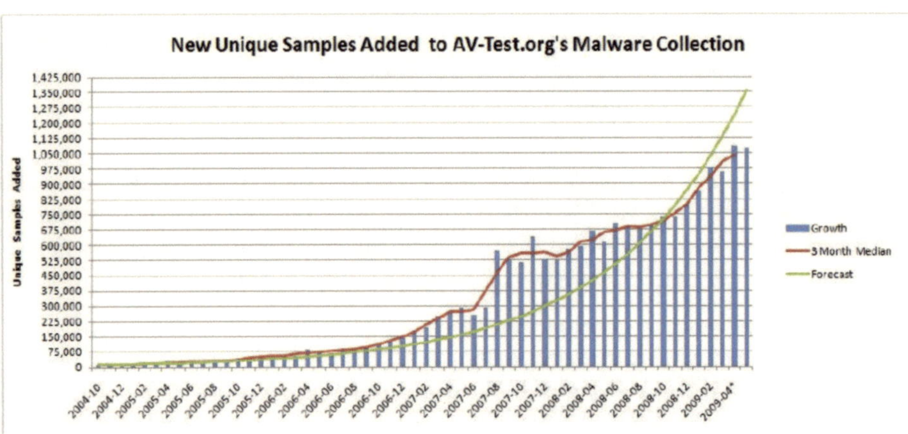

Fig. 17. A graph showing the number of unique new identified malware signatures over time.

The two last figures illustrate the potentially rapid increasing cyber number of attacks in general and more precisely the attacks on Smart grid infrastructures manifested by APT attacks based on RAT. Figure 17 illustrates also the increasing difficulties in pursuing threat analysis and/or attack detections by firewalls.

New technologies supporting increasing resilience of Smart grids include methodologies supporting *analysis and design of in-depth-defense* such as *Attack/Consequence Funnel* and *Last Line of Defense*.

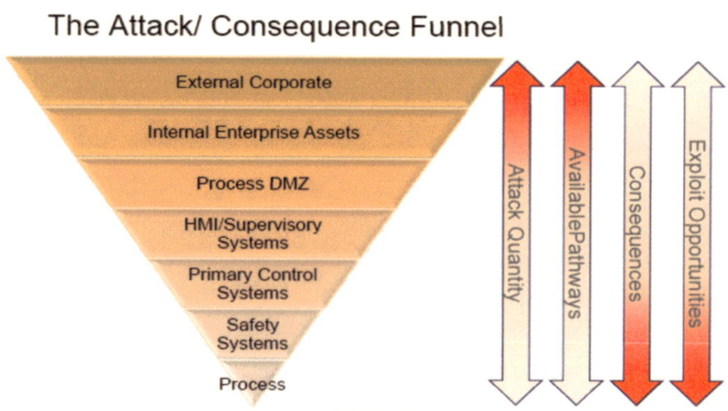

Fig. 18. The Attack/Consequence Funnel where the reddish arrow ends indicate increasing numbers.

The following Figure 19 illustrates means and reasons for securing the Last -Line-of-Defense of critical systems such as Smart Grids.

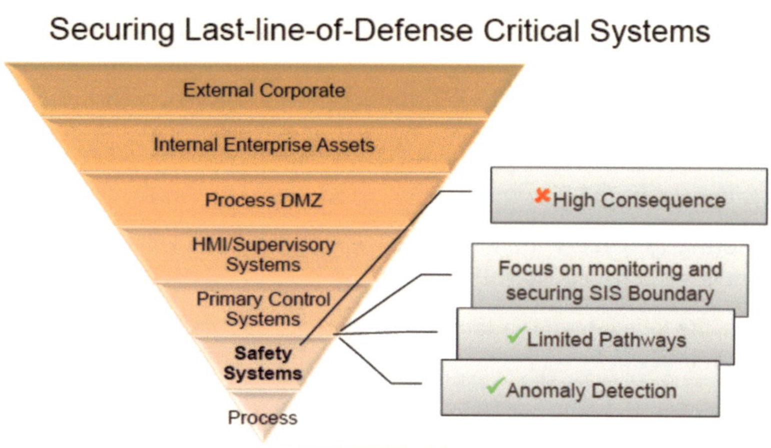

Fig. 19. Position and mechanisms enabling Last-line-of-Defense.

Figure 18 and Figure 19 gives a layered view of defense of critical systems. To implement those layers we could use the architecture given in the *standard ANSI/ISA-99* (now IEC 6222443.02.01 security standard of *"Zones and Conduits"*. The standard offers:

- A *level of segmentation* and *traffic control* inside the control system

- Control networks divided into *layers or zones based on control function*
- *Multiple separated zones* support *implementation* of a *defense in depth strategy*

Figure 20 illustrates the *overlay architecture of zones connected by conduits*. The information flows of the conduits are protected by carefully designed and implemented *firewalls*.

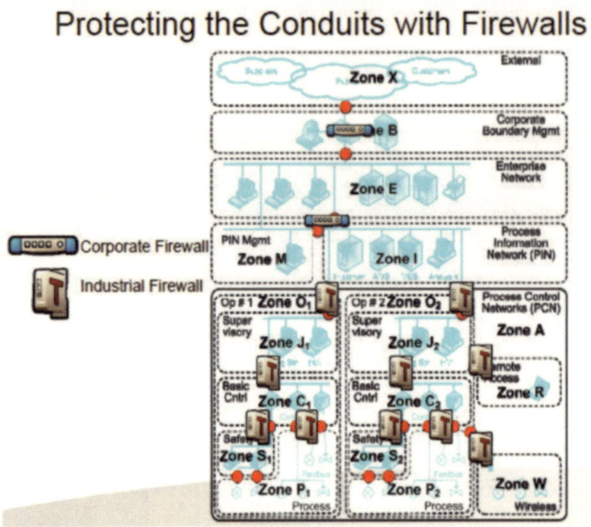

Fig. 20. The overlay security architecture of zones, conduits and firewalls.

Traditional firewalls are controlling the information flows at *system boundaries*. IT-based firewalls could handle IP-based traffic but not OPC based traffic that is the standard control system protocol, e.g., SCADA. Moreover, the more severe attacks are aiming at the *runtime environment* after the IT-based firewall. To remedy those shortcomings there are now products coming that could inspect OPC traffic. For instance *Hirschmann OPC Enforcer*[22] allows deep packet inspection of OPC. Stuxnet made extensive use of RPC (Remote Procedure Call) protocol, which is the basis for OPC. Also we have the *ECAT*[23] tool supporting *on-line monitoring of system memories* to address *APT threats*.

On-line monitoring as a basis for *system hardening* have been investigated in two thesis works [e, g] and several papers [8, 9, 10, 15, 25].

Finally, here are the following important recent reports addressing cyber security and Smart Grids by NIST Smart Grid[24]

- *Regulatory Recommendations* for Data Safety, Data Handling and Data Protection, issued by Expert Group 2 on February 16, 2011 [31].
- *Introduction to NISTIR 2628 Guidelines for Smart Grid Cyber Security* issued by Cyber Security Working Group in September 2010.

---

[22] Home page: http://www.hirschmann.com/en/Hirschmann_Produkte/Industrial_Ethernet/security-firewall/EAGLE20_Tofino_OPC/index.phtml

[23] Home page: http://www.siliciumsecurity.com/

[24] Home page: http://www.nist.gov/smartgrid/

Implementing defense-in-depth solutions, using architectures and standards mentioned above to monitor and protect data flows could be a way forward towards increased resilience of Smart grids. However, we need to investigate selected use cases and to develop suitable tool to pursue those efforts. This is the topic of next Sections.

7. Use cases

Use cases plays an important role in setting up and evaluate Smart grid scenarios. We have focused on use cases involving customers as stakeholders. Firstly, because *empowerment* of customers has been singled out as an important *enabler aiming at increased Energy Efficiency* (Section 1). Secondly, customers must be supported to *trustworthy collaborate* with other stakeholders of Smart grids (Figure 7), Thirdly, empowered *customers will require trusted cyber security and privacy* to accept Smart grid services.

We have advocated that setting up and monitoring Service Level Agreements (SLAs) provide a *structured methodology* to ensure Interoperability between stakeholders cooperating in delivering the services needed to fulfil the objectives and concerns related to the use cases at hand (Section 5). Typically, a *business case consists of related use cases*.

We have developed demonstrators of two customer empowerment related use cases in the areas of:

1. Smart homes
2. Green energy

The use cases are described in more detail in [12, 13, 26, 27, i]. The first use case aims at empowerment of customers to *change their profiles* to include more Renewable Energy Sources (RES), However, inclusion of RES will affect the energy balance of the (sub)grid to which the customer is connected. To have a minimal model of this sub grid we take into account the following three main stakeholders (Figure 3):

- A Distribution Grid Operator (DSO
- A Mediator/Facilitator (MF)
- Customers (C)

The Stakeholders participate in several SLAs coordinating the Business case based on changing the DER part of the power delivery to customer that is *empowered to change this amount*. In order to have a *separation of concerns* between the *business view* (buying power from RES) and the *grid view* (keeping energy balance of Voltage and Frequency) we model two types of SLAs (Figure 5):

- $SLA_{MF\text{-}DSO}$: Specifically addressing the coordination between MF and DSO maintaining the electricity balance within pre-set values.
- $SLA_{MF\text{-}C}$: Specifically addressing the requirements from C to change the amounts of RES to be delivered.

The two SLAs corporate in the following cycle:

1. Customer asks to change the DER amount with ΔDR during an interval Δ t.
2. The MF checks firstly if this allowable according to the $SLA_{MF\text{-}C}$, *if NO,* the request is *denied.*

3. If YES, the change is allowed if the energy balance could be maintained, The SLA_{MF-C} sends a request to the SLA_{MF-DSO} with that question. If NO, the request is denied.
4. If YES, the request from Customer C is *granted.*

It should be noted that the reasons behind the denials or granting could be given according to the setting up of the SLAs. The message exchanges, driven by SLAs, are recorded in databases for billing, accountability and traceability.

Our next demonstrator focus on issues related to *building and maintaining trust* between stakeholders. The use case is curtailment of service offerings.

This use case has the following stakeholders; *RES* (Renewable Resources), *DSO* and *C*. The market is defined as follows:

1. The DSO provides, monitor and bills energy to C (the present situation).
2. RES can and will occasionally generate and distribute energy ΔDR to the DSO.

The use case is defined as follows:

1. C *asks* for ΔDR during an interval Δ t.
2. DSO *asks* RES if ΔDR can be delivered during Δ t.
3. RES *confirms* this and DSO *inform* C that the *request is granted.*

Normally, the agreement between RES, DSO, and C is settled with a SLA. However, we can *encounter a breakdown* leading to that DSO *curtails* the delivery of ΔDR! The reasons could be either of or a combination of the following three events:

- DSO *identifies* that inclusion of ΔDR will *unbalance the grid* (voltage). Since DSO is responsible for the proper functioning of the grid we have *curtailment.*
- C *discovers* that he *cannot receive* ΔDR . DSO is informed and we have curtailment.
- RES discovers that ΔDR *cannot be produced* during Δ t.

Obviously the curtailment could eventually generate *losses of revenue,* and/or *in trust* and/or *willingness to invest* in supporting infrastructures. The overall consequence can be *resistance to implement parts of Smart grids* unless *proper regulatory frameworks and trustworthy SLAs are in place.* In short, *Risk assessments and mitigation* of technical and economic nature have to be in place to enable acceptance of Smart grid solutions.

Our simple use cases illustrates that design, implementation, maintenance, and acceptance of Smart grids have to be *carefully engineered* along the lines outlined in this chapter. This in turn *requires suitable tools.* That is the topic of next section.

8. Tools and environments

We have implemented and assessed two types of tools related to design and development of Smart grid pilots and Field tests, that is tools supporting *SLA procurement* and tools supporting *configuration and implementation of experimental environments* [8, 9, 10, 11, 12, 13, 15, 22, 24, 25, 27, e, g, i].

The following Figure 21 gives a architecture of SLA design environment.

The SLA agreement is based on a *selected use case* (goal, stakeholders, goal architecture, tasks, non-functional requirements, KPIs, exception procedures, *etc.*). Functional and non-

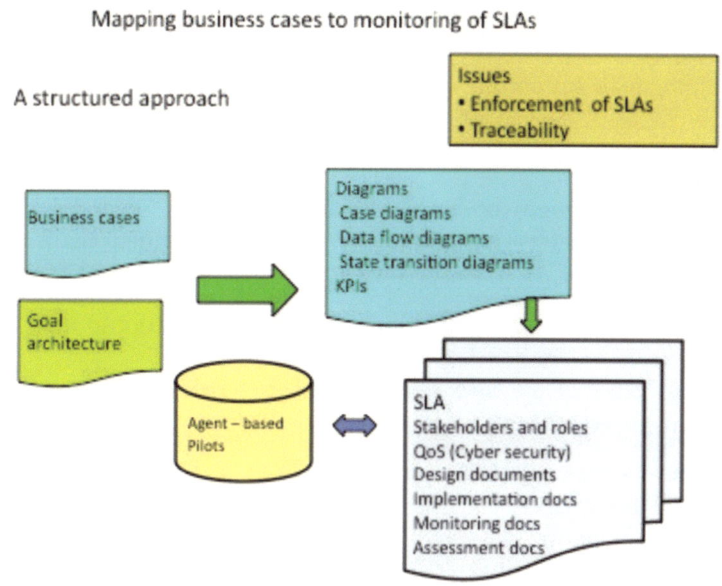

Fig. 21. Structured approach of setting up SLAs.

functional concerns are selected from *Interoperability Framework*, processed using our Framework supporting *Engineering trustworthiness* (Figure 10) and resulting in a *SLA footprint on the Interoperability Framework* (Figure 11).

During the SLA negotiation process, the stakeholders have *agreed on the terminology and meaning (semantics)*, the *KPIs and the conditions of the SLA*. Furthermore, issues related to *monitoring and data management* (including cyber security and privacy) have also been resolved. The SLA agreement is also a basis for assessing interoperability and simulations/pilots.

The *role of SLA agreements* is to map *use cases* into a *control and management structure* that *monitor the activities* (KPIs) *related to the use case*. Those activities are stored to enable billing, traceability and accountability. Furthermore, in cases of breakdowns, activities are recorded and eventually self-healing actions are activated according to the SLA. Furthermore a high-level *meta information* of the SLA agreement is kept to allow *reuse and adaptation* of use cases [6, 21, 23, a].

From Figure 3, Figure 4, Figure 5, Figure 6, and Figure 11 it follows that the *data flows* corresponding to the SLAs, firstly are *distributed and secondly have potentially large volumes.* Since the sets of SLAs are *overlays* of the infrastructure *into cells* of stakeholders and infrastructure components, we can form a *SLA-based data overlay* to address the challenges of data management in Smart grids [Figure 20, i].

Our *configurable experimental environments* have been evaluated in pilots aiming at:

- Hardening the execution environments (Software security)

- Designing experimental Smart grid environments by configuration of selected standard platforms.
- Evaluating tools supporting remote configuration and monitoring of pilots (France - Sweden)
- Evaluating tools supporting monitoring of data flows.

9. Conclusions and future work

Addressing the challenges of Future Smart grids poses several known and unknown challenges. In the Chapter we give a overview on identified challenges and promising routes toward solutions. We focus on the concept of Empowered customer of three reasons:

- *Active customers* have been identified as a key stakeholder to meet the expectations of the *EU 20-20-20 Energy Package.*
- *Other stakeholders in new business processes* of the Smart grid *must support active customers.*
- Active customers *will only accept and take up trustworthy services.*

In the Chapter we make a selection of key issues to address towards those ends. Firstly we emphasise the importance of *ensuring interoperability* of Smart grids. To that end, we advocate the use of the *Interoperability Frameworks* by NIST and GWAC, specifically addressing *interoperability between the Technical* (Syntax), *Informational* (Semantics), and *Organizational* (Pragmatics) Categories (Levels).

Interoperability assures that stakeholders can *coordinate their activities towards a common goal.* Our starting point of customer empowerment is that the customer is *empowered to change his/her energy-based services to meet individual goals.* To address this we propose *Service Level Agreements* (SLAs) as a mean to *coordinate different sets of stakeholders towards common goals that can be trusted, monitored, maintained, and billed.*

Information management (collect, store, access, process, distribute) is a key enabler of interoperability. Since the meaning of a given data item might be different to different stakeholders (system components) at different times we address these challenges briefly in the Chapter. Related to information processing is information sharing and information protection (e.g., privacy). Those important aspects of Cyber security and Privacy are outlined in the Chapter. We also illustrate some threats towards those non-functional concerns with some novel methods and techniques to implement a more resilient future Smart grid.

We also illustrate some use and shortcomings of present technologies in two use cases.

The work reported is promising but still in its infancy. We will continue our investigations and explorations is some on going and planned projects with involvement of KTH. Examples include:

- KIC InnoEnergy[25]
- Stockholm Royal Seaport[26]
- EU Grid4EU (new)[27]

---

[25] Home page: http://www.kic-innoenergy.com/

[26] Home page: http://www.stockholmroyalseaport.com/

10. Acknowledgements

The author acknowledge the importance of fruitful discussions and work in national and international collaborative projects and within the research groups at BTH and KTH as reported in the references.

11. References

[1] Energy Efficiency in SmartGrids, (2011) Download from http://seesgen-ict.rse-web.it.

[2] Akkermans, J.M., Ygge, F., Gustavsson, R.: Homebots: Intelligent decentralized services for energy management. In Proceedings of The Fourth International Symposium on the Management of Industrial and Corporate Knowledge, ISMICK '96, Rotterdam, The Netherlands, 21-22 October 1996. (1996).

[3] Gustavsson, R.: Agents with power. Communications of the ACM. March 99/Vol. 42, No. 3, pp. 41–47 (1999).

[4] Gustavsson, R.: Ensuring dependability in service oriented computing. Proceedings of The 2006 International Conference on Security & Management (SAM06) at The 2006 World Congress in Computer Science, Computer Engineering, and Applied Computing, Las Vegas (2006).

[5] Gustavsson, R.: Proper use of Agent Technologies in Design and Implementation of Software Intensive Systems. The Second International Workshop on Integration of Software Engineering and Agent Technology (ISEAT 2006) at The Sixth International Conference on Quality Software (QSIC 2006) (2006)

[6] Gustavsson, R., Fredriksson, M.: Process algebras as support for sustainable systems of services. Applicable Algebra in Engineering, Communication and Computing. 16, 179–203 (2005).

[7] Gustavsson, R., Fredriksson, M., Meilstrand, P.: The proper role of agent technologies in design and implementation of dependable network enabled systems. Weyns, D. and Holovet, T. (eds.) Multiagent Systems and Software Architecture, Proceedings of the Special Track at Net.ObjectDays, Erfurt, Germany, September 19, 2006, pp. 49-58. (2006)

[8] Mellstrand, P., Gustavsson, R.: An Experiment Driven Approach Towards Dependable and Sustainable Future Energy Systems. Proceedings of the 3rd International Conference on Critical Infrastructures (2006).

[9] Mellstrand, P., Gustavsson, R.: Experiment based validation of Critical Information Infrastructure Protection (CIIP). Critical Information Infrastructures Security, First Inter-national Workshop, CRITIS 2006, Samos, Greece, August 31 - September 1, 2006. Revised Papers in Lecture Notes in Computer Science, Springer Verlag, Volume 4347/ 2006, 15-29, DOI: 10.1007/11962977_2 (2006).

[10] Ståhl, B., Le Thanh, L., Caire, R., Gustavsson, R.: Experimenting with Infrastructures. Proceedings of 5th International Conference on Critical Infrastructure (CRIS 2010). Beijing, pp. 1 - 7, (2010).

[11] Pepink, G., Kok, K., Dimeas, E., Hatzpargyrious, N., Hadjsaid, N., Caire, R., Gustavsson, R., Salass, R., Niesing, J., Hamilton, L., others: ICT-platform based

[27] Home page: http://www.enel.com/en-GB/innovation/smart_grids/european_initiatives/grid4eu/

Distributed Control in electricity grids with a large share of Distributed Energy Resources and Renewable Energy Sources. Proceedings of the First International ICST Conference on E-Energy–E-Energy 2010. Athens, October 14–15, 2010. (2010).

[12] Hussain, S., Gustavsson, R.: Coordinating Energy Business Models and Customer Empowerment in Future Smart Grids. ICST Conference on E-Energy. E-Energy, 2010. Proceedings of the First International ICST Conference on E-Energy–E-Energy 2010. Athens, October 14–15, 2010. (2010).

[13] Gustavsson, R., Stahl, B.: The empowered user - The critical interface to critical infrastructures. Proceedings of 5th International Conference on Critical Infrastructure (CRIS 2010). Beijing, pp. 1 - 7, (2010).

[14] Kok, K., Karnouskos, S., Nestle, D., Dimeas, A., Weidlich, A., Warmer, C., Strauss, P., Buchholz, B., Drenkard, S., Hatziargyriou, N., others: Smart houses for a smart grid. Electricity Distribution-Part 1, 2009. CIRED 2009. 20th International Conference and Exhibition on. pp. 1–4 (2009).

[15] Gustavsson, R., Ståhl, B.: Self-healing and resilient critical infrastructures. Proceedings of 3rd International Workshop on Critical Information Infrastructures Security Rome, (2008).

[16] Törnqvist, B. and Gustavsson, R.: On Adaptive Aspect-Oriented Coordination for Critical Infrastructures. In Proceedings of the First International Workshop on Coordination and Adaptation Techniques for Software Entities, Oslo, 2004, (2004)

[17] Van Craenenbroeck, T., De Wispelaere Vreg, B.: Service level agreements and regulatory aspects of data communication between DGO's and suppliers. Electricity Distribution, 2005. CIRED 2005. 18th International Conference and Exhibition on. p. 1–4 (2005).

[18] Vrba, P.: Java-based agent platform evaluation. Holonic and Multi-Agent Systems for Manufacturing. 1086–1087 (2004).

[19] Barwise, J., Perry, J.: Situations and attitudes. CSLI Publications MIT Press 1983 (reprint 1999).

[20] Devlin, K.: Infosense: Turning Information Into Knowledge. W.H. Freeman (2001).

[21] Malik, K., Choudhary, P.: Business Organizations and Collaborative Web: Practices, Strategies and Patterns. Igi Global (2011).

[22] Lundberg, J., Gustavsson, R.: Challenges and opportunities of sensor based User empowerment. Engineering principles for open socio-technical systems. Proceedings of 8th IEEE International Conference on Networking, Sensing and Control, April 11-13, 2011, Delft, the Netherlands (ICNSC2011). (2011).

[23] Fakhfakh, K., Chaari, T., *et.al.*: A Comprehensive Ontology-Based Approach for SLA Obligations Monitoring. The Second International Conference on Advanced Engineering Computing and Applications in Sciences. p. 217–222 (2008).

[24] Gustavsson, R. and Mellstrand, P. : Dependable Virtual Power Plants. In *Proceedings of CRIS Workshop 2006 - Influence of Distributed and Renewable Generation of the Power System Security (DiGeSEC´06),* Magdeburg, Germany (2006).

[25] Gustavsson, R. and Ståhl, B.; Self-Healing and Resilient Critical Infrastructures, In *Proceedings of 3rd International Workshop on Critical Information Infrastructures Security, October 13-15, Rome.* Selected to be published in a special issue on *CRITIS´08 by Springer Verlag (2008).*

[26] Hussain, S., Honeth, N., Gustavsson, R., Sandels, C., and Saleem, A.: Trustworthy Injection/Curtailment of DER in Distribution Networks Maintaining Quality of Service. In *Proceedings of 16th International Conference on Intelligent System Applications to Power Systems ISAP 2011* (2011).

[27] Gustavsson, R., Hussain, S., and Nordström, L.; Engineering of Trustworthy Smart Grids Implementing Service Level Agreements. In *Proceedings of 16th International Conference on Intelligent System Applications to Power Systems ISAP 2011* (2011).

[28] *The Future of Cloud Computing. Opportunities for European Cloud Computing Beyond 2010.* EC Expert Group Report, Public Version 1.0 (Eds . Jefferry, K. (ERCIM) and Neidecker-Lutz, B. (SAP)).

[29] *Introduction to NISTIR 7628 - Guidelines for Smart Grid Cyber Security.* The Smart Grid Interoperability Panel, Cyber Security Working Group. September 2010, NIST.

[30] *NIST Framework and Roadmap for Smart Grid Interoperability Standards*, Release 1.0, NIST Special Publication 1108.

[31] *Regulatory Recommendations for Data Safety, Data Handling and Data Protection.* Task Force Smart Grids. Expert Group 2. Report June 22, 2010.

[32] Ådahl. K. and Gustavsson, R.: Decision Support by Visual Incidence Anamneses for increased Patient Safety. In *Efficient Decision Support Systems: Practice and Challenges – From Current to Future.* InTech – Open Access Publisher (2011). ISBN: 978-953-308-63-9.

Thesis works

a. Fredriksson, M.: *On the nature of open computational systems – Online Engineering.* PhD Thesis, Dissertation Series No. 2004:5, Blekinge Institute of Technology. ISBN 91-7295-045-5.

b. Rindebäck, C.: *Designing and Maintaining Trustworthy Online Services.* PhL Thesis, Licentiate Dissertation Series No. 2007:8, Blekinge Institute of Technology, ISBN 978-91-7295-129-4.

c. Brandt, P.: *Information in Use – Aspects of Information Quality in Workflows.* PhD Thesis, Dissertation Series No. 2007:04, Blekinge Institute of Technology. ISBN 978-91-7295-111-2.

d. Östlund. L.: *Information in Use – In- and Outsourcing Aspects of Dogital Services.* PhD Thesis, Dissertation Series No. 2007:05, Blekinge Institute of Technology. ISBN 978-91-7295-110-2.

e. Mellstrand, P.: *Informed System Protection.* PhD Thesis, Dissertation Series No. 2007:10, Blekinge Institute of Technology. ISBN 978-91-7295-106-8.

f. Lundberg, J.: *Engineering Principles for Open Socio-Technical Systems.* PhD Thesis, Dissertation Series No. 2011:01, Blekinge Institute of Technology. ISBN 978-91-7295-103-9.

g. Ståhl, B.: *Exploring Software Resilience*. PhL Thesis, Licentiate Dissertation Series No. 2011:05, Blekinge Institute of Technology, ISBN 978-91-7295-206-5.
h. Ådahl K.: *On Decision Support in Participatory Medicine supporting Health Care Empowerment*. PhD Thesis, Dissertation Series No. 2011:01, Blekinge Institute of Technology. ISBN 978-91-7295-221-8
i. Hussain, S: *Coordination and Monitoring Servises Based on Service Level Agreements in Smart Grids*. PhL Thesis, Licentiate Dissertation Series No. 2012:01, Blekinge Institute of Technology, ISBN 978-91-7295-224-9.

5

Energy Consumption Inequality and Human Development

Qiaosheng Wu[1], Svetlana Maslyuk[2] and Valerie Clulow[3]
[1]*School of Economics and Management, China University of Geosciences, Wuhan*
[2]*School of Business and Economics, Monash University, Victoria,*
[3]*College of Business, RMIT University, Melbourne,*
[1]*China*
[2,3]*Australia*

1. Introduction

Empirical evidence shows that growing energy consumption leads to a rapid increase in global greenhouse gases emissions (henceforth GHG). As the largest market failure ever experienced, diffusion of GHG in the global atmosphere happens quickly, regardless ofwhere the GHG is emitted (Sinn, 2007). Evidently, by century's end, energy-related carbon dioxide emissions would, at current rates, more than double, putting the world onto a potentially catastrophic trajectory, which could lead to warming of 5°C or more compared with preindustrial times (IEA, 2009). The existing energy system with most of the energy consumed by the developed nations, has underpinned and constructed deeply unequal social relations, as well as imbalanced nature-society relations. At present given current resource constraints, developing nations cannot follow the path previously chosen by the developed nations to achieve economic growth.

Following Jacobson et al. (2005), the distribution of and access to energy resources may result in significant social, environmental and economic inequalities. To date, inequality in energy consumption across countries has received very limited analytical attention. In the recent literature devoted to climate change, there have been several attempts to use the tools of conventional income distribution analysis to measure inequality in carbon dioxide (CO_2) emissions across countries and changes in inequality over time (see Heil & Wodon, 1997, 2000; Hedenus & Azar, 2005; Duro & Padilla, 2006; Padilla & Serrana, 2006; Groot, 2010). Yet, very few studies in the energy literature apart from Jacmart et al. (1979), Jaconson et al. (2005) and Rosas-Flores et al. (2010) have analysed inequality in energy consumption for a large sample of countries.

One of the first to notice the correlation between per capita energy consumption, standard of living and the degree of a country's development and to use the Lorenz curve to measure energy consumption inequality for 1950, 1969 and 1975 was Jacmart et al. (1979). They proposed that changes in the distribution of energy among countries provides another measure of trends in world's inequality and reported a decline in energy consumption inequality over time. In the analysis of the distribution of residential energy consumption in Norway, USA, El Salvador, Thailand and Kenya, Jacobson et al. (2005) found dramatic differences between energy use of developed and developing nations with Kenya, El Salvador and Thailand having the highest inequality in energy consumption respectively. These differences can be explained by the differences in a nation's wealth, income distribution and government infrastructure as well as climatic conditions, energy efficiency measures and size and geographic distribution of the rural population. In the analysis of inequality in the distribution of expenses associated with main energy fuels in Mexico, Rosas-Flores et al. (2010) found that natural gas, electricity and gasoline were consumed mainly by the higher income earners, while firewood and kerosene were the main fuels for the lower income consumers.

In the past, the improvements in the human quality of life meant greater use of energy, however it is no longer possible under the current supply contraints and climate change conditions. In fact the literature shows that good quality of life can be achieved on much lower energy consumption levels (Pasternak, 2000, Pachari and Spreng, 2003, Spreng, 2005). According to the United Nations (UN) 2007/2008 Human Development Report, under the energy supply constraints and the constant necessity to improve energy efficiency, when energy use is associated with human development, it is possible to find opportunities for the synergetic development of energy and society, by shifting the focus of the economy to satisfying basic human needs. It is possible to introduce a sufficientarian 'development threshold' attributed to global energy consumption, by the use of the nationally-weighted human development indicators such as the United Nations Development Program (UNDP) Human Development Index (HDI).

The purpose of the study reported in this chapter, is to measure energy consumption inequality by using the standard tools of economic analysis - the Lorenz curve and Gini coefficient. These inequality measures also provide critical insights into the temporal evolution of energy management in different states and nations, and allow us to visualise the impact of factors such as new technologies, government policies, etc (Jacobson et al., 2005). In this chapter, four Lorenz curves were generated based on the four equity criterions namely production-based, energy consumption-based, human development and economic activity equity criterions.

The list of 129 countreis analyzed in this study is given in Table 1 below. To calculate energy consumption inequality measures we use UNDP HDI and the International Energy Agency (IEA) data on per capita energy consumption. HDI is composed of three elements including longevity (L), as proxied by the life expectancy at birth, education index (E, a combination of adult literacy and gross enrollment indeces) and income as measured by the GDP per capita PPP USD index. Because they are equally important, HDI components are weighted equally. The following equations represent how the HDI components are calculated:

$$L = \frac{Life\ Expectancy - 25}{85 - 25} \tag{1a}$$

$$E = \frac{2}{3} * Adult\ Literacy\ Index + \frac{1}{3} * Gross\ Enrollment\ Index \tag{1b}$$

$$GDP = \frac{Log(GDP\ per\ capita) - \log(100)}{\log(40000) - \log(100)} \tag{1c}$$

The 2009 UNDP Human Development Report divided nations into three groups based on their HDI level. High human development economies (HHD) have HDI≥0.85, medium

Country	HD Category	Country	HD Category	Country	HD Category
Albania	M	Gabon	M	Nigeria	L
Algeria	M	Georgia	M	Norway	H
Angola	L	Germany	H	Oman	M
Argentina	H	Ghana	L	Pakistan	L
Armenia	M	Greece	H	Panama	M
Australia	H	Guatemala	M	Paraguay	M
Austria	H	Haiti	L	Peru	M
Azerbaijan	M	Honduras	M	Philippines	M
Bahrain	H	Hungary	H	Poland	H
Bangladesh	L	Iceland	H	Portugal	H
Belarus	M	India	M	Qatar	H
Belgium	H	Indonesia	M	Romania	M
Benin	L	Iran	M	Russian Federation	M
Bolivia	M	Ireland	H	Saudi Arabia	M
Bosnia and Herzegovina	M	Israel	H	Senegal	L
Botswana	M	Italy	H	Singapore	H
Brazil	M	Jamaica	M	Slovakia	H
Brunei Darussalam	H	Japan	H	Slovenia	H
Bulgaria	M	Jordan	M	South Africa	M
Cambodia	M	Kazakhstan	M	Spain	H
Cameroon	L	Kenya	L	Sri Lanka	M
Canada	H	Korea	H	Sudan	L
Chile	H	Kuwait	H	Sweden	H
China	M	Kyrgyzstan	M	Switzerland	H
Colombia	M	Latvia	H	Syrian Arab Republic	M
Congo	L	Lebanon	M	Tajikistan	M
Congo(Democratic Republic)	L	Libyan Arab Jamahiriya	M	Tanzania	L
Costa Rica	H	Lithuania	H	Thailand	M
Côte d'Ivoire	L	Luxembourg	H	Togo	L
Croatia	H	Macedonia	M	Trinidad and Tobago	M
Cuba	H	Malaysia	M	Tunisia	M
Cyprus	H	Malta	H	Turkey	M
Czech Republic	H	Mexico	M	Turkmenistan	M
Denmark	H	Moldova	M	Ukraine	M
Dominican Republic	M	Mongolia	M	United Arab Emirates	H
Ecuador	M	Morocco	M	United Kingdom	H
Egypt	M	Mozambique	L	United States	H
El Salvador	M	Myanmar	M	Uruguay	H
Eritrea	L	Namibia	M	Uzbekistan	M
Estonia	H	Nepal	L	Venezuela	M
Ethiopia	L	Netherlands	H	Viet Nam	M
Finland	H	New Zealand	H	Yemen	L
France	H	Nicaragua	M	Zambia	L

Note: The grouping of the countries is based by the 2009 UNDP Human Development Report. H—high human development countries, M --medium human development countries, L --low human development countries.

Table 1. Countries included in the sample.

human development economies (MHD) have 0.6≤HDI<0.85 and low human development economies (LHD) have HDI<0.6. In 2007, 47 economies corresponded to HHD, 60 to MHD and 22 to LHD nations respectively. The period 1998 to 2007 was chosen for this analysis because it corresponds to comparable metgodology of the HDI calculation used by the UNDP allowing us to compare the inequality measures across a common time period.

Table 2 contains total primary energy supply (TPES) per capita, GDP, population and HDI values for 30 countries with the largest per capita energy consumption in the world.

Country	TPES/pop, toe/capita	GDP, Billion 2000$,PPP	% of World total GDP	GDP/pop, 2000$/capita, PPP	Population, Million	% of World total population	HDI
Qatar	26.5392	29.02	0.047	34548	0.84	0.013	0.901
Iceland	15.7377	10.83	0.018	34935	0.31	0.005	0.968
United Arab Emirates	11.8296	113.85	0.185	26053	4.37	0.066	0.879
Bahrain	11.6523	16.12	0.026	21493	0.75	0.011	0.878
Trinidad and Tobago	11.4646	20.35	0.033	15301	1.33	0.02	0.813
Kuwait	9.4631	70.73	0.115	26590	2.66	0.04	0.893
Luxembourg	8.7901	31.2	0.051	65000	0.48	0.007	0.96
Canada	8.1686	1046.87	1.704	31743	32.98	0.499	0.959
United States	7.7459	11468	18.669	37962	302.09	4.571	0.953
Brunei Darussalam	7.114	6.03	0.01	15462	0.39	0.006	0.866
Finland	6.8962	164.81	0.268	31155	5.29	0.08	0.953
Saudi Arabia	6.2128	360.74	0.587	14907	24.2	0.366	0.819
Oman	5.9536	44.73	0.073	17204	2.6	0.039	0.83
Australia	5.8703	666.78	1.085	31541	21.14	0.32	0.965
Singapore	5.83	135.88	0.221	29603	4.59	0.069	0.928
Norway	5.7075	190.75	0.311	40499	4.71	0.071	0.971
Sweden	5.5118	298.31	0.486	32602	9.15	0.138	0.957
Belgium	5.3683	323.58	0.527	30469	10.62	0.161	0.946
Netherlands	4.9107	534.06	0.869	32604	16.38	0.248	0.955
Russian Federation	4.7455	1603.73	2.611	11323	141.64	2.143	0.803
Korea	4.5855	1065.75	1.735	21992	48.46	0.733	0.931
Czech Republic	4.4324	209.12	0.34	20264	10.32	0.156	0.893
Kazakhstan	4.2931	127.68	0.208	8248	15.48	0.234	0.788
Estonia	4.1972	22.03	0.036	16440	1.34	0.02	0.872
France	4.1483	1737.96	2.829	27339	63.57	0.962	0.949
Germany	4.0268	2315.34	3.769	28147	82.26	1.245	0.936
Japan	4.0195	3620.16	5.893	28336	127.76	1.933	0.951
New Zealand	4.0075	101.07	0.165	24122	4.19	0.063	0.942
Austria	3.99	266.51	0.434	32032	8.32	0.126	0.946
Turkmenistan	3.6416	38.18	0.062	7698	4.96	0.075	0.764
Other countries	1.154	34787.8	56.6	6150	5656.1	85.6	-
World	1.82	61428.02	100	9294	6609.27	100	-

Table 2. Top 30 energy consumers.

Although the majority of these nations are developed economies, the list also contains resource-rich developing nations such as Qatar and Oman. The United States with high human development level (HDI is 0.953 in 2007) was the largest energy consumer in the world, consuming 20 percent of the world's total energy. Other nations with relatively high levels of energy use are Qatar, Iceland, United Arab Emirates, Bahrain, Trinidad and Tobago, Kuwait, Luxembourg and Canada. Norway has the highest human development level due to the highest HDI value.

In this study we found that inequality of energy consumption has been decreasing over the entire time period of analysis. This can be attributed to several factors including globalization and improved access to energy and infrastructure in some developed countries (e.g. China and India). We suggest that concerns to do with inequality of energy consumption must be incorporated and integrated into the development strategies for all countries irrespective of their human development level.

The chapter is structured as follows. Section 2 describes inequality measures used in this chapter. Section 3 discusses energy consumption inequality using four equaity criteria. Section 4 provides an overview of inequality in time from 1998 to 2007 and Section 5 concludes the chapter by analysing policy implications of our findings.

2. Measuring energy consumption inequality

In order to visualize HHD-MHD/-LHD energy consumption inequality between countries this chapter uses the Lorenz curve and the Gini coefficient. In traditional economics, the Lorenz curve shows what percentage of the total income is held by the corresponding percentage of households, where households are ranked by level of income. Applying the Lorenz curve in the context of energy consumption, means replacing households by countries, and ranking by income is replaced by ranking by energy consumption per capita across countries. Doing so results in a Lorenz curve that depicts distribution of cumulative percentage of world population on the abscissa axis versus the cumulative percentage of the energy consumption distributed along the ordinate axis.

Mathematically Lorenz curve can be represented as

$$y = f(p), \tag{2a}$$

where p is the cumulative population share of persons earning income equal to or below income level x, y is the cumulative income share of population subgroup p. Any Lorenz curve must have the following properties,

$$\frac{dy}{dp} > 0, \frac{d^2y}{dp^2} > 0, y(0) = 0, y(1) = 1, \tag{2b}$$

and is defined on the domain $0 \le p \le 1$.

Applying the Lorenz curve in the context of energy consumption, means replacing households by countries, and ranking by income is replaced by ranking by energy consumption per capita across countries. Doing so results in a Lorenz curve that depicts distribution of cumulative percentage of world population on the abscissa axis versus the cumulative percentage of the energy consumption distributed along the ordinate axis (Jacobson et al., 2005). In fact, the criterion to rank countries is fully determined by the variables used on the coordinate axes in a Lorenz diagram (Groot, 2010). Therefore, one can also construct a Lorenz curve where the horizontal axis measures cumulative world GDP shares instead of cumulative world population shares (Groot, 2010).

Figure 1 shows an energy consumption Lorenz curve in 2007 for countries sorted by per capita GDP PPP. The 45 degree line represents the line of perfect equality, where national energy consumption is equalized globally on a per capita basis. The area between the perfect equity line and the actual distribution (Lorenz) curve is given by the Gini coefficient wich is calculated as

$$Gini = \sum_{i=0}^{n} \frac{100^2 - [(P_{i+1} - P_i)(E_i + E_{i+1})]}{100^2}, \tag{3}$$

where P_i is the population share of country i and E_i is its energy consumption share in world population and in total world energy consumption respectively. In this case the Gini coefficient indicates the degree of global inequality in per capita energy consumption. A Gini coefficient of zero corresponds to perfect equality in per capita energy consumption among all countries in the sample (every country consumes the same amount of energy and the Lorenz curve corresponds to the 45-degree line), while a Gini coefficient of one would indicate perfect inequality in energy consumption, arising due to all the world's energy being consumed by one nation. For the year 2007, Gini coefficient corresponding to Lorenz curve shown on Figure 1 is 0.47, implying that distribution of energy consumption in 2007 between the richest and the poorest nations that was not equal.

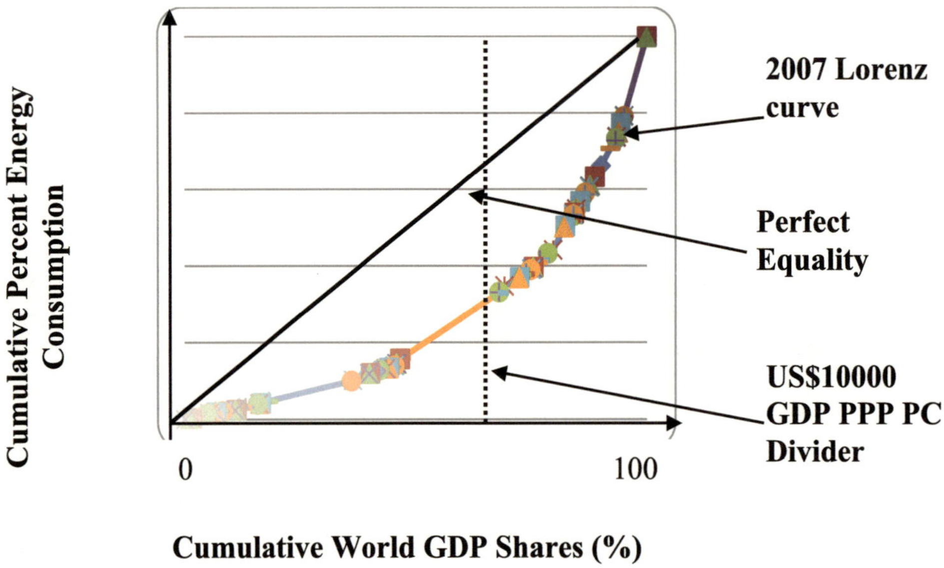

Fig. 1. The Lorenz curve for energy consumption in 2007 for countries sorted by per capita GDP PPP.

A potentially more intuitive way to interpret Figure 1 is by using GDP of US$ 10000 PPP as a divider between lower and higher income countries. Then, in 2007, 75% of the world's population with per capita GDP of less than US$10000 accounted for 40% of global energy consumption. The remaining 25% of population with GDP PPP per capita of more than US$10000 accounted for 60% of global energy consumption.

3. Energy consumption inequality criterions

By ranking countries in a different way it is possible to construct a different Lorenz curve, and it will be shown that the criterion to rank countries is fully determined by the variables used on the coordinate axes in a Lorenz diagram (Groot, 2010). In this chapter we generate four Lorenz curves based on four equity criterions. The first is an energy consumption-based equity criterion which is predicated on the rationale that all countries should have an equal right to use energy for its social and economic development. In this case the Lorenz curve is constructed by plotting per capita energy consumption shares in the cumulative world energy consumption on the vertical axis, and cumulative world population shares (%) on the horizontal axis. Second is an energy production-based sovereignity equity criterion which is connected to a country's capabilities to produce and consume its own energy. In this case, the horizontal axis of the Lorenz curve is found by sorting cumulative world population shares (%) by per capita energy production. Third is an economic activity equity criterion. In this study we use energy intensity or the number of energy units used in the production of a nation's GDP as the proxy for economic activity. High/low energy intensity represents high/low cost of converting energy into GDP. The Lorenz curve is sorted by energy intensity, where cumulative world GDP shares (%) ranked by energy intensity is on the horizontal and cumulative world energy consumption shares (%) are on the vertical axes. Last is a human development equity criterion which is based on the HDI. In this case cumulative world energy consumption shares (%) are on the vertical axis and cumulative world population shares (%) ranked by the HDI are on the horizontal axis. According to the conventional welfare theories, to achieve higher human development, each individual should enjoy development rights, including social, economic, political, as well as the basic survival needs and the provision of non-material services based upon demand for natural resources. Therefore, the concept of human development is important because it is not only concerned with the current state of the human well-being but also with the realization of human potential. This criterion implies that each member of the society is entitled to realize their basic human right to development potential given constrained natural resources.

Figure 2 shows the distribution of 2007 energy consumption under energy consumption-based equity criterion. Based on this criteria, the Gini coefficient was 0.50. Top 10 countries in terms of energy cosumption include: Qatar, Iceland, United Arab Emirates, Bahrain, Trinidad and Tobago, Kuwait, Luxembourg, Canada, United States and Brunei Darussalam. These countries harbour 5.52 % of the world's population, and use 24.06 % of the world's energy.

Figure 3 shows the distribution of 2007 energy consumption under energy production equity criterion. In 2007 the Gini coefficient was 0.39. Per capita energy production in the top 10 countries include: Qatar, Kuwait, Brunei Darussalam, Norway, United Arab Emirates, Trinidad and Tobago, Oman, Saudi Arabia, Bahrain and Libyan Arab Jamahiriya. These nations harbor 0.77 % of the population, and produce 12.23 % of the world's energy, but consume 2.95 % of the world's energy.

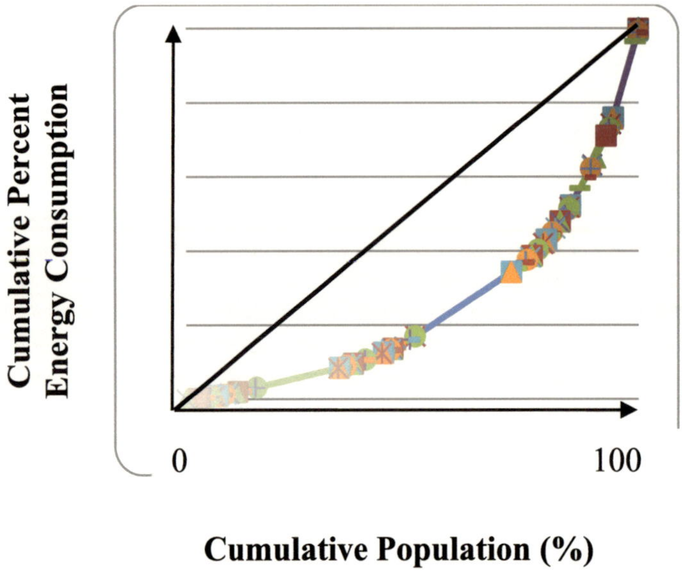

Fig. 2. The Lorenz curve in 2007 for countries sorted by per capita energy consumption.

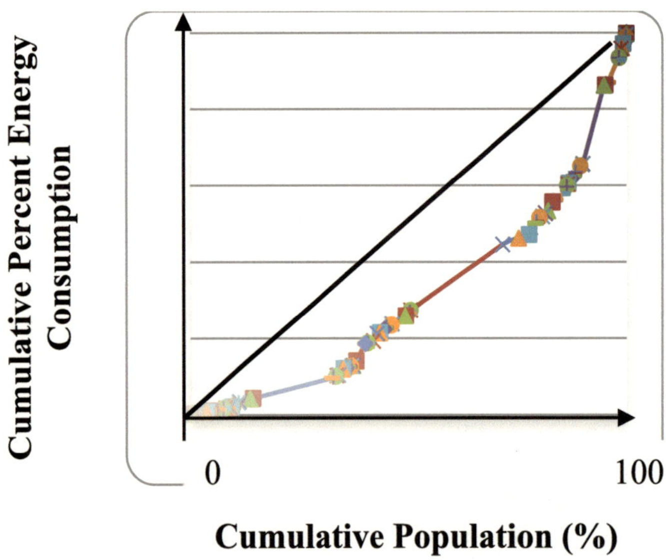

Fig. 3. The Lorenz curve in 2007 for countries sorted by per capita energy production.

Figure 4 shows the distribution of 2007 energy consumption under economic activity equity criterion. In 2007 the Gini coefficient was 0.19. The energy intensity of the top 10 countries namely, Uzbekistan, Qatar, Trinidad and Tobago, Nigeria, Tanzania, Zambia, Bahrain, Kazakhstan, Jamaica and Tajikistan, with GDP of 0.78 % of the 129 countries, indicateduse of 2.65 % of world's energy.

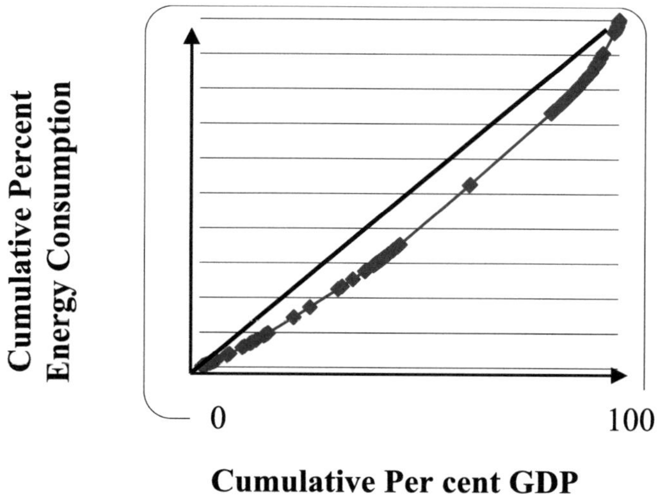

Fig. 4. The Lorenz curve in 2007 for countries sorted by energy intensity.

Figure 5 shows the Lorenz curve sorted by HDI criterion, where cumulative world energy consumption shares (%) are on the vertical axis and cumulative world population shares (%) are on the horizontal axis. The Gini coefficient in 2007 is 0.46. Top 10 HDI nations are Norway, Iceland, Australia, Ireland, Luxembourg, Canada, Sweden,the Netherlands, Finland, United States. Their total GDP accounts for 23.7 % of the world's GDP, 6.3 % of the world's population, and use 25.9 % of the world's energy. The energy use of HHD countries is 48.5 % of the world's total, their GDP accounts for 52.3 % of the world's total, and they are the home countries of 17.8 % of the world's population. MHD countries use 48.1 % of the world's energy, harbor 67.3 % of the world's population and account for 43.1 % of the world's GDP. LHD countries harbor 14.9 % of the world's population and only use 3.4 % of the world's energy.

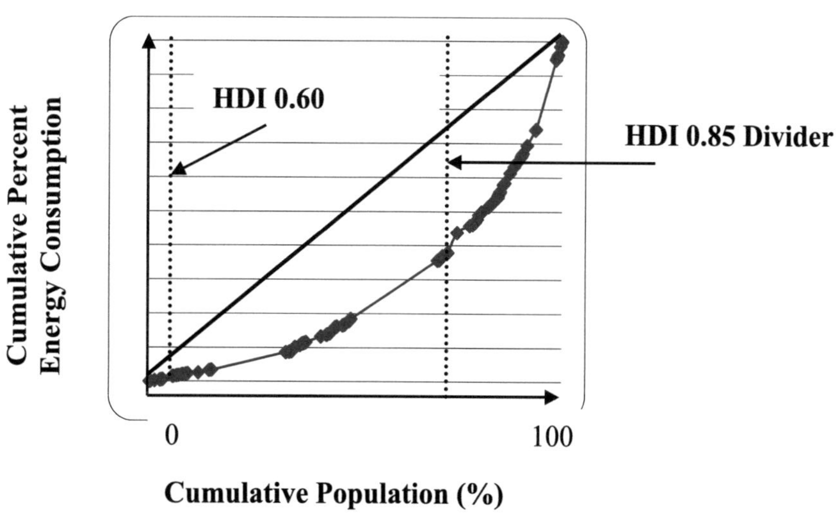

Fig. 5. The Lorenz curve in 2007 for countries sorted by HDI.

4. Energy consumption inequality from 1998 to 2007

Table 3 and Figure 6 present calculated Gini coefficients calculated based on the four equity criterions from 1998 to 2007. One can see that although inequality in energy consumption (shown by the difference between the respective Lorenz curve and the diagonal and decline in the Gini coefficient values) has diminished over the time according to all four criterions analysed, it did not disappear completely.

Year	Gini coefficient			
	Energy production-based criterion	Energy consumption-based criterion	HDI criterion	Economic activity criterion
1998	0.4273	0.5365	0.5052	0.2082
1999	0.4237	0.5356	0.5013	0.2025
2000	0.4262	0.5384	0.5059	0.2018
2001	0.4240	0.5364	0.4956	0.1990
2002	0.4206	0.5323	0.4965	0.1971
2003	0.4129	0.5258	0.4876	0.1939
2004	0.4043	0.5172	0.4781	0.1899
2005	0.3996	0.5125	0.4746	0.1882
2006	0.3951	0.5054	0.4656	0.1876
2007	0.3890	0.5000	0.4572	0.1870

Source: Authors' own calculations based on the UNDP (2000-2009) and IEA (2009).

Table 3. The Gini based on equity criterions from 1998 to 2007.

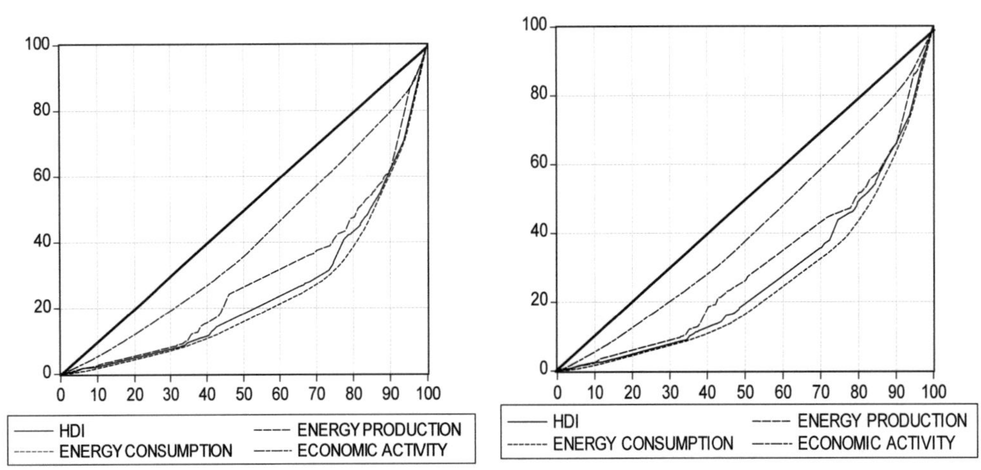

1998 2007

Note: HDI - Human development equality criterion, ENERGY PRODUCTION – Energy production-based equality criterion, ENERGY CONSUMPTION – Energy consumption-based equality criterion, ECONOMIC ACTIVITY – Economic activity equality criterion.

Fig. 6. The Lorenz curve in 1998 and 2007 for different equality criterions.

One can see that the largest inequality is based on the HDI and energy consumption criterions. This finding can be explained by the continued poor access to energy resources by the developing nations, insufficient and in some cases inadequate infrastructure facilities and the use of energy-inefficient technologies. Although over the time, developed nations have improved access to energy resources, on average they are still consuming much less energy on a per capita basis as compared to the developed nations.

5. Conclusion

The distribution of energy resources may result in significant social, environmental and economic inequalities (Jacobson et al., 2005). A critical issue faced by policy makers across the world is how to distribute the costs and benefits through policies designed to address such problems. This chapter argues that energy consumption has a distinct and critical social dimension. Based on the UN Human Development Index, it analyses the energy consumption equality problem involving the different HDI groups. Although energy consumption inequality has been declining over time, it is not yet on a dissapearing trend. Economic growth, as well other socio-economic factors such as urbanisation and population increases are unbalanced globally, meaning that the contributions of developed and developing countries to climate change are changing. Therefore, compared with developed countries (which typically have high levels of energy consumption and corresponding high HDI and are aiming to keep a high standard of living), developing countries (usually they have lower HDI) have different tasks concerning energy consumption and human development. If the goal of low and medium HDI nation is to achieve improvement in its HDI, the goal of the high HDI nation is its maintenance.

In this study, we consider world energy consumption inequality from 1998 to 2007 and found that all of the conventional income inequality approaches can also be applied to the distribution of per capita energy consumption provided appropriate adjustments are made. We have chosen to apply the Lorenz curve and Gini coefficient to examine the inequality of per capita energy consumption across countries under different equality criteria. As stated earlier 1998 to 2007 was chosen as a sample period because it corresponded to the same methodology of HDI calculation used by the UNDP. In 2010 the UNDP has changed the HDI calculation methodology and approach to country classification. Therefore the calculation of inequality measures based on the new HDI definition is left to the future, but these measures will not be strictly comparable with the past.

Energy consumption inequality, as measured by the divergence of Lorenz curve from the diagonal and by the Gini coefficient, was found to be different based on different equity criterions. In particular, Gini coefficient was much lower when energy consumption shares are pictured against world GDP shares rather than world population shares. Irrespective of the equity criterion used, energy consumption inequality was found to be diminishing over time. These are the reasons that could have lead to a reduction in energy consumption inequality:

a. Globalization or the international integration of markets for goods, services and capital (Brune and Garrett, 2005). Globalization for developing countries often leads to an increase in the energy consumption as developed countries shift production and technologies to developing countries.

b. Creation of essential infrastructure and establishing access to electricity in developing countries. In 2009 the number of people without access to electricity was 1.3 billion or almost 20% of the world's population (IEA, 2011). The speed of electrification in developing countries is still relatively slow, but it is happaning.
c. Changes in the energy consumption mix towards more efficient energy use and a shift towards alternative energy in developed countries and some developing. For example, in 2009 more than 84% of energy produced in Brazil was due to alternative energy sources, the largest of which was sugar cane ethanol. Although the shift towards alternative energy resources is still in the introductory stages, there is a lot of research underway in terms of solar energy, algae and wave energy. At the same time, technology for some energy sources, such as direct geothermal, has been already established.
d. Introduction of the climate change mitigation policies in both developed and developing nations in order to prevent dangerous anthropogenic interference with the climate system. Such policies target reduction in GHG gases, which can be achieved due to a reduction in energy consumption and more efficient energy use. Examples of such policies are carbon taxes and emissions trading schemes (ETS). While ETS are more recent instrument (e.g. ETS to control GHG in European Union have been operational since 2005), carbon taxes have been used since 1990s. ETS have been proposed to be introduced in Australia, Japan, US, Canada, Korea, India and China in the near future.

Figure 7 below shows that relationship between the HDI and energy consumption per capita (in tonnes of oil equivalent) is not linear. This means that at low human development levels, increase in energy consumption will lead to large increases in a country's HDI. This is supported by Martinez and Ebenhack (2008), who calculated that addition of 400 kg of oil-equivalent per capita in the poorest nations with HDI values less than 0.4 will support a doubling of their HDI. However, as a country develops, the importance of energy in establishing higher HDI diminishes. Therefore for high and medium human development levels, simply increasing energy consumption is not enough to maintain its human development progress. In this case, a combination of factors such as more efficient energy use, development of energy-saving technologies, establishing appropriate social welfare systems and others are necessary to achieve and maintain high HDI.

Maintenance of high HDI would require policies targeting efficient energy use both on personal and company-based level and promoting energy-efficient technologies. Such policies should be country-specific and reflect current energy mix, industrial structure, potential fossill fuel and alternative energy resources, exisiting climate change mitigation policies (e.g. environmental taxes, subsidies for clean energy initiatives, creating a market for pollution, etc) and global action in climate change mitigation. For example, Canada and Germany are the world leaders in terms of direct geothermal energy and solar power respectively.

At the same time, low HDI countries should reduce energy poverty by creating essential infrastrure, changing their energy consumption mix and establishing access to modern energy sources. For instance, low HDI nations such as Nairobi and Gabon are largely dependent on biomass (firewood, charcoal or dung) as the primary energy source, which is not efficient energy source and highly GHG pollusive. Effors targeting establishing access to

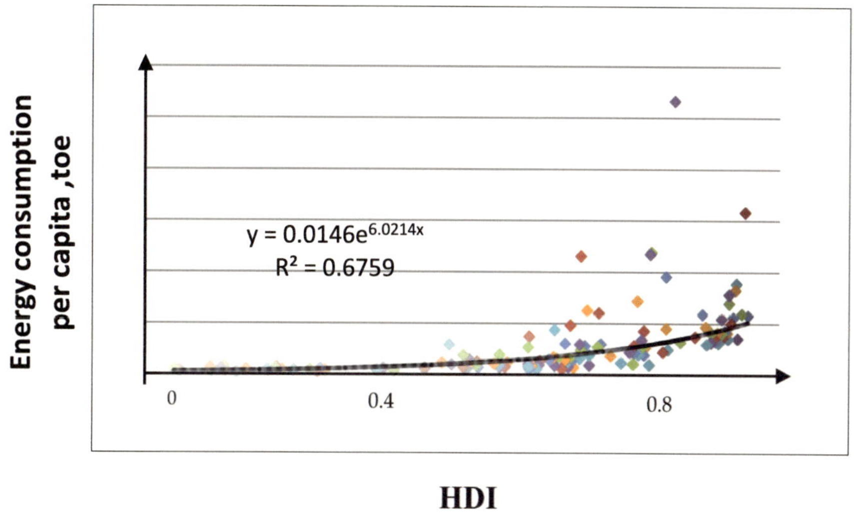

Fig. 7. Energy consumption per capita and the HDI (2007).

modern energy, electrification and creation of essential infrastrucuture are more likely to achieve improvement in HDI. However, global efforts together with inidividual low HDI country efforts might be necessary in order to achieve improvement in human development.

For example, lets consider Australia and Kenya as HHD (HDI = 0.970) and MHD (HDI = 0.541) nations in 2007 respectively. It should be noted that in the beginning of the sample, Kenya, which is now largest economy in East Africa, had lower HDI value. However in a less than decade government policies on improving human development with the help of international organizations (for example, UN World Food Programm since 2004 was installing energy-efficient stoves in Kenyan schools) have been relatively successful, although a lot of challeneges still remain. While primary energy sources in Australia are brown and black coal and natural gas, Kenya is largely dependent on biomass (wood), imported crude oil and electricity with respective shares 70 per cent, 21 per cent, and 9 per cent of total energy use (UNEP, 2006). While in Australia, major electricity source is coal, in Kenya major sources of electricity are hydro, geothermal and thermal power (UNEP, 2006). Governments of these two countries face different challenges, namely maintaining already high HDI (Australia) and achieving improvement in HDI (Kenya). In both cases, this would require efficient use of energy resources, but for Australia this would also mean significant climate change mitigation policy constraints. For example, Australia has pledged to reduce its GHG emissions (the primary means of achieving is goal is transitional carbon tax on producers and introduction of a national mandatory emissions trading scheme in 2015) and increase investment in alternative energy such as direct geothermal and wave energy. For Kenya, where 80% of population depends on biomass as the primary source of energy, the challenges lie in improving electricity generation and distribution, creating essential transmission and distribution infrastructure, reducing the cost of electricity, reducing its dependence on crude oil imports and investing in green energy sources (UNEP, 2006). However, the poverty still remains acute in Kenya due to high income inequality, disproportionate access to essential resources including land, susceptibility to natural disasters such as floods and still inadequate access to basic social services including education (Hendriks, 2010, p.99).

Since HDI is composed of three elements (longevity as proxied by the life expectancy at birth, education as proxied by the gross enrollment and adult litarcy indeces, standard of living as proxied by the GDP per capita), its improvement or maitenance would require achieving progress in all of them. For instance, achieving higher economic growth is no possible without energy use, but the strategies will be different for HHD, MHD and LHD because of different energy mix, different demographic characteristics and different techological levels (more pollusive in developing nations). Due to existing infrastrucuture faciliting as well as technologies, high HDI countries should take a leadership role on reducing energy consumption, reducing emissions and improving energy efficiency measures that could help fostering replicable models of the good quality life that are based on much lower energy consumption levels.

Potential direction for future research would be analysis of causality between energy consumption and human development. Causal relationship (i.e. lead-lag relationsip) between energy consumption and GDP has been examined in the literature at the country-specific level, as well as based on the panel data analysis. In summary, the findings of these research efforts are mixed and largely depend on the time period of analysis, energy mix and level of country's economic development. However, current literature gap lies in stydying causality between energy consumption, including different enenrgy sources, and development indicators other than GDP, such as for example Human Development Index.

6. References

Atkinson, A.B., 1970. On the measurement of inequality. Journal of Economic Theory, 2, 244–263.

Australian Government, 2011. Strong Growth, Low pollution- modelling a carbon price. Available online at: www.treasury.gov.au/carbonpricemodelling/content/report.asp

Banerjee, A., Yakovenko, V. M., 2010. Universal patterns on inequality, New Journal of Physics, 12, 075032.

Birol, F., 2007. Energy economics: a place for energy poverty in the agenda?. The Energy Journal, 28, 01–06.

Blackorby, C., Donaldson, D., 1978. Measures of relative equality and their meaning in terms of social welfare. Journal of Economic Theory, 18, 59–80.

Brune N., Garrett G., 2005. The globalization Rorschach test: international economic integration, inequality, and the role of government. Annual Review of Political Science, 8, 399–423.

Chakravarty, S.R., 1990. Ethical social index numbers. Springer-Verlag, Heidelberg, New York,London, Paris, Tokyo, Hong Kong.

Champernowne, D.G., Cowell, F.A., 1998. Economic inequality and income distribution. Cambridge Univ. Press, United Kingdom.

Cowell, F.A., 1995. Measuring income inequality (2nd ed.), Harvester Wheatsheaf, Hemel Hempstead.

Dalton, H., 1920. The measurement of the inequality of incomes. Economic Journal, 30, 348–361.

Dasgupta, P., Sen, A., Starrett, D., 1973. Notes on the measurement of inequality. Journal of Economic Theory, 6, 180–187.

Druckman A, Jackson T., 2008. Measuring resource inequalities: the concepts and methodology for an area-based Gini coefficient. Ecological Economics, 65, 242-52.

Duro, J.A., Padilla, E., 2006. International inequalities in per capita CO_2 emissions: a decomposition methodology by Kaya factors. Energy Economics, 28, 170-187.

Foster, J.E., 1985. Inequality measurement. Proceedings of Symposia in Applied Mathematics, 33, 31-68.

Foster, J.E., Shorrocks, A.F., 1988. Inequality and poverty orderings. European Economic Review, 32, 654-662.

Giannini Pereira, M., Vasconcelos Freitas, M. A., Fidelis da Silva, N., 2011. The challenge of energy poverty: Brazilian case study. Energy Policy, 39, 167-175.

Groot L., 2010. Carbon Lorenz curves. Resource and Energy Economics, 32, 45-64.

Hedenus, F., Azar , C., 2005. Estimates of trends in global income and resource inequalities. Ecological Economics, 55, 351-364.

Heil, M.T., Wodon , Q.T., 1997. Inequality in CO_2 emissions between poor and rich countries. Journal of Environment and Development, 6, 426-452.

Heil, M.T., Wodon ,Q.T., 2000, Future inequality in CO_2 emissions and the impact of abatement proposals. Environmental and Resource Economics, 17, 163-181.

Hendriks, B., 2010, Urban livelihoods, institutions and inclusive governance in Nairobi. 'Spaces' and their impacts on quality of life, influence and political rights. Amsterdam University Press.

International Energy Agency (IEA), 2011. Access to electricity. Available online at: http://www.iea.org/weo/electricity.asp

Jacobson, A., Milman, A. D., Kammen, D. M., 2005. Letting the (energy) Gini out of the bottle: Lorenz curves of cumulative electricity consumption and Gini coefficients as metrics of energy distribution and equity. Energy Policy, 33, 1825-1832.

Jacmart, M. C., Arditi, M., Arditi, I., 1979. The world distribution of commercial energy consumption. Energy Policy, 7, 199-207.

Kolm, S.-C., 1976a. Unequal inequalities I. Journal of Economic Theory, 12, 416-442.

Kolm, S.-C., 1976b. Unequal inequalities II. Journal of Economic Theory, 13, 82-111.

Lambert, P.J., 1993. The Distribution and redistribution of income: a mathematical analysis. Manchester Univ. Press, Manchester and New York.

Moriarty, P., Honney, D., 2010. A Human needs approach to reducing atmospheric carbon. Energy Policy, 38, 695-700.

Opschoor, H., 2009. Sustainable development and a dwindling carbon space. Environmental and Resources Economics, DOI 10.1007/s10640-009-9332-2.

Pachauri, S., Spreng, D., 2003. Energy use and energy access in relation to poverty. Centre for Energy Policy and Economics—Swiss Federal Institutes of Technology, Working Paper No. 25, Zurich, Switzerland.

Padilla, E., Serrano, A., 2006. Inequality in CO_2 emissions across countries and its relationship with income inequality: a distributive approach. Energy Policy, 34, 1762-1772.

Pasternak, A. D., 2000. Global energy futures and human development: a framework for analysis. US Department of Energy Report UCRL-ID- 140773, Lawrence Livermore National Laboratory, Livermore, CA.

Rosas-Flores, J.A., Galvez, D. M., 2010. What goes up: Recent trends in Mexican residential energy use. Energy, 35, 2596-2602.

Rothschild, M., Stiglitz, J.E., 1973. Some further results on the measurement of inequality. Journal of Economic Theory, 6, 188–204.

Sen, A., 1973. On economic inequality. Oxford Univ. Press, Oxford.

Sinn, H.W., 2007. Public policies against global warming. NBER Working Paper No.W13454, CESifo (Center for Economic Studies and Ifo Institute for Economic Research)/ NBER (National Bureau of Economic Research).

Spreng, D., 2005. Distribution of energy consumption and the 2000 W/capita target. Energy Policy, 33, 1905–1911.

United Nations Development Programme (UNDP)–Human Development Report Office. 1998-2009. "The Human Development Index (HDI)." New York. Available online at: http://hdr.undp.org/ en/statistics/hdi/.

United Nations Environment Project (UNEP), 2006. Kenya: Integrated assessment of the Energy Policy. Nairobi, UNEP. Available online at: www.unep.ch/etb/areas/pdf/Kenya%20ReportFINAL.pdf

Part 2

Energy Efficiency on Demand Side

6

Effect of an Electric Motor on the Energy Efficiency of an Electro-Hydraulic Forklift

Tatiana Minav, Lasse Laurila and Juha Pyrhönen
Lappeenranta University of Technology/LUT Energy
Finland

1. Introduction

Mobile working machines play an important role in modern industry. These machines are widely used for instance in the mining, process and goods manufacturing industry, forest harvesting and harbour terminal work. Figure 1. illustrates typical examples of mobile working machines.

(a) (b) (c)

photo: Cargotec photo: Rocla photo: Sandvik

Fig. 1. Examples of mobile working machines: a) straddle carrier, b) forklift truck and c) mine underground truck.(Minav, 2011d)

Mobile working machines can be classified as light mobile machines that operate by battery power and heavy machines that work using a diesel engine. Both of these types include mechanical structures, which are driven by hydraulics. With the rising concern in global scale environmental issues, energy saving in vehicles and mobile machines is an important subject and reduction of fuel consumption is strongly required (Petrone, 2010 ; Saber, 2010 ; Bhattacharya, 2009 ; Montazeri-Gh, 2010 ; Mapelli, 2010, Liu, 2010). Hybrid propulsion concepts in working machines are emerging to improve their fuel economy and reduce CO_2 emissions (Fakham, 2011; Hui, 2010, Paulides, 2008). In addition, now there are government mandated Tier IV reduction regulations for harmful exhaust gases for diesel powered equipment (Wagner, 2010; EPA, 2011). According to (Kunze, 2010), a hydraulic hybrid can be consider the greatest innovation potential for the industrial sector. Traditional hydraulic systems' control methods are giving way to direct electric–drives-based control of hydraulic force (Grbovic, 2011; Berkner, 2008;). Hydraulics is one of the widely adopted engineering approaches as it provides high force densities (Burrows, 2005), but its efficiency is often limited by lossy control methods but nowadays we have an opportunity to reduce the power losses by using an efficient electric drive system together with the hydraulic part of a mobile machine(Ahn, 2008; Iannuzzi, 2008). Improving working machines' efficiencies has attracted a lot of attention among researchers and manufacturers all over the world (Yang, 2007; Liang, 2001; Rahmfeld, 2001; Rydberg, 2005; Innoe, 2008; Mattila, 2000). A great number of different types of non-road machines are manufactured for different purposes. Typically, non-road vehicles can be classified as construction machines, transportation of goods or material handling equipment, and janitorial and agricultural machines. Regeneration-capable hydraulic systems that are based on combinations of an electromechanical unit and a reversible hydraulic machine have the potential of improving the energy efficiency by operating the system components within their optimum efficiency ranges and, especially, by making use of the regenerative processes in all the above-mentioned machines (Yoon, 2009). Forklift is one of the machine types that can be modified for energy recovery. Energy recovery is an efficient way to extend the driving range with limited energy sources (Minav, 2009; Andersen, 2005, Lin, 2010b; Rydberg, 2007).

There is a wealth of literature focused on the energy flow control of hybrid electric vehicles (Moreno, 2006), but publications devoted to the analysis of the electric and hydraulic parts of the vehicles are relatively rare. Energy efficiency and the analysis of losses are, however, gaining importance in all fields. This paper first addresses the idea of direct electric drive control of a hydraulic system and then in more details the effect of the type of the electric machine on the efficiency of an electric energy recovery system of a forklift.

First, the scheme and principle of the novel energy recovery system are described, and then, a theoretical evaluation of the system is performed step by step. A theoretical estimation of the motor efficiency was carried out for two different motor types. The differences in the energy efficiency between the two motors used in the test setup are discussed. Finally, conclusions are given.

2. Overview of the test setup

Traditional light forklifts use accumulators to supply electric energy. The lifting control is often based on valve-controlled hydraulic servo systems (Jelali, 2003). The experimental test setups on which the work is based are illustrated in Fig. 2. There are two setups to evaluate the effect of electric machine itself on the energy efficiency in a forklift. The setups are based on a commercial battery operated forklift, equipped either with a low voltage permanent magnet synchronous motor drive or with a safe voltage induction motor drive. The control of the experimental setup is quite different from traditional forklifts as it uses a speed-controlled electric servo drive rotating a hydraulic machine in both rotating directions to directly control the amount of hydraulic oil flow in the system. (Minav, 2008)

In the first (I) experimental case, the system has a network supply instead of a motor-generator set common in higher power hybrid drives. The rest of the components in the case of network supply are: 400 V electric machine frequency converter ACSM1 by ABB, a 400 V permanent magnet electric machine, a hydraulic pump capable of operating also as a motor,

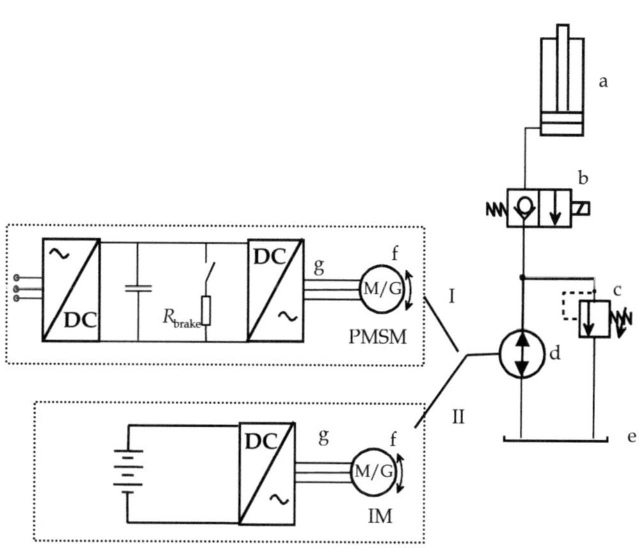

Fig. 2. Structures of the hydraulic and electric system to be tested. The experimental system consists of: a) single-acting cylinder, b) two-way normally closed poppet valve, c) pressure relief valve, d) hydraulic machine acting both as a pump or a motor, e) oil tank, f) electric machine operating both as a motor or a generator, g) frequency converter, and brake resistor R_{brake} (Minav, 2008).

a two-way normally closed poppet valve , an pressure relief valve, and a hydraulic cylinder. In the second (II) safe voltage case the 34 V motor was an induction motor manufactured by Danaher Motion and the converter manufactured by ZAPI. The battery voltage was 48 V. The hydraulic pump with a certain displacement moves a certain amount of oil from the tank to the rest of the hydraulic circuit where the piston should have a corresponding movement. The pump raises the oil pressure to the level required by the load. The flow runs through a two-way, normally closed poppet valve (b), which is here referred to as a control valve. Its role is to prevent accidental uncontrolled fast lowering of the load. The hydraulic pump produces a flow depending on the displacement of the pump and the rotating speed of the servo motor. A pressure relief valve (c) keeps the pressure under a maximum level in the system. The unique feature of the energy recovery system is that the oil travels through the system through the same route when lifting and lowering a load. While lowering a mass, the potential energy of the load produces a flow that rotates the hydraulic machine as a motor, and the mechanically connected electric motor acts as a generator, which is controlled by the frequency converter. Hence, the generator controls the amount of fluid flow and the position of the fork during lowering, instead of traditional valve control without energy recovery. Here, the converter rectifies the generated electric energy to the DC link where an accumulator should be in the case of a hybrid drive.

An upper level controller controls both the electrical and hydraulic parts of the forklift system. The performance of the system can, therefore, be easily controlled at different speeds when lifting or lowering masses. The system recovers as much energy as possible during a lowering movement. The test setup was equipped with pressure, current and voltage measurement sensors to define the efficiencies of the different parts of the system.

An example of the measured data for an internal gear pump is shown in Figures 3–5. In Figure 3 is shown the measured speed and current of the PMSM. The torque is the estimated torque of the PMSM. U_{dc} is the DC voltage in the DC link of the ACSM1, and the RO status shows the position of the relay that controls the two-way normally closed poppet valve.

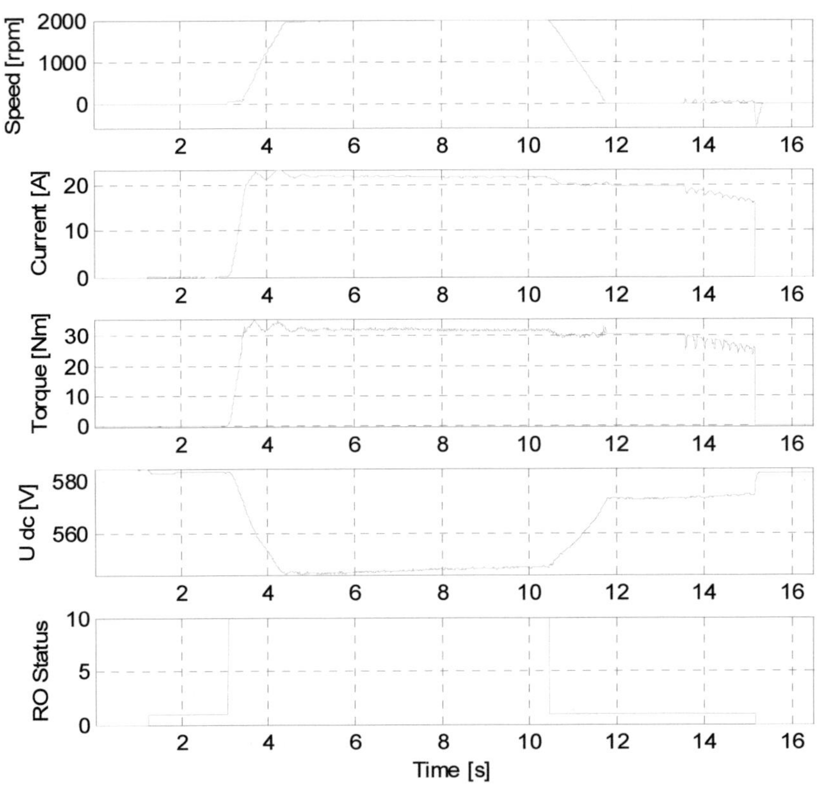

Fig. 3. Example of the data from the ACSM1 frequency converter (Minav, 2011d).

Figure 4 shows the measured phase voltages and currents measured with the Yokogawa PZ4000 power analyzer.

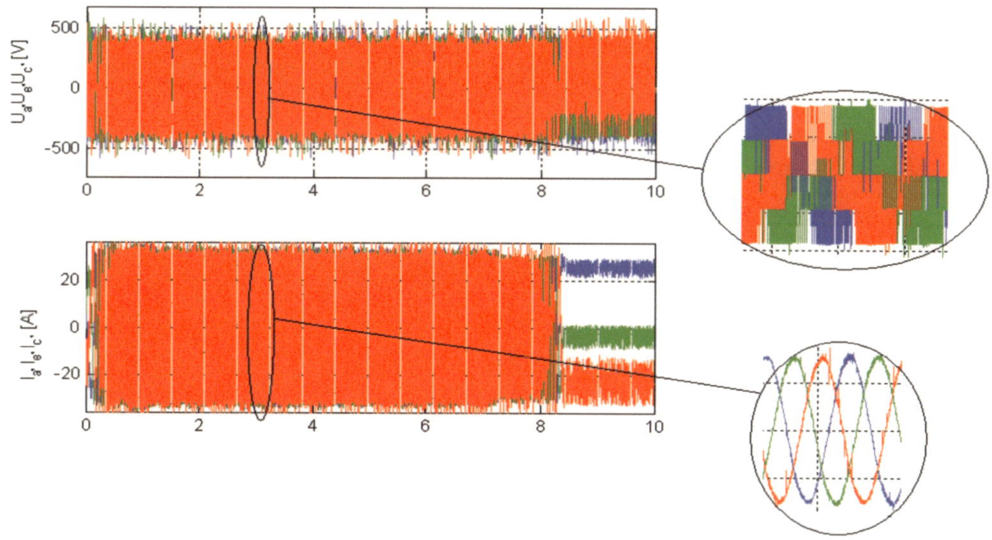

Fig. 4. Example of the data measured with the Yokogawa PZ4000 power analyzer (Minav, 2011d).

Figure 5 shows an example of measured DC current and DC voltage from the DC link with the Yokogawa PZ4000 power analyzer during a lowering motion.

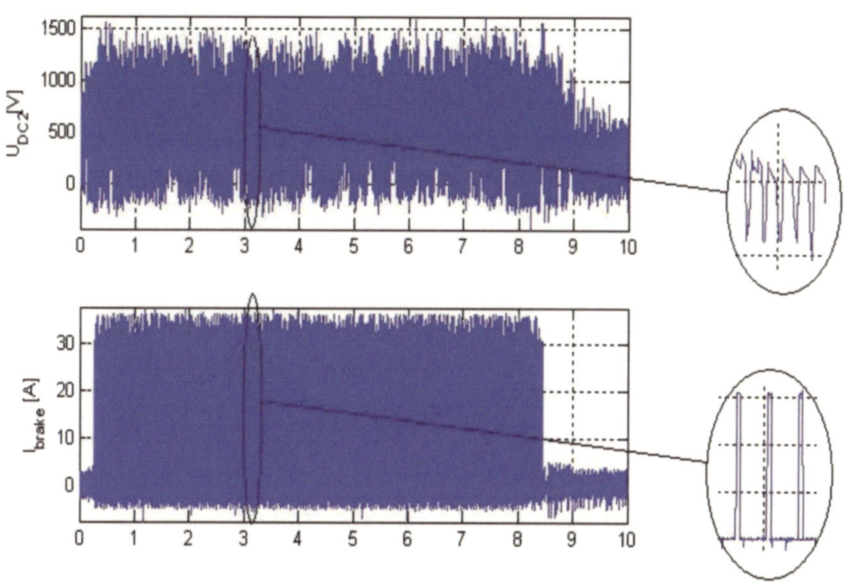

Fig. 5. Example of the measured DC current and DC voltage with the Yokogawa PZ4000 power analyzer during a lowering motion (Minav, 2011d).

3. Theoretical model of the system

The efficiency evaluation of the test setup is based on theoretical information from the literature and the material provided by manufacturers. With a theoretical model, it is possible to track the problem points in a setup and find targets to improve the performance of the system. The energy efficiency of the proposed hydraulic system was calculated as

$$\eta_{SYS} = \eta_{inv} \cdot \eta_{EM} \cdot \eta_{HM} \cdot \eta_{VP} \cdot \eta_{G} \cdot \eta_{C}, \quad (3.1)$$

where the subscript INV denotes the inverter, EM the electric machine, HM the hydraulic machine, VP the valves and pipes, G the mechanical gears, and C the hydraulic cylinder.

3.1 Electric machine

By and large, the most significant contributors to the described hydraulic lifting system efficiency are the pump and the electrical machine. Two different machine types were chosen to be evaluated. The first is a permanent magnet synchronous servo motor that was chosen for the application because of its high efficiency and high overload capability. The servo motor used in the tests was a 10 kW PMSM (CFM112M) motor by SEW-Eurodrive with 30 Nm nominal torque, 10 A nominal current, 108 Nm maximum torque, and 80 A maximum current (SEW, 2007). The second machine is an induction motor Danaher Motion TSP112/4-150-T 3-Phase AC with 56 Nm nominal torque, 88 Nm maximum torque, and 241 A rated current (Danaher, 2011). The different machine types with different voltages can be fairly compared as the voltage level itself does not affect the efficiency of the drive. Actually, the MOSFET based ZAPI converter has a slightly higher efficiency (c. 99 % at the rated point) than the 400 V IGBT converter (c. 98 % at the rated point). The electric power loss analysis is based on the testing and redesign of the PMSM and the IM (Pyrhönen, 2008). A detailed description and calculations using the machine parameters measured in a fixed temperature for the PMSM are given in (Minav, 2011a). The PMSM data including the manufacturer's data, measured and calculated motor parameters are shown in Table 3.1.

Parameter	Value
Rated power, kW	10
Rated voltage, V	400
Rated speed, rpm	3000
Rated current, A	10
Number of stator slots	36
Air gap, m	0.0012
Rotor diameter, m	0.0882
Outer diameter, m	0.17
Rotor length, m	0.1193
Number of pole pairs	3

Table 3.1. PMSM motor parameters.

Figure 6 shows the calculated efficiency of the PMSM CFM112M with some measured efficiency points.

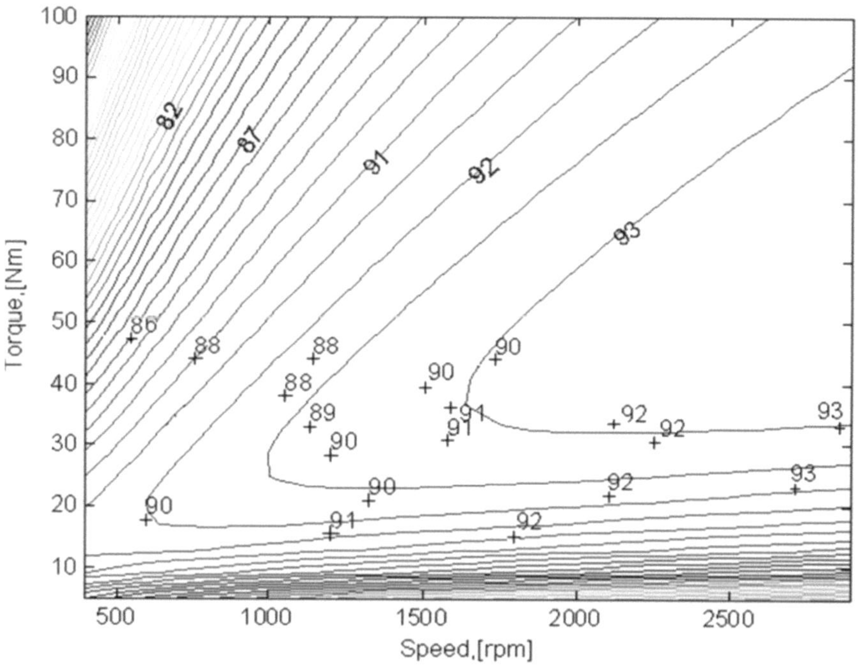

Fig. 6. Efficiency of the 10 kW CFM112M PMSM (solid line). Some measured efficiency points in lifting are indicated by + signs (Minav, 2011a).

A procedure similar to (Minav, 2011a) to produce theoretical system efficiency was performed for the induction machine test setup. This induction machine is the original safe voltage induction motor with which the lifting function of the original battery operated forklift was equipped.The electric power loss analysis for the IM is based on testing and redesign of the machine based on (Pyrhönen, 2008). The IM motor parameters are shown in Table 3.2.

Parameter	Value
Rated power, kW	10
Rated voltage, V	34
Rated speed, rpm	2300
Synchronous speed, rpm	2400
Rated current, A	241
Number of stator slots	36
Number of rotor slots	46
Air gap diameter, m	0.1101
Rotor diameter, m	0.1096
Outer diameter, m	0.204
Rotor length, m	0.15
Number of pole pairs	2

Table 3.2. IM motor parameters (Galkina, 2008; Danaher, 2011)

The resistive losses of the stator, also known as copper losses, are calculated as

$$P_{Cus} = 3 \cdot R_s \cdot I_s^2 , \tag{3.2}$$

where P_{Cus} is the stator winding copper loss (W), R_s is the stator AC resistance (Ω) in the average operating temperature, and I_S the stator current (A). The resistive losses of the rotor are calculated as follows:

$$P_{Cur} = 3 \cdot R'_r \cdot I'^2_r , \tag{3.3}$$

where P_{Cur} is the rotor winding copper loss (W), R'_r is the rotor AC resistance (Ω) referred to the stator in the average operating temperature, and I_r' is the rotor current (A) referred to the stator.

The secondary losses comprise the core losses and the additional losses. The largest contribution to the secondary losses is due to the harmonic energies generated when the motor operates under load. These energies are dissipated as currents in the copper windings, harmonic flux components in the iron parts, and as leakages in the laminate core. The core losses at the rated point are:

$$P_{Fe} = 3 \cdot \frac{|\underline{E}_m|^2}{R_{Fe}(f)}, \tag{3.4}$$

where R_{Fe} is the core loss resistance (Ω) and E_m is the air gap voltage (V), calculated by

$$\underline{E}_m = \underline{U}_{sph} - \underline{I}_s \cdot \underline{Z}_s \tag{3.5}$$

Where $\underline{U}_{sph}$ is the phase voltage (V) and $\underline{Z}_s$ is the stator circuit impedance (Ω) including the stator resistance and leakage. The Iron loss is frequency dependent as a function of f^2 which is taken into account in the efficiency calculations as a function of speed.

The additional losses in the IM were calculated as

$$P_{ad} = 3 \cdot |\underline{I}_s| \cdot |\underline{U}| \cdot \cos\varphi_x \cdot 0.5 \cdot 10^{-2}, \tag{3.6}$$

The sum of losses P_{Los} in the IM machine are calculated as

$$P_{Los} = P_{Cus} + P_{Cur} + P_{Fe} + P_{ad}, \tag{3.7}$$

The efficiency η_{IM} of the electric machine is obtained by

$$\eta_{IM} = \frac{P}{P + P_{Los}} \cdot 100 , \tag{3.8}$$

where P is the shaft power (W).

Calculations based on (3.2–3.8) were repeated for a fixed torque for all speeds. The machine parameters, except the iron loss resistance, were kept as constants calculated in the rated operating point.

Figure 7 shows the calculated efficiency of the IM with the measured efficiency points. The measured points indicate that the modelling has been fairly accurate for this purpose

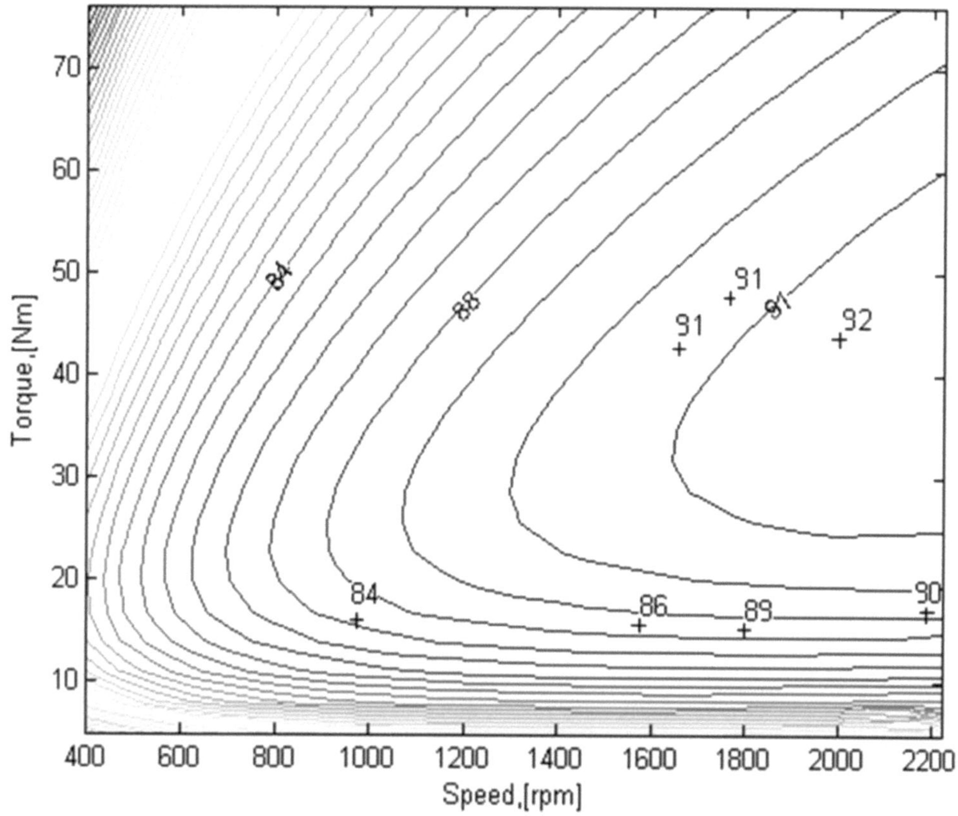

Fig. 7. Efficiency of the 10 kW IM (solid line). Some measured efficiency points in lifting are indicated by + signs.

3.2 Frequency converter

In the experimental test setup I, a servo motor inverter ACSM1 by ABB was used. The ACSM1 high-performance machinery drive provides speed, torque, and motion control for the PMSM. The ACSM1 can control servo motors with or without feedback of the motor speed. It uses Direct Torque Control (DTC) motor control technology to guarantee high performance (Pyrhönen, 1998). The maximum possible efficiency of the ACSM1 frequency converter is in the range of 98 % (ABB, 2007). In the experimental test setup II, a motor converter with MOSFET switches manufactured by ZAPI was used.

3.3 Hydraulic machine and system

In the experiments, we used an internal gear pump with a fixed displacement, as the hydraulic machine is capable of working also as a motor (Erkerle, 2007). Machines of this type are highly efficient in the tested operation range (Minav, 2011c). Fig. 8 gives the efficiencies of continuous-travelling positive displacement machines. A pressure of 12 MPa corresponds to the maximum tested 920 kg payload. At this point the internal gear pump has the efficiency of 84 %. A pressure of 5 MPa corresponds to 0 kg payload in our test arrangement, and the internal gear pump has the efficiency of 87 % at this point.

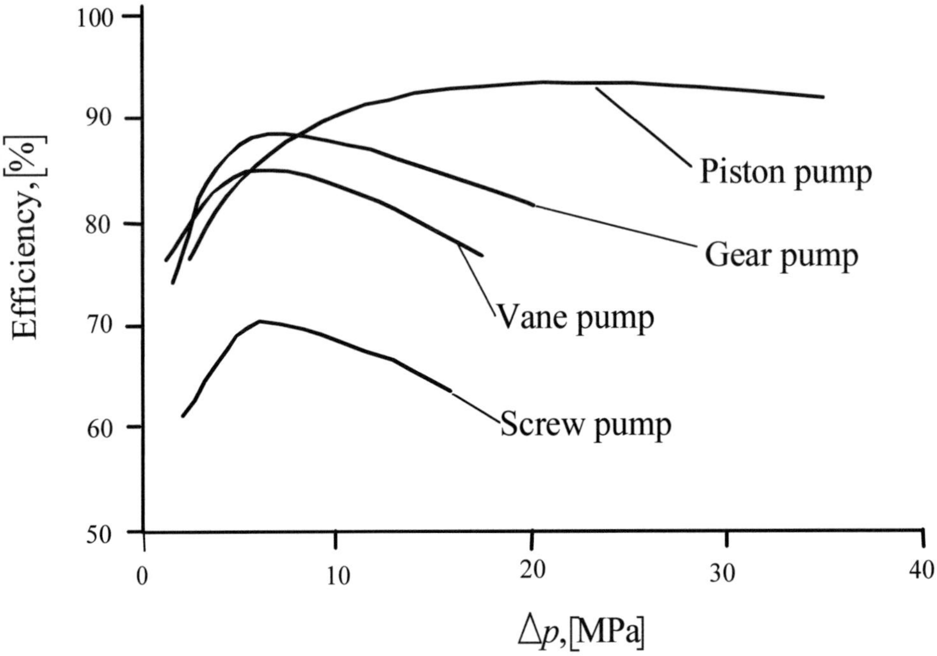

Fig. 8. Comparison the of the efficiencies of different hydraulic pumps operating at constant speed (Minav, 2011c).

The hydraulic losses in the piping systems consist of pipe friction losses in valves, elbows and other fittings, entrance and exit losses, and losses from changes in the pipe size by a reduction in the diameter. In our test setup, the power losses vary between 5–10 % depending on the operation mode. A single-acting cylinder is used in the experimental setup. It is known that the total seal friction of a hydraulic cylinder is 2–5 % of the total cylinder force (Majumdar, 2002). Based on the information from (Minav, 2011b), the cylinder efficiency was assumed to be 95 %. A mechanical chain gear is embedded in the fork construction. Its efficiency varies between 98 % and 99 %. Mechanical gears do not significantly affect the total efficiency, but the efficiency will be taken into account for calculating the system efficiency of the test setup.

The test setups in the two different cases are identical. Only the motor drives were different. Table 3.3 shows the differences in the motor efficiencies at different operating points.

Parameters		PMSM efficiency,[%]	IM efficiency, [%]	Difference, [%-units]
Speed, [rpm]	Torque, [Nm]			
500	15	91	80	11
	30	91	79	12
	45	86	76	10
1000	15	92	86	6
	30	92	88	4
	45	91	87	4
1500	15	90	87	3
	30	92	90	2
	45	92	90	2
2000	15	91	87	4
	30	92	91	1
	45	93	91	2

Table 3.3. Comparison of the efficiencies of the IM and the PMSM.

4. Empirical results

The model shows that the IM is not very efficient compared with the PMSM, and the same result is observed in the figure below. Also, there exists a difference in the performance of the IM in motoring and generating modes (Fig. 9). Figure 9 shows an example of the measurement results of the system efficiency.

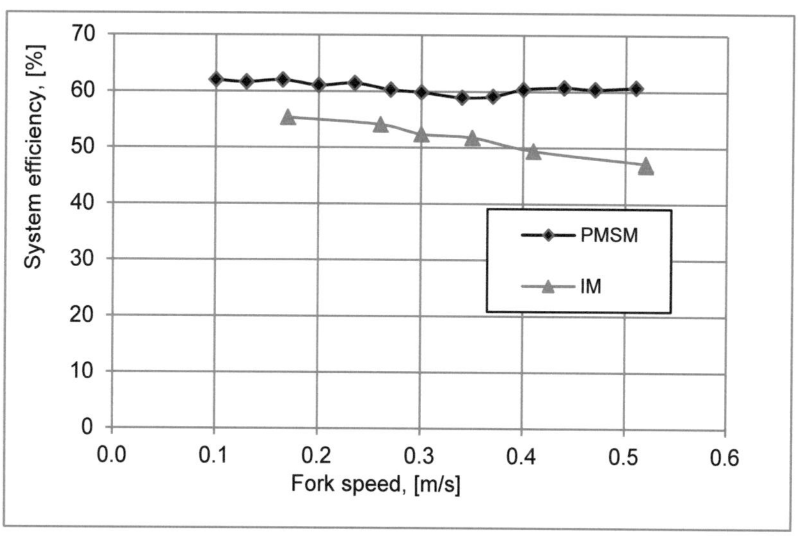

Fig. 9. System efficiency for lifting with 0 kg payload corresponding to 15 Nm shaft torque. In this case the PMSM brings clearly better results compared to the IM.

Figure 9 shows that the effect of the PMSM is significant. In lifting, the efficiency of the system can be improved from 50 % up to 60 %.

Figure 10 shows an example of the measurements of the system efficiency for lowering for different test setups.

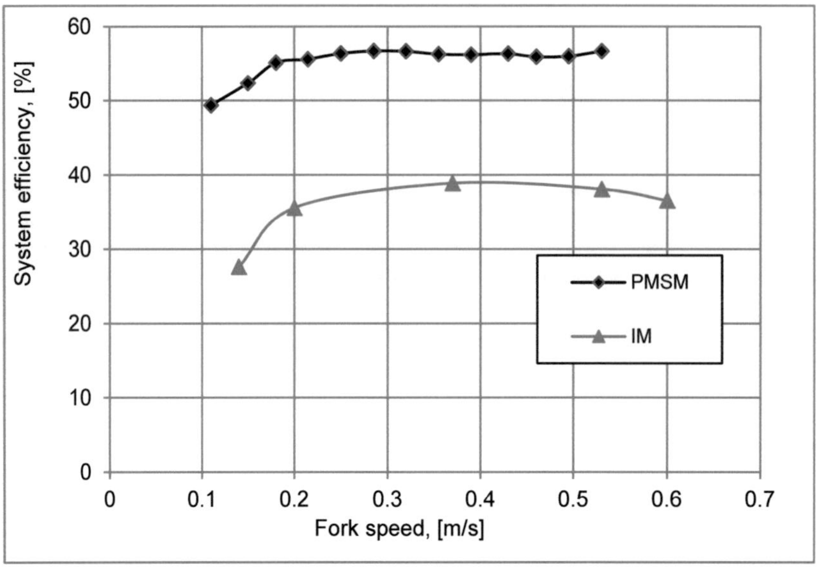

Fig. 10. System efficiency for lowering with 920 kg payload corresponding to 20 Nm shaft torque.

In lowering, the efficiency of the system can be improved from 40 % up to 60 %.It can be seen that in both cases (lifting and lowering), the PMSM setup has a significantly higher efficiency. Partly this high efficiency difference is based on a better matching of the PMSM properties to the load than the corresponding IM properties. The IM is selected according to S3 15 % operating principle and obviously suffers from under dimensioning in this case. However, in low power systems the permanent magnet excitation brings a big benefit from the energy efficiency point of view.

5. Conclusion

The energy efficiency of an electro-hydraulic forklift was studied. The hydraulic system and the electric parts of the working machine were evaluated and the theoretical approach was verified by practical experiments in order to determine the effect of the PMSM and the IM on the system efficiency. The energy recovered in the tests showed that the PMSM has a significant impact on the efficiency of the system. Our theoretical investigation predicted a possible improvement, which was then shown empirically. By choosing an appropriate motor, in this case the 10 kW PMSM, the total system efficiency can be improved during lifting even 14 percentage units and during lowering even 16 percentage units compared to the 10 kW IM. In lifting the best efficiency for IM was 58 % at low speed and zero payload. For PMSM, in lifting, the efficiency is higher, being around 60 % at all range of measured speeds. In lowering the system efficiency with PMSM varies from 49 % to 59 % with increasing speed with 920 kg payload, being clearly higher than for the system with IM, where the system efficiency varies from 27 % to 39 %.

6. Acknowledgments

The research was enabled by the financial support of Tekes, the Finnish Funding Agency for Technology and Innovation, European Union, European Regional Development Fund and Regional council of South Karelia and FIMA (Forum for Intelligent Machines) at the Department of Electrical Engineering, the Institute of Energy Technology, Lappeenranta University of Technology.

7. References

ABB: *Motion Control program Firmware*, 2007, http://www.abb.com, accessed November 13, 2007

Ahn, K. K., Ho, T. H. and Dinh, Q. T. (2008). A study on energy saving potential of hydraulic control system using switching type closed loop constant pressure system. *In Proceedings of the 7th JFPS International Symposium on Fluid Power*, TOYAMA 2008 September 15-18, 2008.

Andersen T. O., Hansen M. R., Pedersen H. C. and Conrad F., (2005), *Regeneration of potential energy in hydraulic forklift truck*, Proc. of the 6th International Conference on Fluid Power Transmission and Control, pp. 302-306, Hangzhou.

Berkner, P. (2008). How, Why, and When to apply electric motors to mobile hydraulic systems. In Parker Hannifin 2008 Global Mobile Sales Meeting & Symposium Whitepaper # 0001., [Online], [Accessed January 10, 2010]. Available at: http://www.allaboutmotion.com

Bhattacharya T., Giri V. S., Mathew V., Umanand L., (2009), Multiphase Bidirectional Flyback Converter Topology for Hybrid Electric Vehicles, *IEEE Trans. Ind. Electron.*, vol. 56, no. 1, pp. 78-84, Jan. 2009.

Burrows, C. R. (2005). Control and condition monitoring in electro-hydraulic systems. *In UKACC Control 2004 Mini Symposia*. [Online]. Available at: http: //www.ieee.com.

Danaher Motor, TSP112/4-150-T3 product datasheet, (2011) http://www.danahermotion.co.jp/products/files/AC_motor_sec.pdf, accessed September 10, 2011.

EPA (2011). *Clean Air Nonroad Diesel - Tier 4 Final Rule*, [Online], [Accessed 14 January 2010] available at: http://www.epa.gov/nonroaddiesel/2004fr.htm.

Fakham H., Lu D., Francois B., (2011),Power Control Design of a Battery Charger in a Hybrid Active PV Generator for Load-Following Applications, *IEEE Trans. on Industrial Electronics*, vol. 58, no. 1, pp.85 - 94, January 2011

Galkina A., (2008), *System onvestigation for hybrid electric vehicle*, Master's thesis, Dept. Electrical Eng., Lappeenranta University of Technology. , 80 pages.

Grbovic P. J., Delarue P., Le Moigne P., Bartholomeus P., (2011),The Ultracapacitor-Based Controlled Electric Drives With Braking and Ride-Through Capability: Overview and Analysis, *IEEE Trans. on Industrial Electronics*, vol. 58, no. 3, pp.925 - 936, March 2011

Hui, S. and Junqing, J. (2010). Research on the system configuration and energy control strategy for parallel hydraulic hybrid loader. *Journal Automation in Construction*, Vol. 19, Issue 2, March 2010, pp. 213–220.

Iannuzzi, D. (2008). Improvement of the Energy Recovery of Traction electrical drives using supercapacitors. *In Proceedings of the 13th International Power Electronics and Motion Control Conference.*

Inoue, H. (2008). *Introduction of PC200-8 Hybrid Hydraulic Excavators.* [Online]. Available at: http://www.komatsu.com/CompanyInfo/profile/report/pdf/161-E-05.pdf.

Jelali M., (2003,)*Hydraulic servo systems Modeling, identification and control*, Springer, 2003,p. 355.

Kunze, G. (2010). *Mobile construction machinery-trends and new developments.* ATZonline. April 2010. [Online]. available at: http: //www.atzonline.com.

Liang X. and Virvalo T., (2001) *What's wrong with energy utilization in hydraulic cranes*, Proc. 5th international conference on fluid power transmission and control, Hangzhou, China.

Lin, T., Wang, Q., Hu, B. and Gong W. (2010a). Development of hydraulic powered hydraulic construction machinery. *Journal Automation in construction.* Vol. 19, issue 1, January 2010, pp. 11–19.

Lin, T., Wang, Q., Hu, B. and Gong, W. (2010b). Research on the energy regeneration systems for hydraulic excavators. *Journal Automation in construction.* Vol. 19, issue 8, December 2010, pp. 1016–1026.

Liu C., Chau K.T., Jiang J.Z.,(2010), A Permanent-Magnet Hybrid Brushless Integrated Starter-Generator for Hybrid Electric Vehicles , *IEEE Trans. Ind. Electron.*, vol. 57, no.12, Dec. 2010.

Majumdar, S. R. (2002). *Oil Hydraulic Systems principles and maintenance.* New York: McGraw-Hill.

Mapelli F. L., Tarsitano D., Mauri M., (2010), Plug-In Hybrid Electric Vehicle: Modeling, Prototype Realization, and Inverter Losses Reduction Analysis , IEEE Trans. Ind. Electron., vol. 57, no. 2, Feb. 2010.

Minav T., (2008), *Electric energy recovery system for a hydraulic forklift*, Master's thesis, Dept. Electrical Eng., Lappeenranta University of Technology, 70 pages.

Minav T., Laurila L., and Pyrhönen J. ,(2011a), *Permanent Magnet Synchronous Machine Sizing: Effect on the Energy Efficiency of an Electro-hydraulic Forklift*, Transactions on Industrial Electronics, in press.

Minav T., Laurila L., Immonen P., Pyrhönen J., Vtorov V. and Niemelä M.,(2011b) *Electric energy recovery system for a hydraulic forklift – theoretical and experimental evaluation*, IET Electric Power Applications, Vol.5, issue 4, Apr. 2011. pp. 377–386.

Minav T., Laurila L., Pyrhönen, J. and Vtorov V., (2011c), *Direct pump control effects on the energy efficiency in an electro-hydraulic lifting system*, Journal International Review of Automatic Control, Vol.4, n.2, March 2011.

Minav T.A., (2011d), *Electric-drive-based control and electric energy regeneration in a hydraulic system*, Ph.D. dissertation, Dept. Elect. Eng, Lappeenranta University of Technology, Acta Universitatis Lappeenrantaensis 436, Lappeenranta, Finland, 2011.

Minav T.A., Laurila L. I. E., Immonen P., Haapala M. and Pyrhönen J. J., (2009). Electric energy recovery system efficiency in a hydraulic forklift, *Proceedings of the* EUROCON 2009, St.Petersburg, Russia, 2009.

Montazeri-Gh, M. Soleymani, M., (2010) Investigation of the Energy Regeneration of Active Suspension System in Hybrid Electric Vehicles , *IEEE Trans. Ind. Electron.*, vol. 57, no. 3, March 2010.

Moreno J., Ortúzar M. E. and Dixon J. W., (2006), *Energy-Management System for a Hybrid ElectricVehicle*, Using Ultracapacitors and Neural Networks, IEEE Trans. Ind. Electron., vol. 53, no. 2, April 2006.

Petrone G. A., Spagnuolo G., Vitelli M.,(2010), Low-Frequency Current Oscillations and Maximum Power Point Tracking in Grid-Connected Fuel-Cell-Based Systems, *IEEE Trans. Ind. Electron.*, vol. 57, no. 6, pp. 2042, June 2010.

Pyrhonen J., Jokinen T. and Hrábovcová V., (2008), *Design of rotating electrical machines* , John Wiley & Sons, 2008, p.538 .

Pyrhonen J., Niemela M., Kaukonen J., Luukko J. and Pyrhonen O., (1998), *Test results with the Direct Flux Linkage Control of Synchronous machine*, Aerospace and Electronic Systems Magazine, IEEE, vol. 4, pp. 23-27, Apr. 1998.

Rahmfeld, R. and Ivantysynova, M., (2001), *Displacement Controlled Linear Actuator with Differential Cylinder - A Way to Save Primary Energy in Mobile Machines.* In Proceedings of 5th International Conference on Fluid Power Transmission and Control (ICFP'2001), Hangzhou, China, pp. 316–322.

Rydberg, K.-E. (2005). Energy Efficient Hydraulic Systems and Regenerative Capabilities. *In Proceedings of the Ninth Scandinavian International Conference on Fluid Power SICFP'05*. Linköping.

Rydberg, K.-E. (2007). Design of Energy Efficient Hydraulic Systems - System Concepts and Control Aspects. *In Proceedings of the 5th International Symposium on Fluid Power Transmission and Control*. Beidaihe, China.

Saber A. Y., and Venayagamoorthy G. K.,(2010) Plug-in Vehicles and Renewable Energy Sources for Cost and Emission Reductions , *IEEE Trans. Ind. Electron.*, in press, Dec. 2010.

SEW Drive Engineering-Practical Implementation. http://www.sew-eurodrive.com, accessed November 13, 2007.

Wagner. (2010). *Tier IV emissions regulations: Knowing the facts.* [Online]. Available at: http: //www.wagnerequipment.com.

Yang H., Sun W., and Xu B., (2007), New Investigation in Energy Regeneration of Hydraulic Elevators, *IEEE/ASME Transactions on Mechatronics*, vol. 12, n. 5, October 2007

Yoon J.I., Kwan A. K., Truong D. Q., (2009), *A Study on An Energy Saving Electro-Hydraulic Excavator*, Proc. ICROS-SICE International Joint Conference 2009, Fukuoka, Japan, August 18-21.

Mattila, J. (2000). *On Energy-efficient motion control of hydraulic manipulator.* Doctoral thesis, Publications 312, Tampere University of Technology.

Paulides, J. J. H., Kazmin, E. V., Gysen, B. L. J. and Lomonova, E. A.(2008). Series Hybrid Vehicle System Analysis Using an In-Wheel Motor Design. In Proceedings of the

IEEE Vehicle Power and Propulsion Conference (VPPC), September 3-5, Harbin, China.

Eckerle (2007) Hydraulic division. EIPS 2 Internal gear pumps.[Online]. [Accessed 20 April 2008], available at: http: //www.eckerle.com.

7

Energy Efficiency Analysis in Agricultural Productions: Parametric and Non-Parametric Approaches

S. H. Mousavi Avval, Sh. Rafiee and A. Keyhani
Department of Agricultural Machinery Engineering
Faculty of Agricultural Engineering and Technology
University of Tehran
Karaj,
Iran

1. Introduction

The relation between agriculture and energy is very close. Agricultural sector itself is an energy user and energy supplier in the form of bio-energy (Alam et al., 2005). It uses large quantities of locally available non-commercial energies, such as seed, farmyard manure and animate energy, and commercial energies directly and indirectly in the form of electricity, diesel fuel, chemical fertilizers, plant protections, irrigation water and farm machinery (Kizilaslan, 2009).

Nowadays, energy usage in agricultural activities has been intensified in response to continued growth of human population, tendency for an overall improved standard of living and limited supply of arable land. Consequently, additional use of energy causes problems threatening public health and environment (Rafiee et al., 2010). However, increased energy use in order to obtain maximum yields may not bring maximum profits due to increasing production costs. In addition, both the natural resources are rapidly decreasing and the amount of contaminants on the environment is considerably increasing (Esengun et al., 2007).

Efficiency is defined as the ability to produce the outputs with a minimum resource level required (Sherman, 1988). In production, efficiency is a normative measure and is defined as the ratio of weighted sum of outputs to inputs or as the actual output to the optimal output ratio. The weights for inputs and outputs are estimated to the best advantage for each unit so as to maximize its relative efficiency. In order to measure the optimal input or output, it is necessary to first specify the production frontier (Mukherjee, 2008).

Efficient use of energy resources in agriculture is one of the principal requirements for sustainable agricultural productions; it provides financial savings, fossil resources preservation and air pollution reduction; for enhancing the energy efficiency it must be attempted to increase the production yield or to conserve the energy input without affecting the output (Singh et al., 2004). Therefore, energy saving has been a crucial issue for sustainable development in agricultural systems. Development of energy efficient agricultural systems with low input energy compared to the output of food can reduces the greenhouse gas emissions from agricultural production systems.

Improvements in the efficiency of resource use in agriculture require not only the definition of spatial and temporal use of current resources but also the development of tightly defined and broadly applicable indices (Topp et al., 2007).

In some studies the indicators of output energy to input energy ratio and energy productivity (i.e. yield to input energy ratio) in crop production systems have been used to evaluate the performance of farmers (Mohammadi et al., 2010; Unakitan et al., 2010). Energy productivity is an important indicator for more efficient use of energy although higher energy productivity does not mean in general, more economic feasibility (Mohammadi et al., 2010). The energy input-output analysis is usually made to measure the energy efficiency and environmental aspects. This analysis will determine how efficient the energy is used. In current years, several researches have been conducted on energy use for production of different agricultural crops (Jianbo, 2006; Meul et al., 2007; Kizilaslan, 2009).

Moreover, in some studies the parametric and non-parametric approaches have been used to analyze the efficiency of farmers in agricultural productions. In parametric approach, an econometric model is used to identify the relationship between energy inputs and yield values of crop productions. In this method, the parameters of the production or cost functions are estimated statistically.

Establishing the functional forms between energy inputs and output for agricultural crops are very useful in terms of determining elasticity of different energy inputs on yield (Turhan et al., 2008). Development of a model consists of several logical steps; one of them is the sensitivity analysis to ascertain how a given model depends on its input factors (Hamby, 1994). Sensitivity analysis quantifies the sensitivity of a model's state variables to the parameters defining the model. It refers to changes in the response of each of the state variables which result from small changes in the parameter values. Sensitivity analysis for the parameters of a developed model is valuable because it identifies those parameters which have most influence on the response of the model (Chalabi and Bailey, 1991). The sensitivity analysis of energy inputs on crop production is essential because it revealed what changes in energy inputs cause greater impacts on the output. Furthermore, it is of especial importance for the policy-makers to frame suitable policies for improving energy use efficiency (Lamoureux et al., 2006).

In recent years, many researchers have developed econometric models between energy inputs and output for different agricultural crops (Banaeian et al., 2010; Mohammadi and Omid, 2010). Singh et al. (2004) and Rafiee et al. (2010) investigated energy inputs and crop yield relationship to develop an econometric model for wheat and apple productions, respectively. Moreover, they applied the marginal physical productivity (MPP) technique to analyze the sensitivity of energy inputs on yield. Kulekci (2010) applied the stochastic frontier analysis technique in the Cobb-Douglas form to determine the technical efficiency for a sample of 117 randomly selected sunflower farms in Erzurum, Turkey. This method is parametric and uses statistical techniques to estimate the parameters of the function; however, this approach requires a pre-specification of the functional form and an explicit distributional assumption for the technical inefficiency term.

Mousavi-Avval et al. (2011b) analyzed the energy efficiency of soybean producers using parametric approach. In this study the Cobb-Douglas production function was applied to develop an aconometric model between inputs and output. The inputs were human labor, machinery, diesel fuel, chemicals, fertilizers, water for irrigation, electricity and seed energies; while, the soybean yield was the single output.

Data Envelopment Analysis (DEA) is a non-parametric linear programming (LP) based technique of frontier estimation for measuring the relative efficiency of a number of decision making units (DMUs) on the basis of multiple inputs and outputs (Zhang et al., 2009). The main advantage of non-parametric method of DEA compared to parametric approaches is that it does not require any prior assumption on the underlying functional relationship between inputs and outputs. It is, therefore, a non-parametric approach. In addition, DEA is a data-driven frontier analysis technique that floats a piecewise linear surface to rest on top of the observations (Zhou et al., 2008).

Due to the high advantages of DEA, there are a large number of its applications for evaluating the performances of DMUs in different issues. Also, it currently has been employed in some agricultural enterprises. In an earlier and related study, DEA was utilized to evaluate the technical efficiency of irrigated dairy farms in Australia. The results from this study proposed that DEA was a useful tool in helping to benchmark the dairy industry, which is continually striving to improve the productive efficiency of farms (Fraser and Cordina, 1999). Also, Dawson et al. (2000) presented technical and overall economic efficiencies for 22 rice farms in Philippines, using a frontier production function approach. The results of their study showed that the overall efficiency was changed from 84% to 95% between the farms.

In another study, DEA was applied to investigate the efficiency of individual farmers and to identify the efficient units for citrus farming in Spain (Reig-Martínez and Picazo-Tadeo, 2004); also, Barnes (2006) identified the technical efficiency scores of Scottish dairy farms by applying the DEA approach. Malana and Malano (2006) employed the DEA technique to benchmark the productive efficiency of irrigated wheat area in Pakistan and India.

In another study by Nassiri and Singh (2009), the DEA technique was subjected to the data of energy use from different inputs by individual paddy producers and the technical, pure technical and scale efficiencies of farmers were estimated. The results showed that, there was high correlation between technical efficiency and energy-ratio, however comparison between correlation coefficient of farmers in different farm categories and different zones showed that energy-ratio and specific energy were not enhanced indices for explaining of all kinds of the technical, pure technical and scale efficiency of farmers. Omid et al. (2010) employed this technique to analyze technical and scale efficiencies of cucumber producers.

Oilseed sunflower (*Helianthus annuus L.*) is one of the most widely cultivated and important oilseed crops in the world (Latif and Anwar, 2009). Sunflower seeds are mainly used for the production of oil for human consumptions. They contain a high amount of oil (26%) which is an important source of polyunsaturated fatty acid (linoleic acid) of potential health benefits (Pimentel and Patzek, 2005). The seeds are also used as a protein source for non-ruminant and ruminant animals. Iran produced more than 43000 tones of sunflower seeds from about 67000 ha harvested land area, in 2008 (FAO, 2008). In Iran, sunflower is mainly grown in Golestan province in the north-east of the country (Anonymous, 2010).

This study focuses on the capability of parametric and non-parametric approaches in energy efficiency analysis for agricultural crop productions. For this purpose the energy balance for sunflower production was investigated and the efficiency of energy use was analyzed using the parametric Cobb-Douglass production function and non-parametric data envelopment analysis techniques.

2. Materials and methods

2.1 Data collection

In this study Golestan province was chosen as a representative of the Iranian sunflower production enterprises since it is the main center of oilseed productions, especially sunflower crop, in the country, mainly due to the very favorable ecological conditions. Other oilseed crops cultivated in this province are soybean and canola.

Golestan province is located in the north-east of Iran, within 36° 30' and 38° 08' north latitude and 53° 57' and 56° 22' east longitude. A survey approach was used to collect the quantitative information on energy inputs used for the production of sunflower in the production period of 2009/10. The required sample size was determined using simple random sampling method as below (Cochran, 1977):

$$n = \frac{N \times s^2 \times t^2}{(N-1)d^2 + (s^2 \times t^2)} \tag{1}$$

where n is the required sample size, N is the number of oilseed sunflower producers in target population, s is the standard deviation, t is the t-value at 95% confidence limit (1.96), and d is the acceptable error. The permissible error in the sample size was defined to be 5% for 95% confidence. Thus the sample size was found to be 95, and then, 95 farmers from the population were randomly selected. Before collecting data, the survey form was pre-tested by a group of randomly selected farmers and these pre-tested surveys were not included in the final data set. The reliability of the questionnaires was tested using *Cronbach's alpha*. So, the *Cronbach's alpha* level of 0.7 demonstrated adequate construct reliability.

2.2 Energy balance analysis method

A standard procedure was used to convert each agricultural input and output into energy equivalents. The inputs used in sunflower production were in the form of chemical fertilizers (nitrogen, phosphate, potassium and sulfur), chemical biocides (herbicides, fungicides and insecticides), diesel fuel, electricity, farmyard manure, water for irrigation, human labor and machine power. Also, the grain yield was considered as output.

The energy equivalent may thus be defined as the energy input taking into account all forms of energy in agricultural productions. The energy equivalents were computed for all inputs and outputs using the conversion factors indicated in Table 1. Multiplying the quantity of the inputs used per hectare with their conversion factors gave the energy equivalents.

The energy equivalent associated with labor vary considerably, depending on the approach chosen; it must be adapted to the actual living conditions in the target region (Moore, 2010). In this study the energy coefficient of 1.96 MJ h^{-1} was applied. It means only the muscle power used in different field operations of crop production. Total energy embodied in machinery included energy for raw materials, manufacturing, repairs and maintenance, and energy for transportation. Taking into account the total weight and the life of machinery as used in practice, the energy required for each operation was calculated assuming that the embodied energy of tractors and agricultural machinery be depreciated during their economical life time (Beheshti Tabar et al., 2010); so, the embodied energy in machinery was calculated by multiplying the depreciated weight of machinery (kg ha^{-1}) with their energy equivalents (MJ kg^{-1}). Also, the weight of machinery depreciated per hectare of sunflower production during the production period was calculated as follows (Mousavi Avval et al., 2011):

$$TW = \frac{G \times W_h}{T} \tag{2}$$

where TW is the depreciated machinery weight (kg ha^{-1}); G is the total machine weight (kg); W_h is the time that machine used per unit area (h ha^{-1}) and T is the economical life time of machine (h).

Depending on the context, manure may be considered either a valuable source of nutrients replacing synthetic fertilizers, a waste product from livestock production, or a potential energy source, e.g. for biogas production. In this study, we regarded manure as a source of nutrients, and the substitution method was used to calculate the energy input of animal manure, which equates the energy equivalent of farmyard manure with that of mineral fertilizer equivalents corresponding to its fertilization effect (Liu et al., 2010).

The energy equivalent of water for irrigation input means indirect energy of irrigation consist of the energy consumed for manufacturing the materials for the dams, canals, pipes, pumps, and equipments as well as the energy for constructing the works and building the on-farm irrigation systems (Khan et al., 2009).

Following the calculation of energy input and output equivalents, the indices of energy ratio (energy use efficiency), energy productivity, specific energy (energy intensity) and net energy were calculated as follow (Rafiee et al., 2010):

$$Energy\ ratio = \frac{Energy\ output\ (MJ\,ha^{-1})}{Energy\ input\ (MJ\,ha^{-1})} \tag{3}$$

$$Energy\,productivity = \frac{Sunflower\ yield(kg\,ha^{-1})}{Energy\ input(MJ\,ha^{-1})} \tag{4}$$

$$Specific\ energy = \frac{Energy\ input(MJ\,ha^{-1})}{Sunflower\ yield(kg\,ha^{-1})} \tag{5}$$

$$Net\,energy = Energy\,output(MJ\,ha^{-1}) - Energy\,input(MJ\,ha^{-1}) \tag{6}$$

Energy ratio index is the ratio between the caloric heat of the output products and the total sequestered energy in the production factors. This index allows us to know the influence of inputs expressed in energy units in obtaining output energy. To improve energy ratio in a

Inputs	Unit	Energy equivalent (MJ unit^{-1})	Reference
A. Inputs			
1. Human labor	h	1.96	(Rafiee et al., 2010)
2. Machinery	kg		
a. Tractor		93.61	(Canakci et al., 2005)
b. Self propelled combine		87.63	(Canakci et al., 2005)
c. Other machinery		62.70	(Canakci et al., 2005)
3. Diesel fuel	L	47.80	(Canakci et al., 2005)
4. Chemicals	kg		
a. Herbicides		238.00	(Erdal et al., 2007)
b. Fungicides		216.00	(Erdal et al., 2007)
c. Insecticides		101.20	(Erdal et al., 2007)
5. Fertilizer	kg		
a. Nitrogen		66.14	(Rafiee et al., 2010)
b. Phosphate (P_2O_5)		12.44	(Rafiee et al., 2010)
c. Potassium (K_2O)		11.15	(Rafiee et al., 2010)
d. Sulfur (S)		1.12	(Rafiee et al., 2010)
e. Farmyard manure		0.30	(Rafiee et al., 2010)
6. Water for irrigation	m^3	1.02	(Rafiee et al., 2010)
7. Seed	kg	3.60	(Beheshti Tabar et al., 2010)
8. Electricity	kWh	11.93	(Mobtaker et al., 2010)
B. Output			
1. Sunflower	kg	25.00	(Beheshti Tabar et al., 2010)

Table 1. Energy equivalent of inputs and output in sunflower production.

process, it is possible either to reduce the energy sequestered in the inputs by optimization of energy use or to increase the yield of product by reducing the losses (Kitani, 1999). Energy productivity is the measure of the amount of a product obtained per unit of input energy. Also, net energy gain is the difference between the gross energy output produced and the total energy required for obtaining it.

The energy associated with inputs comes from different sources which classified as renewable and non-renewable energy forms. Renewable energy (RE) consists of water for irrigation, human labor and seeds, whereas non-renewable energy (NRE) includes machinery, diesel fuel, electricity, fertilizers and chemicals energy inputs. On the other hand, energy demand in agricultural productions can be classified in two main groups including direct and indirect forms. Direct energy (DE) form covers human labor, diesel fuel, water for irrigation and electrical energy; while, indirect energy (IDE) includes energy embodied in fertilizers, chemicals, seeds and machinery (Mobtaker et al., 2010).

2.3 Parametric approach

2.3.1 Model development

There are several parametric and non-parametric techniques to measure productive efficiency. Parametric methods assume a particular functional form (e. g., Cobb-Douglas production function or a Translog function) between inputs and output and estimate the parameters of the production or cost functions statistically; however, in this approach it is difficult to separate inefficiency from random error.

In this study following the estimation of energy balance, the relation between energy inputs and output was investigated using a prior mathematical function relation. In specifying a fit relation, the Cobb-Douglass production function was selected as the best function in terms of statistical significance and expected signs of parameters.

The Cobb-Douglass function is a power relation has been used by several authors to investigate the relationship between energy inputs and output in agricultural crop productions (Singh et al., 2004; Mobtaker et al., 2010).

It can be specified in a mathematical form as follows (Singh et al., 2004):

$$Y_i = a_0 \prod_{j=1}^{k} X_{ij}^{a_j} e^{u_i} \qquad (i = 1,2,\ldots,n; j = 1,2,\ldots,\mathbf{k}) \tag{7}$$

Using a linear presentation, the function to be estimated could be written as:

$$\ln Y_i = a_0 + \sum_{j=1}^{k} a_j \ln\left(X_{ij}\right) + u_i \tag{8}$$

where: Y_i, denotes the yield of the ith farmer, X_{ij}, is the jth input used by the ith farmer for the cultivation of crop, α_0, is a constant term, α_j, represent the regression coefficients of jth input, which is estimated from the model and u_i, is the error term.

In this functional form the parameters to be estimated α_j represent the elasticity of output with respect to each input j which implies the percent change in output augmentation from a 1% increase in the jth input.

Assuming that when the energy input is zero, the crop production is also zero, Eq. (8) reduces to (Mousavi-Avval et al., 2011b; Samavatean et al., 2011):

$$\ln Y_i = \sum_{j=1}^{k} a_j \ln\left(X_{ij}\right) + u_i \tag{9}$$

Assuming that yield is a function of energy inputs, for investigating the impact of each input energy on sunflower yield, the Eq. (9) can be expanded in the following form;

$$\begin{aligned} \ln Y_i = {} & a_1 \ln X_1 + a_2 \ln X_2 + a_3 \ln X_3 + a_4 \ln X_4 + a_5 \ln X_5 \\ & + a_6 \ln X + a_7 \ln X_7 + a_8 \ln X_8 + u_i \end{aligned} \tag{10}$$

where X_j (i=1, 2, ..., 8) stand for energy inputs of human labor (X_1), machinery (X_2),diesel fuel (X_3), chemicals (X_4), total fertilizers (X_5), water for irrigation (X_6), electricity (X_7) and seed (X_8) in MJ per hectare unit.

With respect to this pattern, the impacts of DE and IDE, and the effect of RE and NRE forms on the production yield were investigated. So, the Cobb-Douglass function was selected and specified as the following forms:

$$\ln Y_i = \beta_1 \ln DE + \beta_2 \ln IDE + u_i \tag{11}$$

$$\ln Y_i = \gamma_1 \ln RE + \gamma_2 \ln NRE + u_i \tag{12}$$

where Y_i is the ith farmer's yield (kg ha^{-1}), β and γ are coefficient of exogenous variables. Eqns. (10) to (12) were estimated using Ordinary Least Square (OLS) technique.

2.3.2 Returns to scale

Following the estimation of econometric model, in order to describe the changes in output subsequent to a proportional change in all the inputs (when all inputs change by a constant factor) the return to scale index was investigated (Rafiee et al., 2010). In a Cobb-Douglas production function, the sum of elasticity values derived in the form of regression coefficients represent the degree of returns to scale. If the sum of coefficients is more than, equal to, or less than one, implying that there is increasing, constant, or decreasing returns to scale, respectively (Rafiee et al., 2010). Increasing (decreasing) returns to scale indicate that an increase in the input resources produces more (less) than proportionate increase in outputs. In this study, an increasing, constant and decreasing return to scale indicate that when all of the energy inputs are increased by X value, then the sunflower yield increases by more than, exactly and less than X value, respectively.

2.3.3 Sensitivity analysis

Since several parameters affect the model output, the sensitivity analysis of energy inputs on yield was used to identify which factors had a greater effect on the production yield. For this purpose, the marginal physical productivity (MPP) method, based on the response coefficients of the inputs was utilized. The MPP of a factor implies the change in the total output with a unit change in the factor input, assuming all other factors are fixed at their geometric mean level.

A positive value of MPP of any input variable identifies that the total output is increasing with an increase in input; so, one should not stop increasing the use of variable inputs so long as the fixed resource is not fully utilized. A negative value of MPP of any variable input indicates that every additional unit of input starts to diminish the total output of previous units; therefore, it is better to keep the variable resource in surplus rather than utilizing it as a fixed resource.

The MPP of the various inputs was calculated using the a_j of the various energy inputs as follows (Singh et al., 2004; Rafiee et al., 2010):

$$MPP_{xj} = \frac{GM(Y)}{GM(X_j)} \times a_j \tag{13}$$

where MPP_{xj} is the marginal physical productivity of jth input, a_j, the regression coefficient of jth input, $GM(Y)$, geometric mean of yield, and $GM(X_j)$, the geometric mean of jth input energy on per hectare basis.

2.4 Non-parametric approach

Apart from the parametric approach, in this study a non-parametric method of DEA was employed to evaluate the technical, pure technical and scale efficiencies of individual farmers which use similar inputs, produce the same product (sunflower) and operate in a relatively homogenous region (e. g., topography, soil type, climatic conditions, etc.). So, energy consumptions from different inputs including human labor, machinery, diesel fuel, chemicals, total fertilizers, water for irrigation, electricity and seed energies in terms of MJ ha^{-1} were considered as inputs; while, the grain yield of sunflower was the single output; also, each farmer called a DMU.

DEA technique builds a linear piece-wise function from empirical observations of inputs and outputs. DEA is a nonparametric approach for estimating productive efficiency based on mathematical linear programming techniques. Unlike parametric methods, DEA does not require a function to relate inputs and outputs. The DEA envelops the data in such a way that all observed data points lie on or below the efficient frontier (Coelli, 1996). The efficient frontier is established by efficient units from a group of observed units. Efficient units are those with the highest level of productive efficiency.

In DEA an inefficient DMU can be made efficient either by minimizing the input levels while maintaining the same level of outputs (input oriented), or, symmetrically, by maximizing the output levels while holding the inputs constant (output oriented). Sunflower production similar to wheat (Malana and Malano, 2006), paddy (Chauhan et al., 2006) and greenhouse cucumber (Omid et al., 2010) productions relies on finite and scarce resources; therefore the use of input-oriented DEA models is more appropriate to reduce inputs consumed in the production process.

In this study, the technical, pure technical and scale efficiencies of farmers were analyzed.

2.4.1 Technical efficiency (TE)

TE can be defined as the ability of a DMU (e.g. a farm) to produce maximum output given a set of inputs and technology level. The TE score (θ) in the presence of multiple-input and output factor can be calculated by the ratio of sum of weighted outputs to the sum of weighted inputs or in a mathematical expression as follows (Cooper et al., 2004):

$$\theta = \frac{u_1 y_{1j} + u_2 y_{2j} + \ldots + u_s y_{sj}}{v_1 x_{1j} + v_2 x_{2j} + \ldots + v_m x_{mj}} = \frac{\sum_{r=1}^{s} u_r y_{rj}}{\sum_{i=1}^{m} v_i x_{ij}} \tag{14}$$

Let the DMU_j to be evaluated on any trial be designated as DMU_o (o = 1, 2, . . ., n). To measure the relative efficiency of a DMUo based on a series of n DMUs, the model is structured as a fractional programming problem as follows (Cooper et al., 2006):

$$Max: \ \theta = \frac{\sum_{r=1}^{s} u_r y_{ro}}{\sum_{i=1}^{m} v_i x_{io}} \tag{15}$$

S.t:

$$\frac{\sum_{r=1}^{s} u_r y_{rj}}{\sum_{i=1}^{m} v_i x_{ij}} \leq 1 \qquad j = 1, 2, \ldots, n$$

$$u_r \geq 0, \; v_i \geq 0$$

where n is the number of DMUs in the comparison, s the number of outputs, m the number of inputs, u_r ($r = 1, 2, \ldots, s$) the weighting of output y_r in the comparison, v_i ($i = 1, 2, \ldots, m$) the weighting of input x_i, and y_{rj} and x_{ij} represent the values of the outputs and inputs y_j and x_i for DMU$_j$, respectively. Eq. (15) can equivalently be written as a linear programming (LP) problem as follows (Cooper et al., 2006):

$$\text{Max: } \theta = \sum_{r=1}^{s} u_r y_{ro} \tag{16}$$

S.t:

$$\sum_{r=1}^{s} u_r y_{rj} - \sum_{i=1}^{m} v_i x_{ij} \leq 0 \quad \mathrm{j} = 1, 2, \ldots, \mathrm{n}$$

$$\sum_{i=1}^{m} v_i x_{io} = 1$$

$$u_r \geq 0, \; v_i \geq 0$$

The dual linear programming (DLP) problem is simpler to solve than Eq. (16) due to fewer constraints. Mathematically, the DLP problem is written in vector-matrix notation as follows (Cooper et al., 2006):

$$Min : \theta \tag{17}$$

S.t:

$$\mathrm{Y}\lambda \geq \mathrm{y_o}$$

$$X\lambda - \theta x_o \leq 0$$

$$\lambda \geq 0$$

where y_o is the $s \times 1$ vector of the value of original outputs produced and x_o is the $m \times 1$ vector of the value of original inputs used by the oth DMU. Y is the $s \times n$ matrix of outputs and X is the $m \times n$ matrix of inputs of all n units included in the sample. λ is a $n \times 1$ vector of weights and θ is a scalar with boundaries of one and zero which determines the technical efficiency score of each DMU. Model (17) is known as the input-oriented CCR DEA model. It assumes constant returns to scale (CRS), implying that a given increase in inputs would result in a proportionate increase in outputs.

2.4.2 Pure technical efficiency (PTE)

The TE derived from CCR model, comprehend both the technical and scale efficiencies. So, Banker et al. (1984) developed a model in DEA, which was called BCC model to calculate the PTE of DMUs. The BCC model is provided by adding a restriction on λ (λ =1) in the model (17), resulted to no condition on the allowable returns to scale. This model assums variable returns to scale (VRS), indicating that a change in inputs is expected to result in a disproportionate change in outputs.

2.4.3 Scale efficiency (SE)

SE relates to the most efficient scale of operations in the sense of maximizing the average productivity. An scale efficient farmer has the same level of technical and pure technical efficiency scores. It can be calculated as below (Nassiri and Singh, 2009):

$$SE=\frac{TE}{PTE} \tag{18}$$

SE gives the quantitative information of scale characteristics. It is the potential productivity gained from achieving optimum size of a DMU. However, scale inefficiency can be due to the existence of either IRS or DRS. A shortcoming of the SE score is that it does not indicate if a DMU is operating under IRS or DRS conditions. This problem is resolvable by solving a non-increasing returns of scale (NIRS) DEA model, which is obtained by substituting the VRS constraint of λ =1 in the BCC model with $\lambda \le 1$ (Scheel, 2000). IRS and DRS can be determined by comparing the efficiency scores obtained by the BCC and NIRS models; so that, if the two efficiency scores are equal, then DRS apply, else IRS prevail (Omid et al., 2010). The information on whether a farmer operates at IRS, CRS or DRS status is particularly helpful in indicating the potential redistribution of resources between the farmers, and thus, enables them to achieve to the higher output (Chauhan et al., 2006).

The results of standard DEA models divide the DMUs into two sets of efficient and inefficient units. The inefficient units can be ranked according to their efficiency scores; while, DEA lacks the capacity to discriminate between efficient units. A number of methods are in use to enhance the discriminating capacity of DEA (Adler et al., 2002). In this study, the bencmarking method was applied to overcome this problem. In this method, an efficient unit which is chosen as the useful target for many inefficient DMUs and so appears frequently in the referent sets, is highly ranked.

In the analysis of efficient and inefficient DMUs, the energy saving target ratio (ESTR) was used to specify the inefficiency level of energy usage for the DMUs under consideration. The formula is as follows (Hu and Kao, 2007):

$$ESTR(\%)=\frac{(Energy\,saving\,t\arg et)}{(Actual\,energy\,input)}\times 100 \tag{19}$$

where energy saving target is the total reducing amount of energy inputs which could be saved without reducing the output level. A higher ESTR percentage implies higher energy use inefficiency, and thus, a higher energy saving amount (Hu and Kao, 2007). In this study, the Microsoft Excel spreadsheet, SPSS 17.0 software and the DEA software Efficiency Measurement Systems (EMS), Version 1.3 (Scheel, 2000) were employed to analyze the data.

3. Results and discussions

3.1 Energy balance in sunflower production

Amount of inputs and output in sunflower production are given in Table 2. Based on the evaluation of collected data, average human labor required in the study area was 131.7 h ha^{-1}. Approximately 37% of total human labor was used in harvesting, 30% in weeding and 13% in irrigation operations. Also, machine power mainly was used in harvesting and tillage operations. The use of diesel fuel for operating tractors, combine harvesters and water pumping systems was calculated as 72 L ha^{-1}. Moreover, sunflower production used 1 kg of chemicals, 28 kg of nitrogen and 16.6 kg of phosphate per hectare. As tabulated in the third column of Table 2, the total energy consumption during the production period of sunflower was found to be 9600 MJ ha^{-1}. In some related studies total energy input has been reported as 10491 MJ ha^{-1} for sunflower production in Greece (Kallivroussis et al., 2002), 18297.61 MJ ha^{-1} for canola (Unakitan et al., 2010), 14348.9 MJ ha^{-1} for cotton, 11366.2 MJ ha^{-1} for maize and 18680.8 MJ ha^{-1} for wheat production (Canakci et al., 2005). The average yield value of sunflower seed was found to be 1626.5 kg ha^{-1}; accordingly, the total output energy was calculated as 40663 MJ ha^{-1}.

Inputs	Quantity per unit area (ha)	Total energy equivalent (MJ ha^{-1})
A. Inputs		
1. Human labor (h)	131.7	258.1
2. Machinery (kg)	8.1	676.4
a. Tractor	2.6	239.6
b. Self propelled combine	3.7	323.6
c. Other machinery	1.8	113.2
3. Diesel fuel (L)	72.0	3440.4
4. Chemicals (kg)	1.0	205.0
a. Herbicides	0.7	176.6
b. Fungicides	0.1	10.2
c. Insecticides	0.2	18.1
5. Total fertilizer (kg)	2048.8	2696.7
a. Nitrogen	28.0	1852.3
b. Phosphate (P_2O_5)	16.6	205.9
c. Potassium (K_2O)	3.2	36.0
d. Sulfur (S)	2.7	3.1
e. Farmyard manure	1998.3	599.5
6. Water for irrigation (m^3)	644.8	657.7
7. Electricity (kWh)	137.0	1634.2
8. Seed (kg)	8.8	31.6
Total energy input		9600
B. Output		
1. Sunflower seed (kg)	1626.5	40663
Total energy output		40663

Table 2. Amounts of inputs, output and their energy equivalents for sunflower production.

The percentage distribution of the energy associated with the inputs is seen in Fig. 1. It is evident that, the greatest part of total energy input was consumed by diesel fuel (35.8%); followed by total fertilizer (29%). The distribution of total fertilizers energy input was 68.7% nitrogen, 7.7% phosphate, 1.3% potassium, 0.1% sulfur and 22.2% farmyard manure. Similar studies have also reported that diesel fuel and fertilizer were the most intensive energy inputs (Erdal et al., 2007; Kizilaslan, 2009; Mobtaker et al., 2010); Kallivroussis et al. (2002) reported that the main energy consuming inputs for sunflower production in Greece were nitrogen fertilizer (42.4%) and diesel fuel (33.9%). Excessive use of chemical fertilizers energy input in agriculture may create serious environmental consequences such as nitrogen loading in the environment and receiving waters, poor water quality, carbon emissions and contamination of the food chain (Khan et al., 2009). Integrating a legume into the crop rotation, application of composts, chopped residues or other soil amendments may increases soil fertility in the medium term and so reduces the need for chemical fertilizer energy inputs. Moreover, applying a better machinery management technique, employing the conservation tillage methods or technological upgrade to substitute fossil fuels with renewable energy resources may be the pathways to minimize the fossil fuel usage and thus to reduce its environmental footprints.

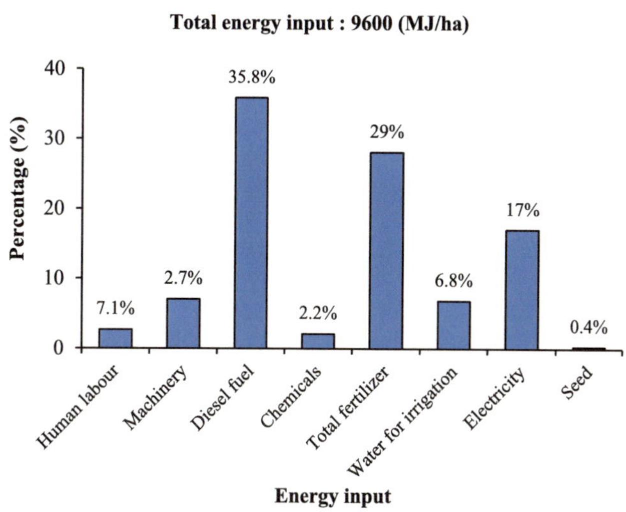

Fig. 1. The shares of energy inputs for sunflower production.

The energy indices including energy output to input ratio, energy productivity, specific energy and net energy gain are presented in Table 3. Energy ratio in sunflower production was found to be 4.24; showing that output energy of sunflower is obtained 4.24 times greater than total input energy. Also, specific energy was accounted as 5.90 MJ kg^{-1}. Energy output to input ratio and specific energy are integrative indices indicating the potential environmental impacts associated with the production of crops (Khan et al., 2009); also, these parameters can be used to determine the optimum intensity of land and crop management from an environmental point of view. Energy ratio in some agricultural crop productions were reported as 1.5 for sesame, 2.8 for wheat, 3.8 for maize, 4.8 for cotton (Canakci et al., 2005) and 4.68 for canola production (Unakitan et al., 2010). The average energy productivity of sunflower production was 0.17 kg MJ^{-1}. This means that 0.17 unit (kg) output is obtained per unit energy (MJ). Calculation of energy productivity for other oilseed crops has been reported as 0.17 kg MJ^{-1} for canola (Unakitan et al., 2010) and 0.18 kg MJ^{-1} for soybean production (De et al., 2001).

Item	Unit	Quantity
Energy ratio	-	4.24
Energy productivity	kg MJ^{-1}	0.17
Specific energy	MJ kg^{-1}	5.90
Net energy	MJ ha^{-1}	31062.7
Direct energy	MJ ha^{-1}	5990.4
Indirect energy	MJ ha^{-1}	3609.6
Renewable energy	MJ ha^{-1}	1546.8
Non-renewable energy	MJ ha^{-1}	8053.3
Total energy input	MJ ha^{-1}	9600

Table 3. Some energy indices in sunflower production.

The input energy classification used for sunflower production according to direct, indirect, renewable and non-renewable energy forms are presented in Table 3 and Fig. 2. It is evident that, the ratios of direct and indirect energy forms are nearly the same; but the ratios of renewable and non-renewable energy forms are fairly different from each other (Fig. 2). The ratio of non-renewable energy is very high (83.9%), indicating that sunflower production in the region depends mainly on fossil fuels. Several researchers have founded the ratio of DE higher than that of IDE, and the rate of NRE much greater than that of RE in production of different agricultural crops (Erdal et al., 2007; Mobtaker et al., 2010).

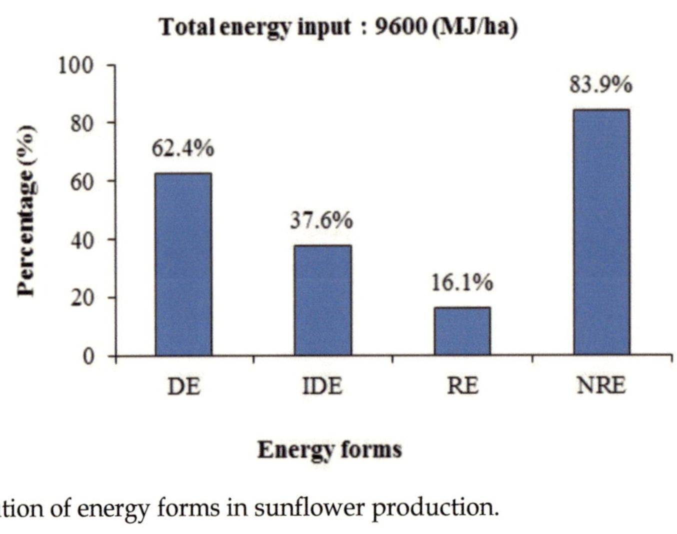

Fig. 2. Distribution of energy forms in sunflower production.

3.2 Econometric model estimation for sunflower production

The results of econometric model estimation of sunflower production are show in Table 4. For the data used in this study presence of autocorrelation in the residuals from the regression analysis was tested using the Durbin-Watson statistical test (Rafiee et al., 2010). The test result revealed that Durbin-Watson value was as 1.66 for Eq. (10); indicating that there was no autocorrelation in the estimated model. The R^2 (coefficient of determination) was as 0.98 for this linear regression model. This implies that all the explanatory variables included in the regression equation had contributed to the yield by 98%.

The estimated regression coefficients for the model 1 are presented in the second column of Table 4. The results revealed that, human labor machinery, diesel fuel, chemicals and seed energy inputs were the most important inputs, significantly contributed to yield. Also, all of the statistically significant inputs showed the positive relationships with output. Moreover, diesel fuel energy input had the highest elasticity on output (0.95). The second and third important energy inputs were machinery and seed with the elasticity values of 0.25 and 0.14, respectively. With respect to the obtained results, increasing 10% in the consumed energy from diesel fuel, machinery and seed energies, would led to 9.5%, 2.5% and 1.4%, increase in sunflower seed yield, respectively. On the other hand, the impacts of electricity, total fertilizer and water for irrigation energies on yield were estimated statistically insignificant and in the cases of total fertilizer and water for irrigation, the coefficients showed the negative relationship with output.

Endogenous variable: yield Exogenous variables	Coefficient	*t*-ratio	MPP
Model 1: $lnY_i = a_1lnX_1 + a_2lnX_2 + a_3lnX_3 + a_4lnX_4 + a_5lnX_5 + a_6lnX_6 + a_7lnX_7 + a_8lnX_8 + e_i$			
Human labor	0.06[c]	1.59	0.52
Machinery	0.25[b]	2.18	0.69
Diesel fuel	0.95[a]	5.58	0.47
Chemicals	0.02[b]	1.98	0.12
Total fertilizer	-0.01	-0.11	-0.01
Water for irrigation	-0.01	-0.71	-0.01
Electricity	0.02	1.42	0.00
Seed	0.14[a]	3.01	7.06
Durbin-Watson	1.66		
R^2	0.98		
Return to scale	1.42		

[a] Indicates significance at 1% level.
[b] Indicates significance at 5% level.
[c] Indicates significance at 10% level.

Table 4. Econometric model estimation for sunflower production.

Mobtaker et al. (2010) developed an econometric model for barley production in Hamedan province of Iran. They reported that human labor, total fertilizer, machinery, diesel fuel, electricity and water for irrigation energies were the important inputs, significantly contributed to yield and machinery energy had highest elasticity. Singh et al. (2004) found that the use of electricity and fertilizers energy inputs in zone 4 of Punjab was inconsistent with output of wheat production.

The degree of returns to scale for the model (1) was calculated by summation of the regression coefficients as 1.42. The value of return to scale more than unity implies increasing return to scale for sunflower production in the region. These results indicate that 1% increase in all the energy inputs would result by 1.42% increase in the sunflower production.

3.3 Sensitivity analysis of energy inputs on sunflower yield

The sensitivity of energy inputs was analyzed using the MPP method and partial regression coefficients on output level. The results are presented in the last column of Table 4. As it is seen, the major MPP value was drown by seed (7.06) and it followed by machinery (0.69) and human labor energies (0.52). This implies that an additional use of 1 MJ ha^{-1} from each of the seed, machinery and human labor energies would lead to an additional increase in yield value of sunflower by 7.06, 0.69 and 0.52 kg ha^{-1}, respectively. In other words, there is a higher potential for increasing output by additional use of these inputs for sunflower production in the surveyed region. On the other hand, the MPP value of total fertilizer and water for irrigation energies were found negative, indicating that use of these inputs is high for sunflower production, resulting in energy dissipation as well as imposing negative effects to environment and human health. The results of sensitivity analysis indicate that which variables should be identified and measured most carefully to assess the state of the environmental system, and which environmental factors should be managed preferentially (Drechsler, 1998). Within this framework, sensitivity analysis of energy inputs is important for improving energy use efficiency and lowering the environmental footprints of energy consumption.

For investigating the relationship between energy forms (i. e. direct, indirect, renewable and non-renewable) and the yield value of sunflower the models (2) and (3) were estimated using Eqs. (11) and (12), respectively. For these models the estimated coefficients, *t*-values, MPP values and validation statistical parameters are presented in Table 5. Durbin–Watson statistical test revealed that Durbin–Watson values were 1.31 and 1.15 for the models (2) and (3), respectively; indicating that there is no autocorrelation in the estimated models. The R^2 values were 0.98 for both the estimated models.

The results of model development between direct and indirect energies showed that both the forms of energy had the expected sign and the impact of direct energy was statistically significant, with an elasticity value of 0.84; while indirect energy form had no statistically significant impact on yield. These imply that a 10% increase in direct energy inputs would led to 8.4% increase in yield.

Looking at the Table 5 it also can be seen that, the regression coefficients of non-renewable energy forms was positive and significant at 1% level, while the impact of renewable energies was insignificant; also, the elasticity of non-renewable energy was higher than that of renewable energy (0.85 versus 0.02), implying that a 10% increase in non-renewable energy inputs would led to 8.4% increase in yield, while 10% increase in renewable energy resources increases the output by only 0.2%.

In the literature, similar results have been reported. For example, the impact of direct energy was more than the impact of indirect energy on yield (Hatirli et al., 2005), and the impact of non-renewable energy was higher than that of renewable energy (Mousavi-Avval et al., 2011a).

As can be seen from Table 5, the MPP values of direct and indirect energy forms were 0.29 and 0.02, respectively. Moreover, the sensitivity analysis of renewable and non-renewable energy forms showed that additional use of 1 MJ in non-renewable energies would lead to an additional increase in yield by 0.22 kg; while in the case of renewable energy forms only 0.04 kg is obtained by additional use of 1 MJ. Rafiee et al. (2010) reported that sensitivity of

Endogenous variable: yield Exogenous variables	Coefficient	*t*-ratio	MPP
Model 2: $lnY_i = \beta_1 lnDE + \beta_2 lnIDE + e_i$			
Direct energy	0.84[a]	3.01	0.29
Indirect energy	0.03	0.37	0.02
Durbin-Watson	1.31		
R^2	0.98		
Model 3: $lnY_i = \gamma_1 lnRE + \gamma_2 lnNRE + e_i$			
Renewable energy	0.02	0.36	0.04
Non-renewable energy	0.85[a]	3.52	0.22
Durbin-Watson	1.15		
R^2	0.98		

[a] Indicates significant at 1% level.

Table 5. Econometric model estimation of energy forms in sunflower production.

direct energy was higher than that of indirect energy; also it was higher for non-renewable energy compared to renewable energy forms.

These results may be due to the fact that renewable energy forms such as human labor and farmyard manure were used partially by only some of the farmers and its share was very low; while, non-renewable energy forms especially diesel fuel and machinery were used intensively by majority of the farmers. Additional use of non-renewable energy sources to boost agricultural productions in developing countries with low levels of technological knowledge not only results in environmental deterioration, but also confronts us with the dilemma of a rapid rate of depletion of energetic resources; while, renewable energy sources can be used indefinitely with minimal environmental impacts associated with their production and use (Fadai, 2007). Development of renewable energy usage technologies such as farm machinery or water pumping systems using biodiesel or solar power, employing integrated pest management technique and utilization of alternative sources of energy such as organic fertilizers (compost, manure, etc.) may be the pathways to substitute the non-renewable energy forms with renewable resources and to reduce their environmental footprints.

3.4 Measuring the efficiency of farmers

The results of distribution of farmers based on the efficiency score obtained by the application of CCR and BCC DEA models are shown in Fig. 3. As it is evident, about 33% (31 farmers) and 54% (51 farmers) from total farmers were recognized as the efficient farmers under constant and variable returns to scale assumptions, respectively. Moreover, 33% and 32%, with respect, had their technical and pure technical efficiency scores between 0.8 and 1 range. Also, when the BCC model is assumed, only 1% had an efficiency score of less than 0.6; whereas, when the CCR model is applied, 11% had the efficiency scores of less than 0.6. The results of returns to scale estimation indicated that all of the technically efficient farmers (based on the CCR model) were operating at CRS, showing the optimum scale of their practices.

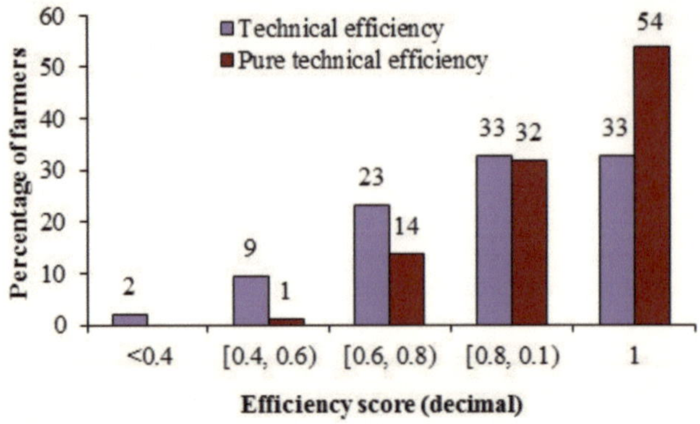

Fig. 3. Distribution of sunflower producers based on efficiency scores

The summarized statistics for the three estimated measures of efficiency are presented in Table 6. The results revealed that the average values of technical and pure technical efficiency scores were 0.83 and 0.93, respectively. Also, the technical efficiency varied from 0.33 to 1 range. The wide variation in the technical efficiency implies that all the farmers were not fully aware of the right production techniques or did not apply them properly. Based on the literature, the technical efficiency scores of 0.77 for paddy production (Chauhan et al., 2006), 0.75 for tomato, 0.81 for asparagus production (Iráizoz et al., 2003) and 0.74 for canola production (Mousavi-Avval et al., 2011c) have been reported.

The average scale efficiency score was relatively low as 0.89, showing the disadvantagiouse conditions of scale size. This indicates that if all of the inefficient farmers operated at the most productive scale size, about 11% savings in energy use from different sources would be possible without affecting the yield level.

Particular	Average	SD	Min	Max
Technical efficiency	0.83	0.17	0.33	1
Pure technical efficiency	0.93	0.10	0.54	1
Scale efficiency	0.89	0.14	0.43	1

Table 6. Average efficiencies of farmers for sunflower production in Golestan, Iran

3.5 Setting realistic input levels for inefficient farmers

A pure technical efficiency score of less than one for a farmer indicates that, at present conditions, he is using higher values of energy than required. Therefore, it is desired to suggest realistic levels of energy to be used from each source for every inefficient farmer in order to avert wastage of energy. The summarized information for setting realistic input levels are given in Table 7. It gives the average energy usage in target conditions (MJ ha^{-1}), possible energy savings and ESTR percentage for different energy sources. It is evident that, total energy input could be reduced to 8028.4 MJ ha^{-1}; while, maintaining the current production level and also assuming no other constraining factors. Diesel fuel, total fertilizer and electricity energies were required as 3311.5, 2145.1 and 987.3 MJ ha^{-1}, respectively. Moreover, machinery, water for irrigation, human labor, chemicals and seeds energy inputs were required as 627.9, 500.9, 248.1, 187.2 and 31.4 MJ ha^{-1}, respectively.

The results of ESTR calculations showed that if all farmers operated efficiently, the reduction of electricity, water for irrigation and total fertilizer energy inputs, with respect, by 39.6%, 23.8% and 20.5% would have been possible without affecting the yield level. These energy inputs had the highest inefficiency which was owing mainly to the excess use of water and also electricity in water pumping systems. High percentage of fertilizer energy inputs can also be interpreted by the low prices and freely availability of these inputs in surveyed region. Accurate fertilizer management by increasing its profitability with the crops and reducing losses by improving management practices can improve energy use. These results are consistent with the results of energy efficiency analysis in parametric approach, in which, the use of these inputs was inconsistent with output (Section 3.3). On the other hand, the ESTR for human labor and seeds energy inputs was found to be 3.9% and 0.5%, respectively; indicating that these inputs were mainly used efficiently by the farmers in the region. This is consistent with the previouse results, in which, the seed and human labor energies had the relatively high MPP values on output of sunflower (Section 3.3). Similar results also had been reported in the literature by Omid et al. (2010).

Moreover, the results revealed that, the ESTR percentage for total energy input was 16.4%, indicating that by adopting the recommendations resulted from this study, on average, about 16.4% (1571.6 MJ ha^{-1}) from total input energy in sunflower production could be saved without affecting the yield level. Singh et al. (2004) reported that 15.9% (11305 MJ ha^{-1}) from total energy input for wheat production could be saved without affecting the yield level. Also, Mousavi-Avval et al. (2011d) found that about 20% of overall resources in soybean production could be reduced if all of the farmers operate efficiently. Using the information of Table 7, it is possible to advise the inefficient farmers regarding the better operating practices followed by his peers in order to reduce the input energy levels to the target values indicated in the analysis while achieving the output level presently achieved by him.

Input	Target use (MJ ha^{-1})	Saving energy (MJ ha^{-1})	ESTR (%)
1. Human labor	248.1	10.0	3.9
2. Machinery	627.9	48.5	7.2
3. Diesel fuel	3300.5	139.9	4.1
4. Chemicals	187.2	17.8	8.7
5. Total fertilizer	2145.1	551.6	20.5
6. Water for irrigation	500.9	156.7	23.8
7. Electricity	987.3	646.9	39.6
8. Seeds	31.4	0.2	0.5
Total energy	8028.4	1571.6	16.4

Table 7. Optimum energy requirement and saving energy for sunflower production.

Fig. 4 shows the distribution of saving energy from different sources for sunflower production. It is evident that the maximum contribution to the total saving energy is 41.2% from electricity. Also, electricity, total fertilizer, water for irrigation and diesel fuel energy inputs contributed to the total saving energy by about 95%. This is consistent with the results of previous studies that diesel fuel and electricity had the highest potential for improving energy productivity in the production of different agricultural crops (Chauhan et al., 2006; Omid et al., 2010). From these results it is strongly suggested that improving the usage pattern of these inputs be considered as priorities providing significant improvement in energy productivity for sunflower production in surveyed region. Improving energy use efficiency of water pumping systems, employing new irrigation systems and leveling farms properly can be suggested to prevent from electrical energy wastage by inefficient farmers. Applying a better machinery management technique, employing the conservation tillage methods and also, controlling input usage by performance monitoring can help to reduce the diesel fuel and fertilizer energy inputs and minimize their environmental impacts. Also, integrating a legume into the crop rotation, application of composts, chopped residues or other soil amendments may increases soil fertility in the medium term and so reduces the need for chemical fertilizer energy inputs.

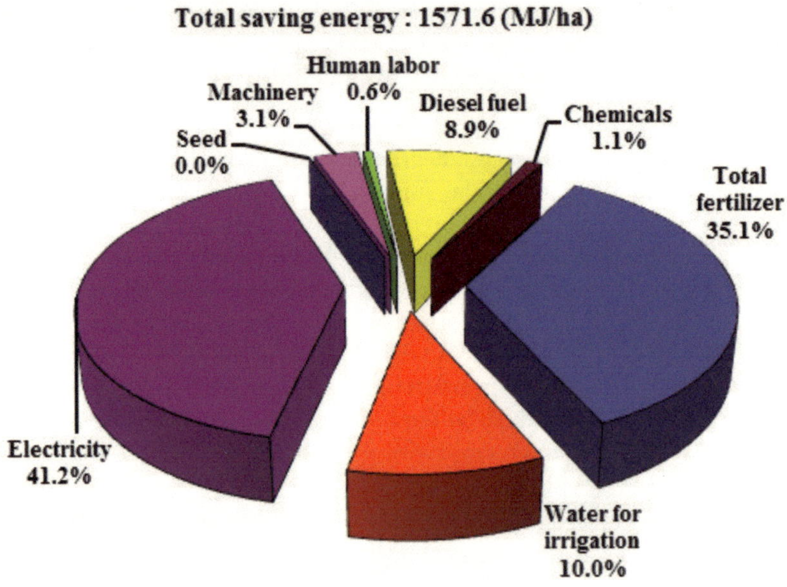

Fig. 4. Distribution of saving energy from different sources for sunflower production.

3.6 Improvement of energy indices

The energy indices for sunflower production in target use of energy are presented in Table 8. It is evident that by optimization of energy use, both the energy ratio and energy productivity indicators can improve by 19.6%. Also, in optimum consumption of energy inputs, the net energy indicator by improvement of 5.1% would increase to 32634.4 MJ ha^{-1}.

Items	Unit	Quantity in optimum use	Change (%)
Energy ratio	-	5.06	19.6
Energy productivity	kg MJ^{-1}	0.20	19.6
Specific energy	MJ kg^{-1}	4.94	-16.4
Net energy	MJ ha^{-1}	32634.4	5.1
Total input energy	MJ ha^{-1}	8028.4	-16.4

Table 8. Improvement of energy indices for sunflower production in Golestan, Iran.

To sum it up, oilseed sunflower is a crop with relatively high requirements on non-renewable energy resources; its fertilizer and electrical energy requirements are high and it needs a high amount of diesel fuel consumption. The farmers mainly don't have enough knowledge on more efficient input use and there is a common belief that increased use of energy resources will increase the yield. The methodologies presented in this study demonstrate how energy use efficiency in sunflower production may improve by applying the operational management tools to assess the performance of farmers. On an average, considerable savings in energy inputs may be obtained by adopting the best practices of high-performing ones in crop production process. Adoption of more energy-efficient cultivation systems would help in energy conservation and better resource allocation.

Some strategies such as providing better extension and training programs for farmers and use of advanced technologies should be developed in order to increase the energy efficiency of agricultural crop productions in the region. The farmers should be trained with regard to the optimal use of inputs, especially, fertilizers and water for irrigation as well as employing the new production technologies. The local agricultural institutes in the region have an important role in these cases to establish the more energy efficient and environmentally healthy sunflower production systems in the region.

4. Conclusion

The study describes the application of parametric and non-parametric approaches to analyze energy efficiency in agricultural production. Therefore, energy use pattern for oilseed sunflower production in Iran was investigated and the parametric method of Cob-Douglas production function and non-parametric method of data envelopment analysis were applied to analyze efficiency of farmers. These methologies helped to identify the impact of energy use from different inputs on output, measure efficiency scores of farmers, segregate efficient farmers from inefficient farmers and to find the wasteful uses of energy by inefficient farmers. The results from both the parametric and non-parametric techniques revealed that the use of machinery, human labor and seed energies had higher impacts on output. In other words, there is higher potential for increasing output by additional use of these inputs for sunflower production. On the other hand, the use of fertilizer, water for irrigation and electrical energy was inconsistent with output, indicating that use of these inputs is high, resulting in energy dissipation as well as imposing negative effects to environment and human health. The results of DEA application further indicated that there are substantial production inefficiencies for farmers; so that, a potential of almost 16% reduction in total energy input use may be achieved if all farmers operated efficiently and assuming no other constraints on this adjustment. Moreover, the results revealed that sunflower production in the region showed a high sensitivity on non-renewable energy sources which may resulted in both the environmental deterioration and rapid rate of depletion of these energetic resources. Therefore, policies should emphasize on development of new technologies to substitute fossil fuels with renewable energy sources aiming efficient use of energy and lowering the environmental footprints. From this study, development of renewable energy usage technologies such as farm machinery or water pumping systems using biodiesel or solar power, applying a better machinery management technique, employing the conservation tillage methods, utilization of alternative sources of energy such as organic fertilizers may be suggested to reduce the environmental footprints of energy inputs and to obtain sustainable food production systems.

5. References

Adler, N.; Friedman, L. & Sinuany-Stern, Z. (2002). Review of ranking methods in the data envelopment analysis context. *European Journal of Operational Research* 140:249-265

Alam, M.S.; Alam, M.R. & Islam, K.K. (2005). Energy Flow in Agriculture: Bangladesh. *American Journal of Environmental Sciences* 1:213-220

Anonymous, (2010). Annual agricultural statistics. Ministry of Jihad-e-Agriculture of Iran. www.maj.ir

Banaeian, N.; Omid, M. & Ahmadi, H. (2010). Energy and economic analysis of greenhouse strawberry production in Tehran province of Iran. *Energy Conversion and Management* 52:1020-1025

Banker, R.; Charnes, A. & Cooper, W. (1984). Some models for estimating technical and scale inefficiencies in data envelopment analysis. Management Science 30:1078–1092

Barnes, A.P. (2006). Does multi-functionality affect technical efficiency? A non-parametric analysis of the Scottish dairy industry. *Journal of Environmental Management* 80:287-294

Barr, S.I.; McCarron, D.A.; Heaney, R.P.; Dawson-Hughes, B.; Berga, S.L.; Stern, J.S. & Oparil, S. (2000). Effects of Increased Consumption of Fluid Milk on Energy and Nutrient Intake, Body Weight, and Cardiovascular Risk Factors in Healthy Older Adults. *Journal of the American Dietetic Association* 100:810-817

Beheshti Tabar, I.; Keyhani, A. & Rafiee, S. (2010). Energy balance in Iran's agronomy (1990-2006). *Renewable and Sustainable Energy Reviews* 14:849-855

Canakci, M.; Topakci, M.; Akinci, I. & Ozmerzi, A. (2005). Energy use pattern of some field crops and vegetable production: Case study for Antalya Region, Turkey. *Energy Conversion and Management* 46:655-666

Chalabi, Z.S. & Bailey, B.J. (1991). Sensitivity analysis of a non-steady state model of the greenhouse microclimate. *Agricultural and Forest Meteorology* 56:111-127

Chauhan, N.S.; Mohapatra, P.K.J. & Pandey, K.P. (2006). Improving energy productivity in paddy production through benchmarking--An application of data envelopment analysis. *Energy Conversion and Management* 47:1063-1085

Cochran, W.G. (1977). Sampling techniques. 3rd Edition, John Wiley & Sons, NY, USA

Coelli, T.J. (1996). A Guide to DEAP Ver. 2.1, A data envelopment analysis (computer)program. Centre for efficiency and productivity analysis. University of New England <www.une.edu.au/econometrics/cepa.htm>

Cooper, L.M.; Seiford, L.M. & Tone, K., (2006). Introduction to data envelopment analysis and its uses. New York: Springer

Cooper, W.; Seiford L, M. & Tone, K. (2004). Data Envelopment Analysis, A comprehensive text with models, applications, references and DEA-solver software. Massachusetts, USA: Kluwer Academic Publishers

De, D.; Singh, R.S. & Chandra, H. (2001). Technological impact on energy consumption in rainfed soybean cultivation in Madhya Pradesh. *Applied Energy* 70:193-213

Drechsler, M. (1998). Sensitivity analysis of complex models. *Biological Conservation* 86:401-412

Erdal, G.; Esengün, K.; Erdal, H. & Gündüz, O. (2007). Energy use and economical analysis of sugar beet production in Tokat province of Turkey. *Energy* 32:35-41

Esengun, K.; Erdal, G.; Gündüz, O. & Erdal, H., (2007). An economic analysis and energy use in stake-tomato production in Tokat province of Turkey. *Renewable Energy* 32:1873-1881

Fadai, D. (2007). Utilization of renewable energy sources for power generation in Iran. *Renewable and Sustainable Energy Reviews* 11:173-181

FAO, (2008). Food and Agricultural Organization. www.fao.org

Fraser, I. & Cordina, D. (1999). An application of data envelopment analysis to irrigated dairy farms in Northern Victoria, Australia. *Agricultural Systems* 59:267-282

Hamby, D.M. (1994). A review of techniques for parameter sensitivity analysis of environmental models. *Environmental Monitoring and Assessment* 32:135-154

Hatirli, S.A.; Ozkan, B. & Fert, C. (2005). An econometric analysis of energy input-output in Turkish agriculture. *Renewable and Sustainable Energy Reviews* 9:608-623

Hu, J.-L. & Kao, C.-H. (2007). Efficient energy-saving targets for APEC economies. *Energy Policy* 35:373-382

Iráizoz, B.; Rapún, M. & Zabaleta, I. (2003). Assessing the technical efficiency of horticultural production in Navarra, Spain. *Agricultural Systems* 78:387-403

Jianbo, L. (2006). Energy balance and economic benefits of two agroforestry systems in northern and southern China. *Agriculture, Ecosystems & Environment* 116:255-262

Kallivroussis, L.; Natsis, A. & Papadakis, G. (2002). The energy balance of sunflower production for biodiesel in Greece. *Biosystems Engineering* 81:347-354

Khan, S.; Khan, M.A.; Hanjra, M.A. & Mu, J. (2009). Pathways to reduce the environmental footprints of water and energy inputs in food production. *Food Policy* 34:141-149

Kitani, O. (1999). CIGR handbook of agricultural engineering, Volume 5: Energy and biomass engineering. ASAE Publications, St Joseph, MI

Kizilaslan, H. (2009). Input-output energy analysis of cherries production in Tokat Province of Turkey. *Applied Energy* 86:1354-1358

Kulekci, M. (2010). Technical efficiency analysis for oilseed sunflower farms: a case study in Erzurum, Turkey. *Journal of the Science of Food and Agriculture* 90:1508-1512

Lamoureux, J., Tiersch, T.R. & Hall, S.G. (2006). Sensitivity analysis of the pond heating and temperature regulation (PHATR) model. *Aquacultural Engineering* 34:117-130

Latif, S. & Anwar, F. (2009). Effect of aqueous enzymatic processes on sunflower oil quality. *Journal of the American Oil Chemists' Society* 86:393-400

Liu, Y.; Høgh-Jensen, H.; Egelyng, H. & Langer, V. (2010). Energy efficiency of organic pear production in greenhouses in China. *Renewable Agriculture and Food Systems* 25:196-203

Malana, N.M. & Malano, H.M. (2006). Benchmarking productive efficiency of selected wheat areas in Pakistan and India using data envelopment analysis. *Irrigation and Drainage* 55:383-394.

Meul, M.; Nevens, F.; Reheul, D. & Hofman, G. (2007). Energy use efficiency of specialised dairy, arable and pig farms in Flanders. *Agriculture, Ecosystems & Environment* 119:135-144

Mobtaker, H.G.; Keyhani, A.; Mohammadi, A.; Rafiee, S. & Akram, A. (2010). Sensitivity analysis of energy inputs for barley production in Hamedan Province of Iran. *Agriculture, Ecosystems & Environment* 137:367-372

Mohammadi, A. & Omid, M. (2010). Economical analysis and relation between energy inputs and yield of greenhouse cucumber production in Iran. *Applied Energy* 87:191-196

Mohammadi, A., Rafiee, S., Mohtasebi, S.S. & Rafiee, H. (2010). Energy inputs - yield relationship and cost analysis of kiwifruit production in Iran. *Renewable Energy* 35:1071-1075

Moore, S.R. (2010). Energy efficiency in small-scale biointensive organic onion production in Pennsylvania, USA. *Renewable Agriculture and Food Systems* 25:181-188

Mousavi-Avval, S.H.; Rafiee, S.; Jafari, A. & Mohammadi, A. (2011a). Energy flow modeling and sensitivity analysis of inputs for canola production in Iran. *Journal of Cleaner Production* 19:1464-1470

Mousavi-Avval, S.H.; Rafiee, S.; Jafari, A. & Mohammadi, A. (2011b). The Functional Relationship Between Energy Inputs and Yield Value of Soybean Production in Iran. *International Journal of Green Energy* 8:398-410

Mousavi-Avval, S.H.; Rafiee, S.; Jafari, A. & Mohammadi, A. (2011c). Improving energy use efficiency of canola production using data envelopment analysis (DEA) approach. *Energy* 36:2765-2772

Mousavi-Avval, S.H.; Rafiee, S.; Jafari, A. & Mohammadi, A. (2011d). Optimization of energy consumption for soybean production using Data Envelopment Analysis (DEA) approach. *Applied Energy* 88:3765-3772

Mousavi Avval, S.H.; Rafiee, S.; Jafari, A. & Mohammadi, A. (2011). Improving energy productivity of sunflower production using data envelopment analysis (DEA) approach. *Journal of the Science of Food and Agriculture* 91:1885-1892

Mukherjee, K. (2008). Energy use efficiency in the Indian manufacturing sector: An interstate analysis. *Energy Policy* 36:662-672

Nassiri, S.M. & Singh, S. (2009). Study on energy use efficiency for paddy crop using data envelopment analysis (DEA) technique. *Applied Energy* 86:1320-1325

Omid, M.; Ghojabeige, F.; Delshad, M. & Ahmadi, H. (2010). Energy use pattern and benchmarking of selected greenhouses in Iran using data envelopment analysis. *Energy Conversion and Management* 52:153-162

Pimentel, D. & Patzek, T.W. (2005). Ethanol Production Using Corn, Switchgrass, and Wood; Biodiesel Production Using Soybean and Sunflower. *Natural Resources Research* 14:65-76

Rafiee, S.; Mousavi Avval, S.H. & Mohammadi, A. (2010). Modeling and sensitivity analysis of energy inputs for apple production in Iran. *Energy* 35:3301-3306

Reig-Martínez, E. & Picazo-Tadeo, A.J. (2004). Analysing farming systems with Data Envelopment Analysis: citrus farming in Spain. *Agricultural Systems* 82:17-30

Samavatean, N.; Rafiee, S.; Mobli, H. & Mohammadi, A. (2011). An analysis of energy use and relation between energy inputs and yield, costs and income of garlic production in Iran. *Renewable Energy* 36:1808-1813

Scheel, H. (2000). EMS: efficiency measurement system users manual, Ver. 1.3, Operations Research and Wirtschaftsinformatik, University of Dortmund. Dortmund, Germany; <http://www.wiso.uni-dortmund.de/lsfg/or/scheel/ems>

Sherman, H.D. (1988). Service organization productivity management. The Society of Management Accountants of Canada, Hamilton, Ontario

Singh, G.; Singh, S. & Singh, J. (2004). Optimization of energy inputs for wheat crop in Punjab. *Energy Conversion and Management* 45:453-465

Topp, C.F.E.; Stockdale, E.A.; Watson, C.A. & Rees, R.M. (2007). Estimating resource use efficiencies in organic agriculture: a review of budgeting approaches used. *Journal of the Science of Food and Agriculture* 87:2782-2790

Turhan, S.; Ozbag, B.C. & Rehber, E. (2008). A comparison of energy use in organic and conventional tomato production. *Journal of Food Agriculture & Environment* 6:318-321

Unakitan, G.; Hurma, H. & Yilmaz, F. (2010). An analysis of energy use efficiency of canola production in Turkey. *Energy* 35:3623-3627

Zhang, X.; Huang, G.H.; Lin, Q. & Yu, H., (2009). Petroleum-contaminated groundwater remediation systems design: A data envelopment analysis based approach. *Expert Systems with Applications* 36:5666-5672

Zhou, P.; Ang, B.W. & Poh, K.L. (2008). A survey of data envelopment analysis in energy and environmental studies. *European Journal of Operational Research* 189:1-18

Part 3

Energy Efficiency in Buildings

8

Energy Consumption Improvement Through a Visualization Software

Benoit Lange, Nancy Rodriguez and William Puech
LIRMM Laboratory, UMR 5506 CNRS,
University of Montpellier II,
France

1. Introduction

In the actual world frame, energy efficiency becomes a necessity. Since 1995, Kyoto protocol has highlighted that humans need to improve their energy consumption and reduce CO_2 signature. Building consumption represent a third of the global energy consumption and Information Technology (IT) equipment is weighing heavily on energy expenses. For example, Google equipment consumes 0,01 percent of the world global Energy, the equivalent of the power usage of a city with 200.000 inhabitants.

Solutions to reduce usage of fossil energy, called "green energies" and supported by standard industries, already exist. For example, faulty processors are removed from production and are re-factored to create solar sensors.The PUE, Power Usage Effectiveness is also a measure created to monitor energy in data centers.

As IT equipment, buildings become greener and smarter. Indeed, buildings are gap of energy and it is necessary to improve them to consume less energy. Governments establish buildings certification to help this development. For example, in France, High Quality Environmental standard has become the actual norm for new or re-factored buildings. Buildings Management Systems (BMS) are widely used to catch data from a set of building sensors. BMS are only reactive systems, there is no prediction. Therefore, it is necessary to propose and develop smarter systems, fully focused in energy improvement.

In this chapter, we present a project called RIDER, for Research for IT Driver EneRgy efficiencies. We are developing a green box composed by a set of IT components; able to generate efficient predictive model and building optimization. The main part of the green box is a pivot model managed by a rule engine and a standard BMS with a specific input/output protocol. Modules can extend the standard green box offering system/user advices, proposing improvements on data model and predicting building behavior.

This chapter introduces the RIDER green box and details the visualization module, which is used to improve model by using user knowledge.

In Section 2, we present the related work on visualization. Section 3 presents a global overview of the RIDER project. Section 4 is dedicated to the visualization method used in this project, whereas Section 5 shows experimental results on real data. Conclusions are presented in Section 6.

2. Related works

Data centers are present in most companies. As they represent a gap of energy, data centers are suitable to be a starting point for energy efficiency research. We work with data extracted from sensors of a dedicated data center located in IBM, Montpellier (France). IBM Montpellier host a specific kind of data center, called "green" (Green Data Center). Effectiveness of a Data center energy is measured with PUE (Power Usage Effectiveness), most of data centers have a high PUE: if value is bigger than two, i.e. a data center consumes twice as much energy as is needed. A data center is told green when its PUE measure is less than 1.5. GDC improvements are developed around management of alleys or usage of a specific kind of server cases. Green Data Center in Montpellier is one of the pilot sites of the RIDER project.

In this chapter, we mainly focus on visualization methods to improve the RIDER model. This section is dedicated to previous visualization methods, level of details and visual analytics domain.

2.1 Visualization

Huge amount of data is produced each day, and visualization is a way to understand this mass. Physics, mechanics, literature and many other domains benefit from visualization.

In physics, scientists use visualization to interpret, understand, render their data, this paradigm is better than spreadsheet. Scientific data contains many dimensions provided by satellites, telescopes, simulations and so on. Scientist's main problem is to give a sense to data, an answer can be given by visualization. Particles are small objects without weight. It is possible to classify particles in three different kinds: static, pre computed or dynamic.

Pre computed particles was used for render explosions in 1960. At each iteration a mathematical function computes location of particles.

Dynamic particles interact with their environment and between them. This method was largely used in fluid simulation, Green (2007) presented results of GPGPU computing [1]: this method use graphic devices as computational device and is based on particles for dynamic fluid simulation. Results are impressive and flow is well rendered. The last kind of particles system is a large static point cloud, they are used to render informations through their colors; cost is low compared to most visualization methods.

Views have to deal with rendering of multivariate and complex data, thus scientists use HPC (High Performance Computing [2]) systems to analyze and produce views from data. These kinds of large computers are gap of energy efficiency and computation time is expensive.

Kapferer & Riser (2008) proposed a method to visualize a large amount of spatial data (clusters of stars) without using expensive HPC. They used a volume rendering method based on particles to render data. They developed their own solution to produce rendering because existent solutions were not efficient. Instead of using traditional HPC method, they used

[1] General-Purpose computation on Graphics Processing Units

[2] Large computer designed to manage large data flow

GPU computing to compute a real time view with a low cost hardware. Authors used GPU computing to apply simulation on data. To display stars, they used a simple point sprite instead of a sphere (spheres are expensive objects in computer graphics). To improve rendering of point cloud, they used lights to create a spherical aspect to these primitives. This solution allows to display more data than full volume representation method. Usage of GPU for astrological data and particle paradigm seems to be well-suited.

Ilcìk (2008) proposed another example of visualization software. In this paper, he analyzed behavior of CAD model within fluid simulation. Particles are used to find weak points of 3D models. When a collision between a particle and fluid occurs, the value of particles change. This method is efficient to analyze CAD models(Indeed, these meshes are well defined and the number of triangles is important). Particles give opportunity to not saturate the object and are an efficient method to understand quickly a problem.

2.2 Level of details

When we focus on 3D visualization, an important question is to fit with hardware's capacities. Current graphic devices are able to render an important quantity of polygons but the number is limited. To solve this problem multi-resolution of meshes appeared. Clark (1976) introduced the concept of level of detail. 3D objects are composed of thousands of polygons. Because a scene is composed of a set of 3D objects, the number of polygons increases quickly. Level of details consists in producing meshes at different resolutions. These different meshes are loaded on the 3D scene depending on camera position. Automatic methods arose in early 90', they were based on usage of different kinds of operators. The first methods were brut forces meshes decimation (Schroeder et al. (1992)). This method use a decimation of vertices as operator but it is not efficient, retriangulation creates some topology mistakes. Luebke (1997; 2001) proposed two overviews of several operators.

But, most of these simplification technics are suitable for triangular meshes and not for volume model. He et al. (1995) proposed a method based on voxel simplification by using a grid for clustering voxels. Lorensen & Cline (1987) have developed marching cube algorithm, used to produce a surface mesh. But this simplification algorithm does not preserve the shape of the mesh. In the work presented in Section 3 and 4, we are looking for a point cloud simplification. Indeed, previous methods dealing with simplification of surface point cloud like Moenning & Dodgson (2004); Pauly et al. (2002); Song & Feng (2009) are not adapted to our case, surface objects do not render inside part of objects.

2.3 Visual analytics

This chapter is focused on visual analytics paradigm proposed by Shneiderman (2001). The goal is to combine two domains: Knowledge Discovery (KD) and Information Visualization (IV). Targets of these domains are similar: extract content from a set of data. Each of these domains uses its own method to reach this goal. KD uses neural network or decisions trees to analyze, simplify and extract interesting features from information. IV is a paradigm where human is the center of the system. Human is used to dig into data instead of an algorithm. Usage of human expertise for digging into data is underlined by Rosen & Popescu (2011). This paper presents an efficient way to count items in a maze; the most important part is to use a smart camera to explore several corridors in the same view. IV and KD have some limits, they can introduce mistakes or misunderstanding on data. Schneiderman had proposed merging of these approaches. Later, Keim et al. (2006) presented a formal definition of this paradigm. In Visual Analytics (VA), automatic methods are used to sort and produce a visualization for user. Then, user can manage data through the visualization and can expose sorting or new content. Often, this data carry different kinds of values, this means that each object can be defined by a set of values: for example, sensor is located by: X, Y and Z coordinates. Visualization tools need to be able to deal with multivariate data, and moreover, it is necessary to give to the user the best method to dig into data: human knowledge can find some unexpected information. To improve its efficiency, user interface needs to be developed around a large set of tools: zooming, panning, scrolling, etc. and also a large set of visualization methods. There are many kinds of visualization to help user's understanding: parallel coordinates, scatter plot, glyph, etc.

Many solutions developed around VA paradigm have emerged from scientific software. Liu et al. (2010) developed Netclinic: a software to find culprit on network. Failure of networks is critical and it is important to find the reason of problems. Netclinic is based on an automatic approach to extract well-known causes of trouble and then software renders view of data designed for non expert users. Human proposals can be analyzed and validated by an automatic method.

Another example of VA paradigm application was proposed by Migut & Worring (2010). Authors explored a method to evaluate risks of a wrong decision from a set of person. They used a machine learning method to extract standard profile of dangerous people. Decision made can be critic and it was necessary to remove false positive. A semantic visualization was given to user to resolve a part of this trouble. The solution in this paper provides a visualization method to understand results from a classifier output using a simple multidimensional space visualization. The software here is composed by a visualization data mining method and a machine learning method. Experts can drive the visualization with a 3D view, based on a scatterplot. This kind of view is presented by Konyha et al. (2009) and a support vector machine provides data filtering. Authors used Voronoï cells to extract borderline of each plot of scatterplot. To improve visualization, pictures and colors were added. Several interaction paradigms (zooming, panning and selection) were implemented to help exploration of data in an efficient way. User case study has been established on Forensic Psychiatry. Experts were able to identify if a patient share same characteristics with a previous patient.

Another data visualization software for multidimensional data is Caleydo. Lex et al. (2010) gave an overview of this software to refine medical diagnostic. Gene exploration is a hard task due to the large amount of multidimensional data. It is not trivial to present a coherent visualization of gene relation. In literature, this problem was apprehended by virtual reality methods. Authors wanted to develop a standard computer method using multiple views and a 2.5D view. The goal was to visualize genes, relations were displayed through the bucket and linked views. Authors used parallel coordinates (Hauser et al. (2003)) or heat map for the visualization of genes. This application was designed around several features: bucket, zooming and links. Their proposal is mainly concentrated on the bucket: it is a special kind of arrangement for multiple views. Each view is mapped to a face of a cube. With this method, links between views can be understood efficiently and interactions give a method to follow the gene path.

3. RIDER project

The RIDER project is a collaborative research project led by IBM France. A consortium around several French industries and universities is in charge of development and research of this project. The goal is to produce a platform, called **green box**, designed to improve energy consumption of a building based on recommend. These are given to monitor, analyze or predict events (temperature increase, crowd, etc). This green box has to be scalable to fit with different building topologies (fit on different kind of buildings: service, industrial or residential buildings).

Buildings become smarter each year. Sets of sensors are used to help monitoring buildings and results are exploited by the building manager. Unfortunately, these results are not easy to understand and they produce huge amount of data. Depending from building, granularity of sensor meshes is different, for example, we can find sensors for: CO_2, temperature, pressure, etc. All these sensors are placed to learn building life, ubiquitous computing is more and more used to help human decisions. But this fast growing brings a large flow of data and actual system does not use all of them. It is necessary to find an efficient way to manage and understand these data. In a second way, it is also necessary to deal with building granularity of sensors. Buildings have not the same number of sensors, some rooms are fully equipped in high resolution, while other rooms are only equipped by a small number of sensors.

This data is managed by a Building Management System (BMS), which only provide analyzes of the current situation and not needs of future. For example, if the temperature increases in a room, the BMS will adapt air conditioner speed depending on the user desire. To improve energy in building, it is necessary to deal with building inertia and adapt air conditioning to the different temperature (external and neighborhood temperature). Improvements on this part are important, because they allow managing building energy in an efficient way. Our green box is designed to understand, anticipate and react to any situation with specific adapted predictive models. This box is composed by a set of tools: a pivot model and recommend modules. Building's model contains by sensors, building's data and external data (from different sources: weather, personal calendar, etc ...).

The global architecture of our system is developed in Figure-1. The heart of our green box is developed around a model driven methodology and a software product line approach for the deployment of "smart buildings" control software. For this purpose, new ways to model and optimize physical infrastructures from an energy-efficiency point-of-view are explored. Also new ways to model software product lines and their variability artifacts are investigated.
The global objective of modeling is to capitalize in a dedicated software product line the architectural knowledge. It aims to study generic model and customizable architecture for a software product line. The idea is to develop modeling tools that will establish new ways to instrument, interconnect and optimize neighborhood or buildings. This method is also developed to facilitate the link between the physical and the computer world. The model will enable creation of real-world aware systems in a context of event-driven computing. New UML models will describe the physical architecture with a higher semantic level to handle the energetic efficiency domain.

A first component is developed around decision making. It aims to build weak dependency solution. This method was developed around learning technics to construct physical and preferences models. The construction of these informations are generated from interviews of final users.

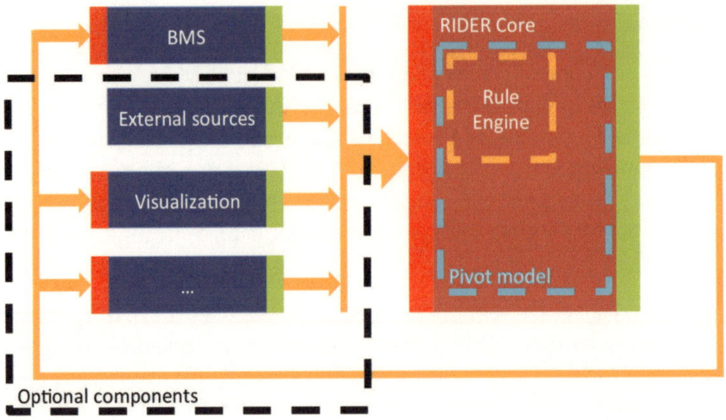

Fig. 1. RIDER architecture.

The main goal of the building manager model is to develop a grey box model for the control and supervision of exchanges and uses of energies between different buildings, including classic and renewable energies, through the use of IT. Energetic problems such as energetic losses will be dealt with information technology in order to bring smarter solutions.

A data mining component has goal to dig into data. It is based on statistical model provided by statistic engines. Informations are extracted from this engine using sensors data. Automatic data mining is done with this component, this allows to validate advices given by the different components.

A visual engine component is also developed. The goal is to propose a mining tool using user knowledge. User is able to feed system when an automatic method fails.

Entire system runs as follow: BMS records data from sensors located in building, then it sends information through a TCP connection with a XML protocol to RIDER Core, this input is the same for each component. RIDER core instantiates the pivot model and applies dedicated rules. Finally it sends results to all subscribers (each component can become a subscriber). If a subscriber wants to propose new data, it can use the same process as the BMS update.

4. Visualization method

In the context of this chapter, interest is on visualization and advices propose to global system. Most of visualization software is developed to render data and not to propose content. Our solution is developed to provide a understandable visualization and also a way to create content. We will present an architecture based on client/server paradigm, different kinds of visualizations used and a digging tool.

4.1 Architecture

Visualization software is based on a client / server paradigm. An HPC (High Performance Computing) is used to produce data, whereas another HPC is used to render this data.

HPC has become a major actor in scientific visualization, unfortunately this kind of computer is expensive and is a gap of energy. HPC are large computers developed for parallel computing, thus it is necessary to use specific algorithms. Instead of using large computer, it is also possible to use an architecture based on graphical engine to execute algorithms. It consists in using a graphic card to compute algorithms in parallel. This section introduces our algorithm used for rendering, and our proposition of a benchmark based on energy consumption of different kinds of devices (a standard computing, parallel computing, ...).

The base of our data is a set of sensors and a coarse building topology. Sensors type, location and value are well known. With all these informations, we want to produce a volumetric view for each room of a building. To produce 3D volume object, voxels and point clouds are the most common methods. Voxels are volumetric pixels, they are also primitives of 3D volume objects. We have adapted the second solution based on point cloud, this method is mostly used in 2D through scatter plot visualization. Piringer et al. (2004) presented a 3D scatter plot method. Primitive used for rendering is a low cost primitive, the goal is to reduce rendering cost (this part will be explained later). This point cloud can be considered as a particle set, as stated previously, particles used in this work are static.

For the generation of the point cloud, we have developed two methods: a regular grid and a concentric pattern. As stated before, our input is a set of sensors and the room's characteristics. Rooms are designed by three measures: length ($l \in \mathbb{R}^+$), width ($w \in \mathbb{R}^+$) and height ($h \in \mathbb{R}^+$). Figure-2 gives a schema of a room with layered sensors. Sensors are located at determinate height in space. Sensors set is defined by: $S = \{s_1, ..., s_{nbS}\}$, where nbS is the number of sensors and $nbS \in \mathbb{N}$. Each sensor is defined by three coordinates: $s_i = \{x_i, y_i, z_i\}$, $i \leq nbS$, where $0 \leq x_i \leq l$, $0 \leq y_i \leq w$, $0 \leq z_i \leq h$. Particles are defined by: $P = \{p_1, ..., p_{nbP}\}$, where nbP is the

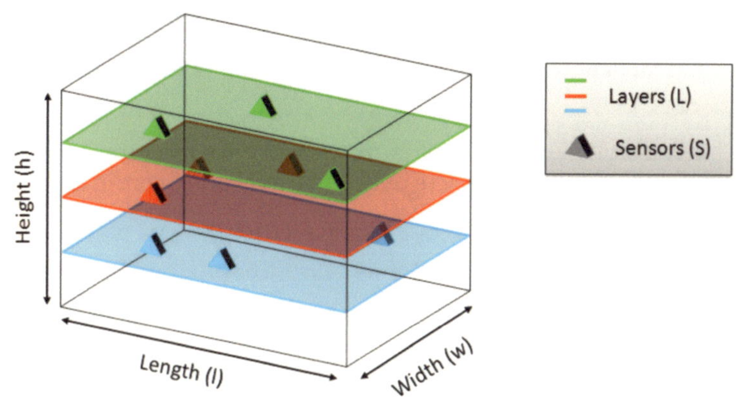

Fig. 2. Topology information of rooms.

number of particles and $nbP \in \mathbb{N}$. Each particle own its coordinates: $p_k = \{x_k, y_k, z_k\}$, $k \leq nbP$, where $0 \leq x_k \leq l$, $0 \leq y_k \leq w$, $0 \leq z_k \leq h$.

For the regular grid, coordinates of each particle can be calculated with:

$$p_k = \begin{cases} x_k = \rho + b \cdot \rho, \\ y_k = \rho + c \cdot \rho, \\ z_k = \rho + d \cdot \rho \end{cases} \tag{1}$$

where b $\in \mathbb{N}$, $0 \leq$ b $< \frac{l}{\rho}$, c $\in \mathbb{N}$, $0 \leq$ c $< \frac{w}{\rho}$, d $\in \mathbb{N}$, $0 \leq$ d $< \frac{h}{\rho}$. ρ represents space between two particles and is defined by depending from resolution desired. We use this measure to produce a multi-resolution mesh designed for different devices. If we want a coarse resolution, we use a high value for ρ, at opposite, we use a low value for ρ to get a high resolution. Each device can visualize a limited number of particles before loosing real time visualization. We can calculate the number of particles for the regular grid by the formula:

$$nbP = \frac{((l-\rho) \times (h-\rho) \times (w-\rho))}{\rho^3}. \tag{2}$$

This grid is regular and fills the entire structure of the room. Figure-3(a) presents a regular grid in a room. Multi resolution can be used to provide the right visualization for each device, simple computer cannot render the same number of particles as a large screen display, see Figure-3(b).

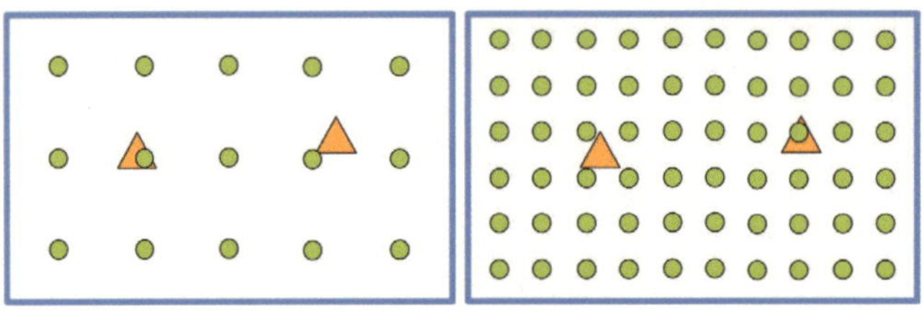

(a) Coarse resolution of a regular grid model. (b) Higher definition of a regular grid model.

Fig. 3. Regular grid model used to render volume of a room. Dots represent particles, triangles represent sensors.

Another possible construction method for particles grid is a concentric pattern. Particles are dropped in rooms with a concentric sphere centered on sensors. Coordinates of these particles can be computed using the following method:

$$p_k = \begin{cases} x_k = S_{xi} + r \cdot \sin\theta \cdot \cos\phi, \\ y_k = S_{yi} + r \cdot \sin\theta \cdot \sin\phi, \\ z_k = S_{li} + r \cdot \cos\theta \end{cases} \tag{3}$$

with

$$\begin{cases} r = k \cdot \rho, & k \in \mathbb{N}, k \leq l, \\ \theta = k \cdot \rho\theta, & k \in \mathbb{R}^+, k \leq w \text{ and } 0 \leq \theta \leq 2\pi, \\ \phi = k, & k \in \mathbb{R}^+, k \leq h \text{ and } 0 \leq \phi \leq \pi, \\ \rho \in \mathbb{R}^+ \end{cases} \tag{4}$$

The number of particles can be estimated using the following formula:

$$\begin{aligned} nbP &\leq nbX \times nbY \times nbZ, \\ &\leq l^3 \times w^3 \times h^2. \end{aligned}$$

To produce a desired point cloud, it is also possible to define the desired number of particles, and algorithm solves this problem. Figures-4 illustrates concentric repartition.

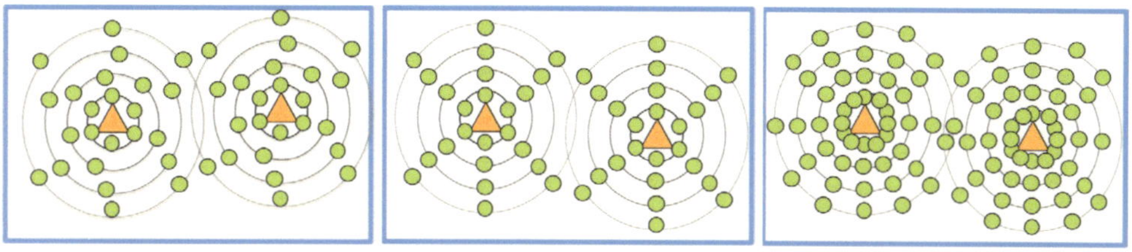

(a) Wiggle model.

(b) Coarse model of concentric representation (angle is the same for each particle).

(c) Higher model of concentric representation (angle is the same for each particle).

Fig. 4. Different concentric methods used. Dots represent particles, triangles represent sensors.

These two kinds of volumetric rendering are suitable for specific visualization. Moreover, particles need to be weighted depending on sensor location. For this, we use two different methods to compute location of particles: a Delaunay triangulation and Voronoï cells. These two methods were presented byAvis & Bhattacharya (1983). The goal of Delaunay triangulation is to produce a set of triangles from a set of vertices; circumcircle of each of these triangles does not own any other point. Thus in 3D, tetrahedrons are extracted using spheres instead of circles. Delaunay tetrahedrons are used to locate particles inside a sensor mesh. The first step is to compute all tetrahedrons, we use the algorithm proposed by Barber et al. (1996). This algorithm is one of the fastest implementation to compute Delaunay triangulation and moreover, it works for any dimensions. We feed its input with coordinates of the sensors and output is a set of tetrahedrons. Then, for each particle, it is necessary to locate it compared to each tetrahedron. To reduce the computation time, we have reduced the number of tested particles using a AABB (axis-aligned bounding box). This method was introduced by Van Den Bergen (1998) to compute hull of mesh using minimum and maximal points coordinates. The final goal is to manipulate only the particle which are inside this box. For each tetrahedron we have four vertices. To compute location of particles, we take three vertices (called a face) from a tetrahedron, then we cast a ray from a particle to the remaining vertex of tetrahedron, and so on. If rays casted from the same particle has collided with a minimum of three faces, this particle is inside tetrahedron. The ray casting method is inspired from works initiated by Snyder & Barr (1987). Authors introduce a method to compute light on a triangular mesh. After this method, some particles have to be located. Most of them are outside the sensor mesh. We use Voronoï cells to compute their location. This method is developed in many algorithms: Qhull by Avis & Bhattacharya (1983) or Vorro++ by Rycroft (2009). Voronoï diagrams are used to partition space in several dimensions. In our case we have used our own method, we locate for each particle the nearest sensors. Figure-5 presents the pipeline of algorithm in 2D. After these steps, particles know their dependency from sensors, this information can be represent by a weight. We compute an invert function based on distance for each particle:

$$\forall t \in T, \forall p \in P^t, W_p = \sum_{s \in S^t} (V_s \cdot (1 - \frac{d(p,s)}{\sum_{s \in S^t} d(p,s)})), \tag{5}$$

with T is the soup of tetrahedrons, P^t represents the set of particles inside the t-indexed tetrahedron and S^t the set of the internal sensors of t-indexed tetrahedron. *d()* is a weight function based on euclidian distance. The entire algorithm is presented in Algorithm-1.

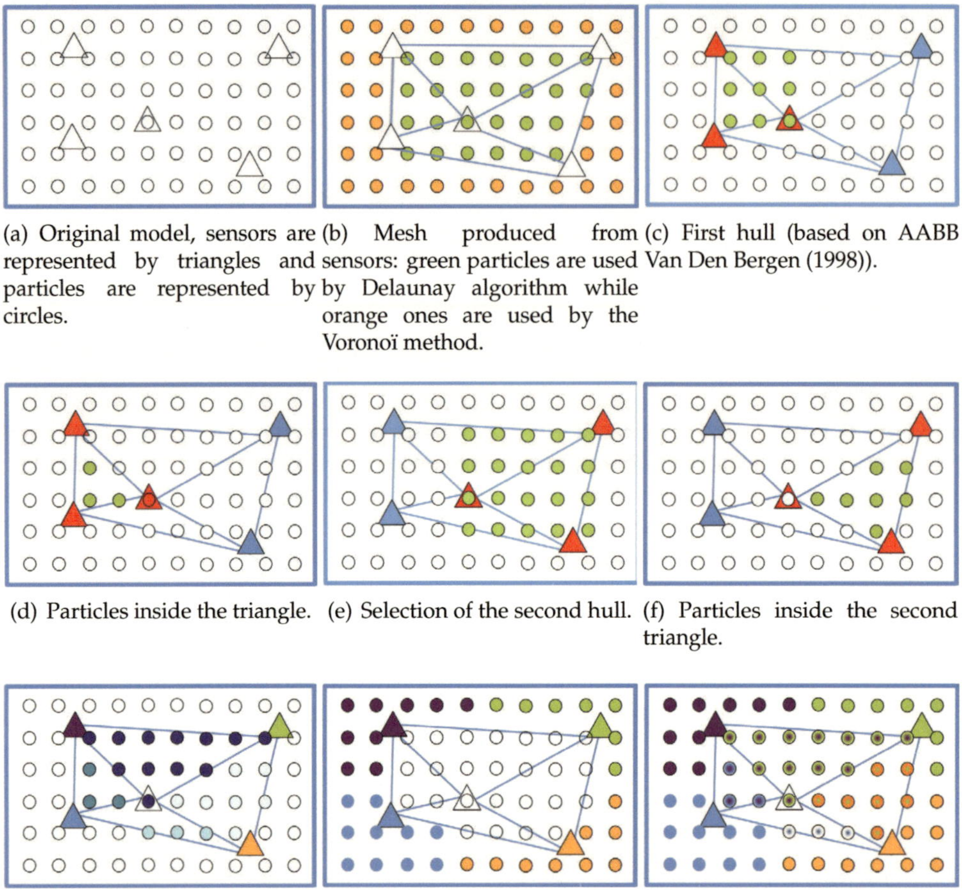

(a) Original model, sensors are represented by triangles and particles are represented by circles.

(b) Mesh produced from sensors: green particles are used by Delaunay algorithm while orange ones are used by the Voronoï method.

(c) First hull (based on AABB Van Den Bergen (1998)).

(d) Particles inside the triangle.

(e) Selection of the second hull.

(f) Particles inside the second triangle.

(g) Final result of Delaunay triangulation.

(h) Final result of Voronoï cells.

(i) Final result after weight function applied on particles.

Fig. 5. These schemas present a graphical representation of the method used to locate particles (representation is in 2D). Sensors are represented by triangles thus particles are dots.

In this chapter, we present a benchmark of different architecture to compute our algorithm. The goal is to find the best energy efficiency hardware. This benchmark is composed from standard computing method and parallel processing method. We present results of this benchmark (execution time) focusing on energy consumption. Parameters are: resolution of the point cloud, particles repartition schema, the compiler used and the method used for parallel processing.

Our visualization method is designed to used in different devices. The first parameter resolution, allows visualization needs to fit with abilities of hardware for rendering. The second parameter is volume creation, we define two kinds of point cloud construction: a regular method and a concentric method. Each of these methods produces a specific kind of render for the visualization and their construction time is also different because the number of particles produced by these two methods is different. Third parameter is compiler: we want to highlight if an important difference between a generic compiler and a dedicated compiler

Algorithm 1 The benchmark algorithm.

S = AddSensors();
P = AddParticles();
T = TetrahedronExtraction(S);
for all T such as t_i **do**
  for all P such as $p_j \in$ AABB(t_i) **do**
    if Locate(p_j, t_i) ≥ 3 **then**
      p_j.sensors.add(t_i)
      $P = P - p_j$
    end if
  end for
end for
for all P such as p_i **do**
  p_i.distance = $+\infty$
  for all S such as s_j **do**
    if p_i.distance $>$ distance(p_i, s_j) **then**
      p_i.distance = distance(p_i, s_j)
      p_i.sensors.add(s_i)
    end if
  end for
end for
ParticlesPonderation();

exists. The fourth and last parameter is based on different parallel processing methods, we have tested standard CPU parallel methods and GPU parallel methods.

This benchmark is focused on construction methods of the point cloud. We change the parameter to produce a multi resolution model of the room: from 0,01 to 5 (0,01 is a high definition mesh and 5 is a coarser mesh). Result is given in Table 1, we can see execution time for creation of the point cloud and number of particles produced. This first experimentation gives better results with concentric models than with a regular grid (for the high resolution mesh), but the number of particles produced by a regular grid is higher than concentric method. For very high resolution, object size does not fit in memory for the regular grid. For coarse resolution, we can see that computation on regular grid is the fastest.

	0,01	0,05	0,25	0,5	1	2	5
RG (time in seconds)			39,288	4,763	0,550	0,034	0,002
RG (mesh size)			6481032	780912	90552	9884	840
CS (time in seconds)	52,395	10,512	2,116	1,073	0,548	0,280	0,071
CS (mesh size)	8330976	1668563	335851	169333	86006	44260	19370

Table 1. Time results and number of particles for two methods, RG means Regular grid and CS is for concentric sphere.

Next benchmark is focused on parallel processing. Parallel processing methods can be implemented with OpenMP, MPI, on HPC, on FPGA, and so on. We have used OpenMP library to compile in parallel with GCC (Gnu Compiler Collection) and with ICC (Intel C++ Compiler), moreover, we have used a method based on GPU with OpenCL. Then, we have used a hybrid parallel method using OpenMP and OpenCL. And finally we have benched our algorithm on a HPC. Results are shown in the Table 2. Lower resolutions of the mesh (step at 0,01 and 0,05) swap memory and software becomes slower, this resolution produce an huge structured data is produce at this resolution. Moreover, the computation time is better compared to a standard computer for high resolution. But if the resolution becomes coarser, the laptop is faster. The results of the HPC in parallel shows that this method is really efficient compared to a standard programming.

	0,25	0,5	1	2	5
GCC 1	39,288	4,763	0,550	0,034	0,002
GCC with OpenMP 2	29,737	3,431	0,412	0,038	0,013
ICC 3	21,938	2,739	0,313	0,012	0,001
ICC with OpenMP 4	18,947	2,341	0,278	0,020	0,013
OpenCL 5		2,466	0,304	0,044	0,020
OpenMP and OpenCL 6		3,256	0,416	0,063	0,037
HPC 7	29,281	3,521	0,394	0,029	0,014

Table 2. Execution time in seconds on a: standard computer using GCC (1), standard computer using GCC and OpenMP (2), standard computer using ICC (3), standard computer using ICC and OpenMP (4), standard computer using OpenCL (5), standard computer using GCC with OpenMP (6) and OpenCL and finally using a HPC with GCC and OpenMP (7).

Next step is focused on energy consumption of each solution, parallel computing is a gap of energy and it is necessary to find the less consuming method. Energy formula is:

$$E = P \times T, \tag{6}$$

where E is the energy in joules, P the power in watts and T is the time in seconds. For this benchmark we have measured the energy consumed by the different devices during execution time. Figure-6 present energy results of the different executions: two main point can be extract, a standard computing is the best solution for a coarse resolution, and HPC is always the worst solution.

We can notice that solution using a spherical method is more efficient than the solution using a regular grid: the number of particles is denser, the resolution of the mesh is more important. None of other method provides bigger resolution for the regular grid on a standard computer. Another important point is that parallel processing is bigger consumer than standard CPU programming. Time of standard programming is longer but the energy used for multi core programming is more important. With a specific compiler: ICC, results are improved for the energy consumption and runtime execution compared to GCC. For the GPU programming we can see that the energy consumption is more efficient than the ICC compiler for high resolution, unless the memory of the graphic card is not enough and does not allow to produce a high-resolution mesh. Finally for the MacBook results, we see that the hybrid solution is not a good choice, exchange between memory and graphic memory produced latency on execution time, and using multi core programming and GPU programming give best results in terms of energy consumption. For the HPC results, energy is higher than the previous solutions. Professional CPUs are less efficient to save energy than standard CPU. The better solution for saving energy is to use a single core application but if we want to have a low computation time, parallel methods using GPU are more efficient for high resolution.

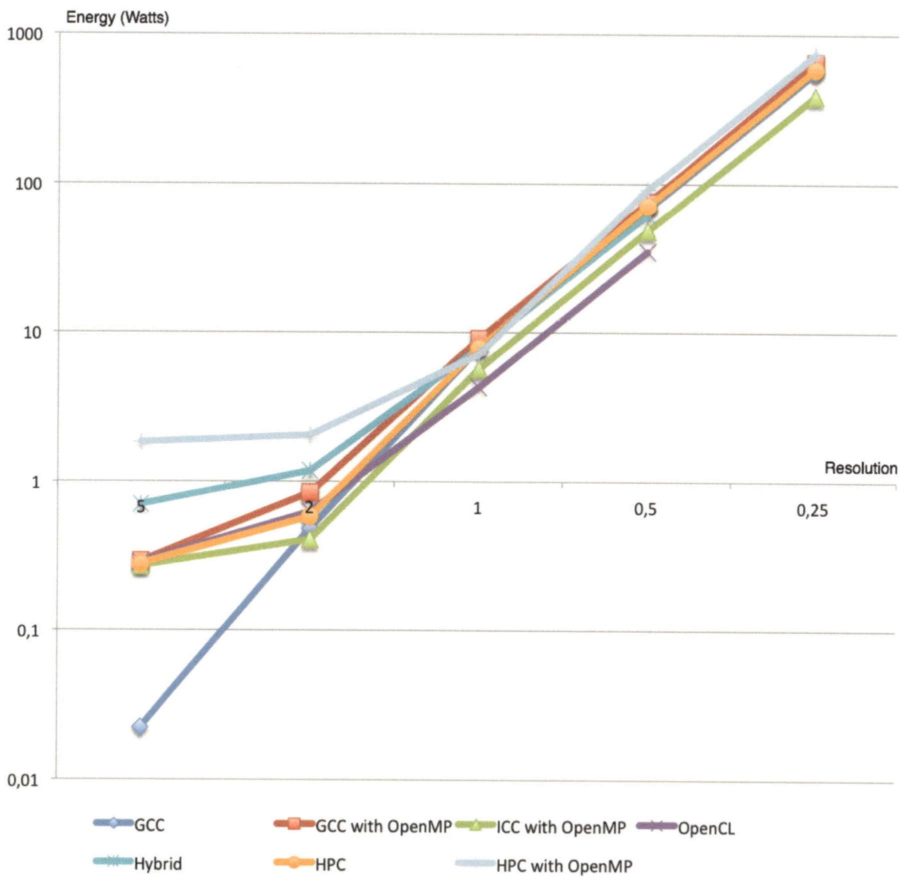

Fig. 6. Results of parallel benchmark.

4.2 One dimension visualization

In this Section, we focus on the rendering engine. It is an important task to render content in an efficient way: data needs to be understandable and their render have to be in real time. In this part we present a method to produce visualization and we explain effectiveness. Moreover, sensors can produce multivariate data, but this part is dedicated to visualization of only one kind.

Data are provided by IBM, Montpellier, France. This data contains value of a set of sensors, located in the IBM Green Data Center (GDC). Sensors are set as layers, and they are multivariate: some of them own temperature, pressure or hygrometry. Another main information is the topology of the rooms. This information can reconstruct virtual rooms with sensors location.

Figure-7 presents the different sensors layers. We use temperature sensors as example, the content produced by this sensors is the most volatile sensors value.

The first step is to view the point cloud produced by the server. This point cloud is based on a simple point sprite primitive, with this method we are able to render a large set of vertices

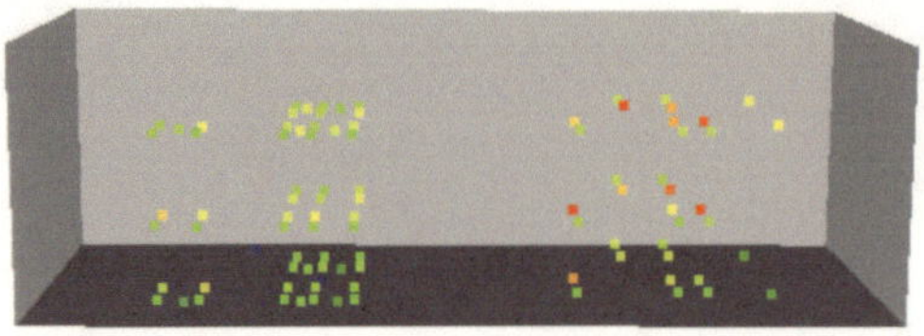

(a) Location of temperature sensors.

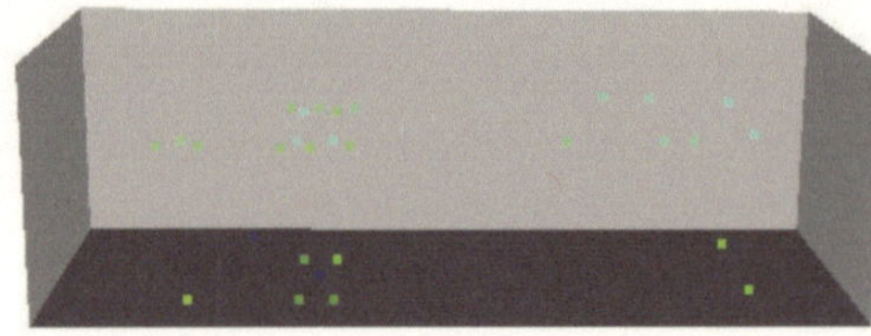

(b) Location of pressure sensors.

(c) Location of hygrometry sensors.

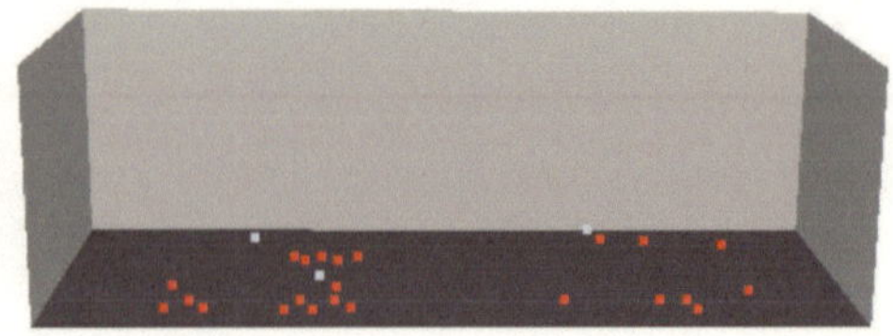

(d) Location of APS sensors.

Fig. 7. Location of the different kind of sensors.

in real time. Figures-8(a) gives a representation of the regular grid pattern and Figure-8(b) is a view of spherical method for the particle representation. This kind of primitive is efficient to navigate into data and understanding is not easy. point clouds have some issues, depth is not easy to catch and moreover aliasing produced by this view is a weak point for the usage of this visualization.

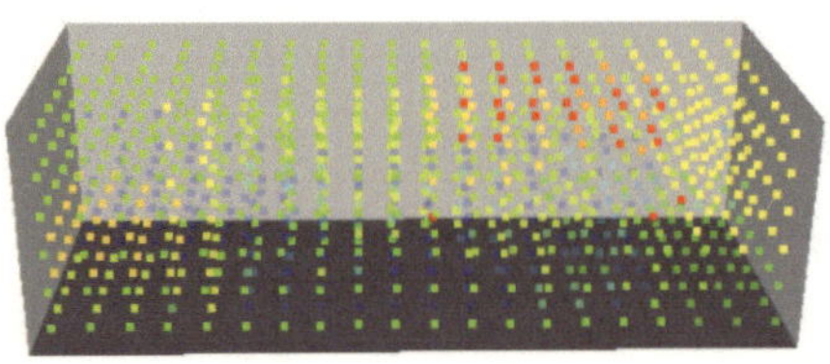

(a) point cloud visualization of a regular grid.

(b) point cloud visualization for a concentric model.

Fig. 8. Raw visualization using point sprite representation.

To smooth result of the previous method, we implement a convolution method. Convolution is used in 2D imaging to smooth an image. It works by using a kernel function applied on each pixel. Results smooth the entire room volume, unfortunately same issues than the previous method persists, visualization is hard to understand and computation time increases. Result seems to be more realistic, without convolution an aliasing color effect is produced by particle, from cells or tetrahedron border, with convolution this brutal change is smoothed.

This kind of view has another important issue, data are dynamic, and it is hard to catch an update. It is necessary to produce efficient method to identify updates. To reduce computation, bandwidth of client, data transmission between server and client is done through LOD paradigm, update is realized on sensors and then on the first neighborhood of it, the graphical result of this method is presented Figure-9.

But the previous method is only a method to understand the update of data, real value is not visible on updated particles. To update and catch this modification, we propose to change

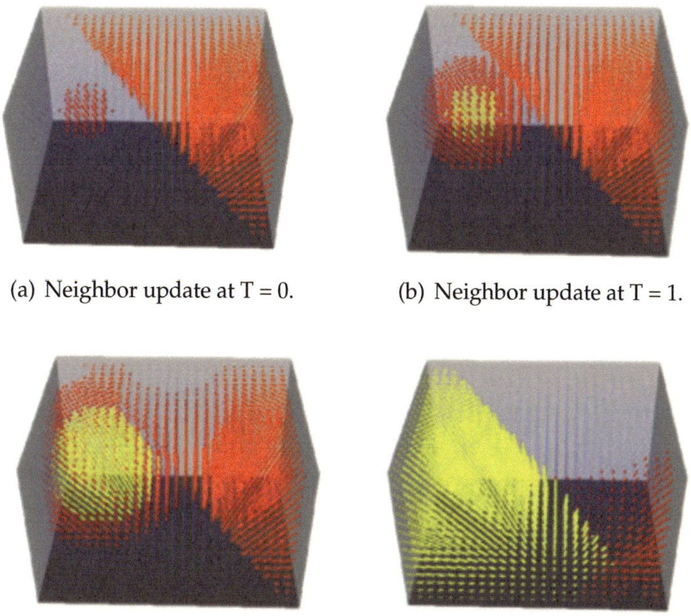

(a) Neighbor update at T = 0. (b) Neighbor update at T = 1.

(c) Neighbor update at T = 2. (d) Neighbor update at T = 4.

Fig. 9. Update processes based on neighbor exploration.

the size of updated particles. Thus, updated particles are smaller than fixed value particles. Updated data is now easier to catch, as shown in Figure-10. The render cost and render issue are the same than a standard point sprite visualization.

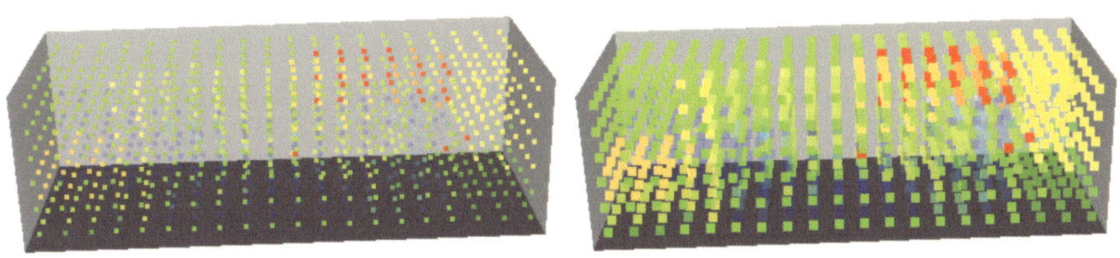

(a) Particle view at T = 0. (b) Particle view at T = 1.

Fig. 10. Updated particles have a bigger size than static value particles.

Another possible solution is to use a method based on convolution algorithm. Updated particles are moved from their original location depending from the result of the convolution kernel. This method gives a simple view of influence of each particle. Airflow of rooms can be model by this method as presented in Figure-11.

To improve the previous solution instead of point cloud, vectors are used to represent results of the convolution mask. Origin of the vector is the location of particle, and the end point is the location of the previous method as illustrated Figure-12. Has we see using vectors is not an efficient method to understand data.

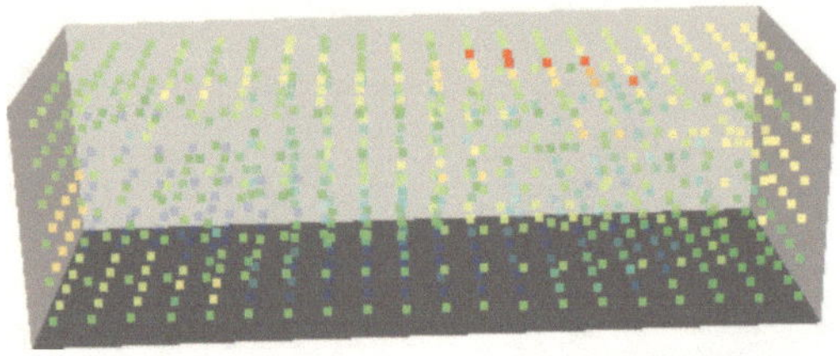

Fig. 11. Particles are moved depending from result of convolution mask.

Fig. 12. V-neck direction of most influent sensors for each particle.

An important point in data visualization is to extract quickly the main information. Main information is variable and depends of the topic. In our case, we produce a thermal visualization and try to extract the relevant temperature area. To do this, we define methods based on boundary extraction or graph theory.

The first method extracts bounding box from a set of particles. The goal is to produce hull of a 3D set of points, we use AABB hull due to the low cost of computation. In our case, data can be aggregated by values, and components are only managed by a delta value. First results are presented in Figure-13.

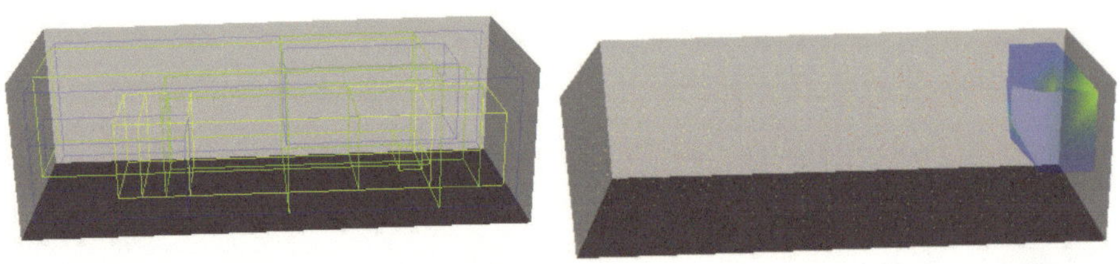

(a) Segmentation of space using extraction of component by value at T = 0.

(b) Triangulation of the biggest component at T = 1.

Fig. 13. Segmentation based on AABB algorithm.

We can also convert our point cloud to a graph. Each neighborhood vertex with a similar value can be connected together. This method creates a set of connected components. Formally, this method is defined by the graph: $G = (V, E)$, E the set of edges and V the set of vertices. G is composed by subgraphes: $G = \{G_1, ..., G_N\}$, G_j where $G_i \cap G_j = \emptyset$, $G_j = (V_j, E_j)$, $V_j \subset V$, $E_j \subset E$ and $\forall j$, G_j is connected. Then, we can use the same bounding method: AABB for each component. Results produce a messy visualization, accuracy of component is more realistic, but the number of boxes is too important. This issue can be visualized on Figure-14, it is also possible to reduce this kind of noise by increasing the delta value of each component.

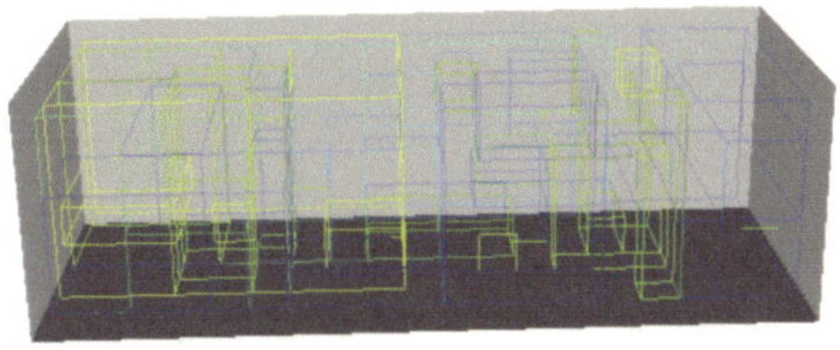

Fig. 14. Segmentation of particle cloud with a simple extraction of connected component.

Finally, to resolve aliasing issue, we focused on rendering of volume. point cloud is not the best method to render a volume, so we have tested several other methods to give user the best point of view.

Our first method is based on triangulation. We take the point cloud and apply a marching cube method, this method is introduced by Lorensen & Cline (1987), the goal is to produce a surface mesh from a point cloud set. Results are given in Figure-15(a). We can see some visual issues due to the triangulation. But the main problem is computation time, indeed, methods used are not in real time for large set of points.

To produce triangulation of the point cloud, we also used the Delaunay Algorithms, Figure-15(b) gives a representation of this method, as we see, this algorithm simplify the point cloud mesh and only extremal points are conserved.

Another idea is based on extraction of the connected component, when we triangulate them, rendering time is too long; result is given in Figure-15(c), some issue appear for color mapping.

This part will present voxel view of rooms. Unfortunately, the rendering cost of these primitives is too important compared to a standard point cloud visualization, the results is presented Figure-15(d). This method is well suited for a volume rendering compared to point cloud, but update of each frame take too mush time.

Last method used is shader programming language. With this method, we get best results and can explore data in an efficient way. Shaders are used after the rendering of the 3D images, computation is done on GPU on a rasterized 2D image. Figure-16 gives some results of this method, different shaders have been used to render each of these figures. Exploring data and understanding is easier than all solutions proposed before, shader methods are mostly used to render flow visualization.

4.3 Data mining through visualization

In this section, we present a method used to dig into data content, the goal is to propose content for the whole system and optimize consumption of a building. We focus on method used for annotate the building model and ways to add content.

We propose some interaction tools to give user abilities to add, update and delete content from current instance of the green box model. Original model extracted from the BMS does not contain enough information, thus it is necessary to add these missing informations. Moreover, buildings have their own life such as the topology and the environment. These informations always changed and BMS are not developed to catch these updates. The interaction tool we propose is composed of two shapes: a sphere and a box. This method gives to the user the

(a) Marching cube method.

(b) Triangulate mesh of point cloud surface.

(c) Triangulate meshes of all component.

(d) Voxel based visualization.

Fig. 15. Severals methods used to produce a surface mesh from our point cloud (visualization not in real time).

(a) Shader visualization, witha fixed shader function.

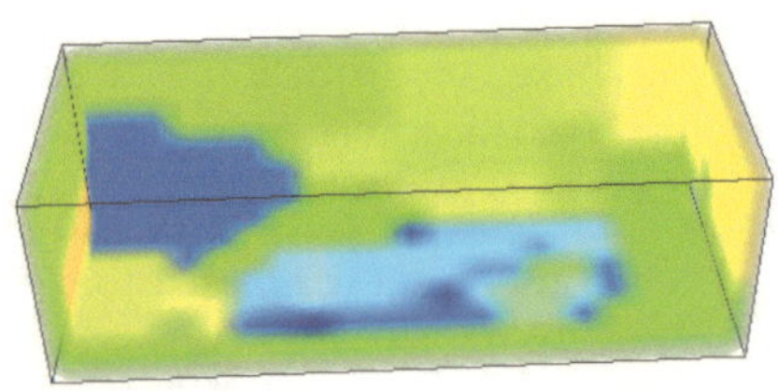

(b) Shader visualization, based on rayTracing method.

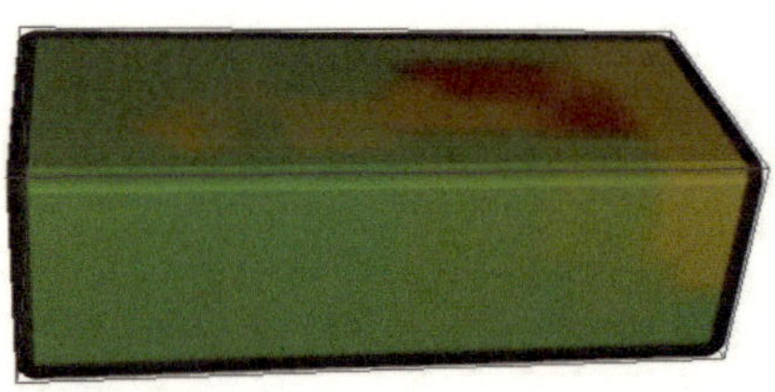

(c) Shader visualization, with isosurface extraction.

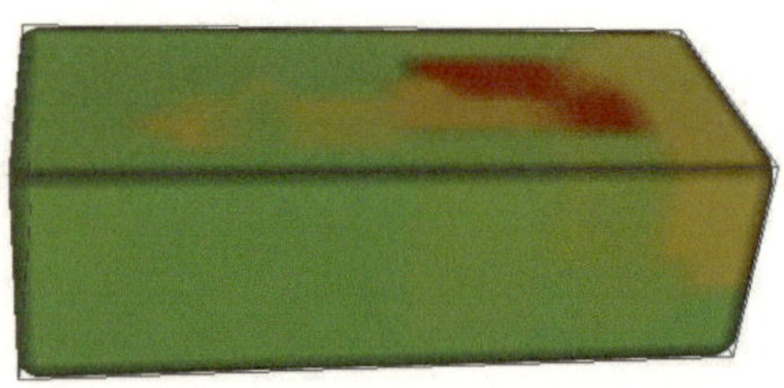

(d) Shader visualization, with isosurface extraction and a transfer function based on light intensity.

Fig. 16. Real time visualization of particles using shader programming language.

possibility to improve building by standard modifications. Another feature allows adding information like influence of sensors, furniture location, split and rooms. All these new contents will introduce new relevant informations for the green box. Figure-17(a) presents two shapes used to dig into data, and manipulator used for placement.

Another content which can be added to the pivot model, is the correlation. Correlations are links between data, when a value of data set changes, each link is influenced. Most modules of RIDER green box are developed to find these correlations, unfortunately some of them can be wrong or missing. But user can discover, update and propose new of them. This tool consists in producing links between data using a simple "Line" primitive with color depending from weight of each node. Correlation can be added to different objects in a room such sensors, furnitures and valve. Figure-17(b) presents an overview of this method.

(a) Digging tools to add, update or remove content. (b) Example of correlation between sensors.

Fig. 17. Tools to dig into data.

To provide new data in the Pivot model, we use a simple XML message. Messages are composed by unique component's id, and new features. For annotation, feature is a text message. For a wall adding, two XML messages are sent: the first one composed by the old room, with topology switch, the second one composed from a new unique ID and shape of this room. These messages are formatted to fit with input of RIDER model, each plug-in sends messages with this format.

5. Case study

In this section, we present results of the Section 4 methods with real data. We use data provided by our pilot: Green Data Center from IBM Montpellier, France. The goal of this data center is to propose efficient methods to reduce energy consumption. This data center is composed by two rooms. The first room is dedicated to servers with a large number of CPU (cooling is done using specific case), called High density. The second room is for low-density servers with a standard air cooling, air management is done through confined alleys. Figure-18 presents a schema of these two rooms. Our data set is composed by shape of merged rooms and sensors data (location, kinds and value). To help other data mining components, it is necessary to provide more content. In this case study, we will extract two spaces (one for each room), then we propose to analyze rooms, extract content and optimizations.

5.1 visualization of the case study

Our first step is to split the two rooms into two spaces. For this, we analyze also sensors location as illustrated Figure-19(a). Obviously, we can find limits of each room, Figure-19(b). It is necessary to identify the content of these rooms. We use the particle view presented Section 4 to identify rooms. We know that the low-density room is organized with area as stated in Figure-18(b). Figure-19(c) and Figure-19(d) present two different views of the room

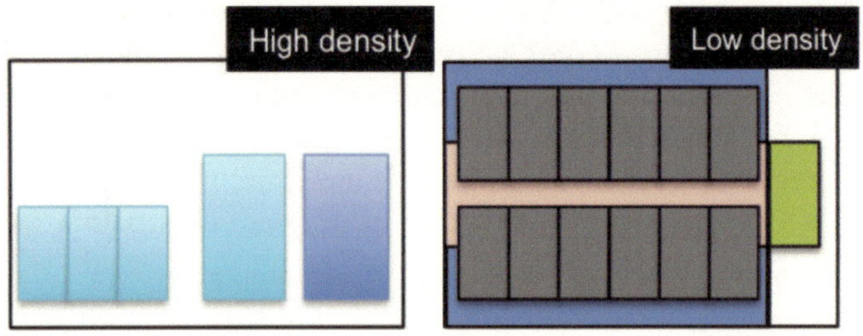

(a) Schema of high-density room (top view). (b) Schema of low-density room (top view).

Fig. 18. User knowledge on GDC, two rooms composed the green data center: the low-density room and the high-density room.

using a volume method. The right room seems to be segmented in different areas. This means that this room is low density.

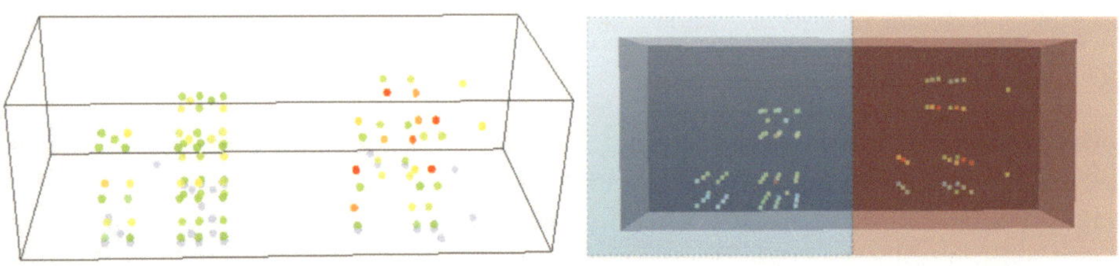

(a) Location of all sensors in GDC. (b) "Building" segmentation.

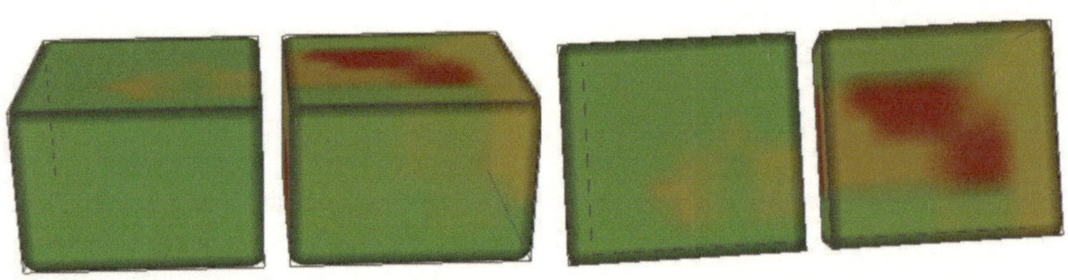

(c) Front view of two rooms, temperature view. (d) Top view of two rooms, temperature view.

Fig. 19. Different steps to identify rooms.

The next goal is to identify content from the high-density room. We analyze data sensors of each kind, using a time function, temperature is most variable value, other kinds are not enough relevant. To find content, we manipulate temperature slider and look for relevant information. Figure-20 gives an overview of different tests done, we found interesting information, for example at center of the room a continuous mass is present. User knowledge is two large cases and a group of servers compose this room, see Figure-18(a). With our flow visualization, we can identify each part of knowledge. But an error occurs on the stack of servers, because they are too low. Another step is to add correlation between stacked sensors. Finally from the green box, we have proposed several contents concerning the location and the size of server (heater part of the room) and moreover some links between sensors.

(a) High density room visualization.

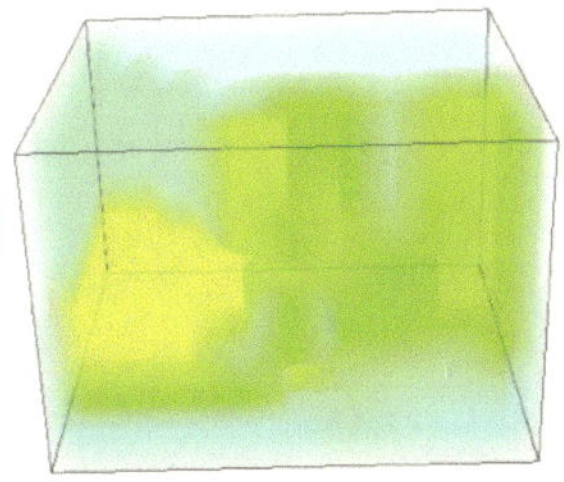

(b) Temperature flow visualization (an alpha function is applied on low temperature).

(c) Extraction of highest temperature, popping of three blobs.

(d) Identification of blobs.

Fig. 20. Visualization mining of high-density room: extraction of servers.

For the low-density room, the goal is to extract alley information, this part of data center is confined to improve energy usage. It is decomposed in three alleys: two cold and a hot alley as stated in 18(b). We use the particle visualization to identify these areas. Figure-21 gives results. Hot alley can be located in center of the room. Another information appears in this visualization, Figure-21(e) presents location of CRAC (CRAC are cooling systems), they are used to extract hot temperature from the alley. As presented before, we can add for each sensor correlation between them (stacked, and alley sensors).

All these new informations are compared to real knowledge from data center and matched. We can now analyze visualization to propose data center optimization. A recurrent issue appears on the low-density room; top servers are too hot during a long while. It is necessary to refactor location of servers to improve their server air flow.

5.2 Case study improvement

IBM Montpellier GDC has a low PUE near 1.5. Using all presented visualization in subsection 5.1 we can propose some improvements for reduce energy consumption of the data center. Firstly, we will focus on energy usage of the high density room, then on the low-density room.

In the high density room we can highlight some points. As stated in Figure-20(d), we can extract location of different servers, the left group is a group of server stacked on a classic server case. If we analyze results from time-lapse visualization, we can see that hottest servers are mostly on the bottom of server case. This part seems to not need changes. For the second air cooled case (center of Figure-20(d)), using a time based visualization, we can

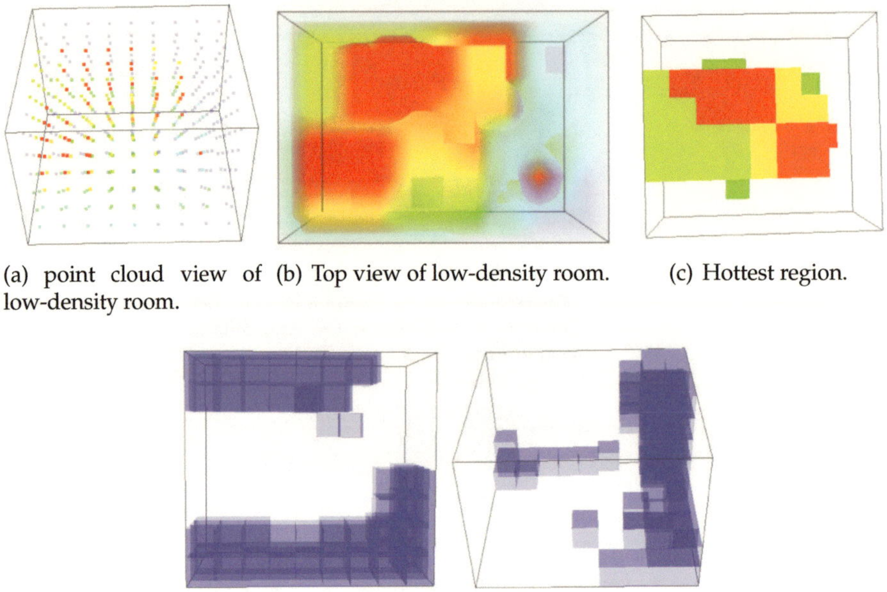

(a) point cloud view of low-density room. (b) Top view of low-density room. (c) Hottest region.

(d) Median temperature, fresh alleys. (e) Coldest temperature, identification of CRAC region .

Fig. 21. Visualization method to extract different alleys from low-density room.

find refactoring optimization in this case. Sometime, hottest servers are placed at the top of the case. To reduce cooling, it is necessary to get down these hot blades. For the last case (the water cooled case), we can find the same pattern as the previous one. Top servers become hotter than down servers. It is necessary to bring them to the bottom of the case to reduce the energy consumption of this case. The cooling system of the global room is located in the left back corner of the room, unfortunately, it is seems to only cool a small part of the room and water cooled server capture an hot air.

The second room visualization present a new way to improve data center. The hot alley of the data center are segmented but we can see that computation of servers is done on the top of the stack. Thus, it is necessary to refactor server blade to improve the airflow of the room and reduce the energy consumption. To propose other improvement, it is necessary to get more data dimensions, for example: energy of each server. Scalability features have to be added to be able to manage buildings and group of buildings in the same time.

6. Conclusions

Specials devices can be used to avoid electrical pic of consumption during journey and smooth usage on the electrical network. Electric usage of consumer can be predict (usage of washing machine) and optimization can be done to reduce the network load. Domotic can be used to automate process, for example window awning can be set depending on sunshine or external temperature, the goal here is to reduce usage of cooling system and thereby reduce electric consumption. Smart buildings are able to react when classic buildings do not.

Buildings consume 38 % of the global energy and produce between 25 % and 40 % of the CO_2 emission. Buildings are expensive investment, not only for their construction cost or buildings fees, but also for their ecological cost.

Our project called RIDER has a goal to develop a new information system to optimize energy efficiency of building or group of buildings. To propose optimizations, energies transfer between buildings can be realized. This transfer can be done on different energies: classic, green (extracted from wind system or photovoltaic) or deadly (produce by industry usage). This project is based on development of new data processing method (models, data mining, visualization, IT architecture and communication).

In this chapter, we present a concept of the green box to dig volumetric data for extracting important features of buildings, the goal is to provide some optimizations. Data are provided from a Green Data Center at IBM Montpellier, France. Our visualization software gives the abilities for users to suggest some topology modifications based on room definition.

Walls have to be added, many informations provided by building manager will be improved. Other stuff needs to be added to improve the building model. It is necessary to set a part of furniture to know the influences of model. To propose this model, we use a particle paradigm to visualize the volume of each room. To dig into data, we give to the user the ability to filter data and submit some updates to IT architect.

First global results of the whole architecture can expect that optimization can reduce energy between 30% to 40 % of a building.

7. Acknowledgments

We would like to thanks the PSSC (Products and Solutions Support Center) team of IBM Montpellier, France for having provided the necessary equipment and data needed for this experimentation. We also thank the FUI (Fonds Unique Interministeriel) for their financial support.

8. References

Avis, D. & Bhattacharya, B. K. (1983). Algorithms for computing D-dimensional voronoï diagrams and their duals, *Advances in Computing Research* 1: 159–180.

Barber, C. B., Dobkin, D. P. & Huhdanpaa, H. (1996). The quickhull algorithm for convex hulls, *ACM Trans. Math. Softw.* 22: 469–483.

Clark, J. H. (1976). Hierarchical geometric models for visible surface algorithms, *Commun. ACM* 19: 547–554.

Green, S. (2007). Cuda particles, *NVIDIA Whitepaper, November* .

Hauser, H., Ledermann, F. & Doleisch, H. (2003). Angular brushing of extended parallel coordinates, *Information Visualization, 2002. INFOVIS 2002. IEEE Symposium on*, IEEE, pp. 127–130.

He, T., Hong, L., Kaufman, A., Varshney, A. & Wang, S. (1995). Voxel based object simplification, *Proc. SIGGRAPH Symposium on Interactive 3D Graphics*, pp. 296–303.

Ilcìk, M. (2008). A framework for global scope interactive visual analysis of large unsteady 3d flow data, *CESCG* .

Kapferer, W. & Riser, T. (2008). Visualization needs and techniques for astrophysical simulations, *New Journal of Physics* 10(12): 125008 (15pp).

Keim, D., Mansmann, F., Schneidewind, J. & Ziegler, H. (2006). Challenges in visual data analysis, *Information Visualization, 2006. IV 2006. Tenth International Conference on*, IEEE, pp. 9–16.

Konyha, Z., Matkovic, K. & Hauser, H. (2009). Interactive visual analysis in engineering: A survey, *Posters at SCCG 2009* pp. 31–38.

Lex, A., Streit, M., Kruijff, E. & Schmalstieg, D. (2010). Caleydo: Design and evaluation of a visual analysis framework for gene expression data in its biological context, *Pacific Visualization Symposium (PacificVis), 2010 IEEE*, IEEE, pp. 57–64.

Liu, Z., Lee, B., Kandula, S. & Mahajan, R. (2010). Netclinic: Interactive visualization to enhance automated fault diagnosis in enterprise networks, *IEEE VAST* pp. 131–138.

Lorensen, W. E. & Cline, H. E. (1987). Marching cubes: A high resolution 3d surface construction algorithm, *SIGGRAPH Comput. Graph.* 21: 163–169.

Luebke, D. (1997). A survey of polygonal simplification algorithms, *IEEE Computer Graphics and Applications* (21): 24–35.

Luebke, D. P. (2001). A developer's survey of polygonal simplification algorithms, *IEEE Computer Graphics and Applications* 21: 24–35.

Migut, G. & Worring, M. (2010). Visual exploration of classification models for risk assessment, *Proceedings of the IEEE Conference on Visual Analytics Science and Technology 2010.*

Moenning, C. & Dodgson, N. (2004). Intrinsic point cloud simplification, *Proc. 14th GrahiCon* 14.

Pauly, M., Gross, M. & Kobbelt, L. (2002). Efficient simplification of point-sampled surfaces, *Visualization, 2002. VIS 2002. IEEE*, IEEE, pp. 163–170.

Piringer, H., Kosara, R. & Hauser, H. (2004). Interactive focus+ context visualization with linked 2d/3d scatterplots, *Coordinated and Multiple Views in Exploratory Visualization, 2004. Proceedings. Second International Conference on*, IEEE, pp. 49–60.

Rosen, P. & Popescu, V. (2011). An evaluation of 3-d scene exploration using a multiperspective image framework, *Vis. Comput.* 27: 623–632.

Rycroft, C. H. (2009). Voro++: a three-dimensional voronoi cell library in c++, *Chaos 19* . Lawrence Berkeley National Laboratory.

Schroeder, W. J., Zarge, J. A. & Lorensen, W. E. (1992). Decimation of triangle meshes, *SIGGRAPH Comput. Graph.* 26: 65–70.

Shneiderman, B. (2001). Inventing discovery tools: Combining information visualization with data mining, *in* K. Jantke & A. Shinohara (eds), *Discovery Science*, Vol. 2226 of *Lecture Notes in Computer Science*, Springer Berlin / Heidelberg, pp. 17–28.

Snyder, J. M. & Barr, A. H. (1987). Ray tracing complex models containing surface tessellations, *SIGGRAPH Comput. Graph.* 21: 119–128.

Song, H. & Feng, H.-Y. (2009). A progressive point cloud simplification algorithm with preserved sharp edge data, *The International Journal of Advanced Manufacturing Technology* 45(5-6): 583–592.

Van Den Bergen, G. (1998). Efficient collision detection of complex deformable models using aabb trees, *J. Graphics Tools.*

9

Succeeding in Implementing Energy Efficiency in Buildings

Mark Richard Wilby, Ana Belén Rodríguez González, Juan José Vinagre Díaz and Francisco Javier Atero Gómez
Department of Signal Processing and Communications
Rey Juan Carlos University
Spain

1. Introduction

In this chapter we address a simply stated problem. *How* do we motivate and enable energy consumption reduction in the building sector? Buildings account for about one third of the world's total energy expenditure (UNEP, 2007). This makes them a primary focus for energy reduction strategies. However, building ownership, management and usage extends across the whole population and is not limited to the control of a few international organizations. This diversity of active participants makes the process for initiating change an extremely complex problem. It requires not only technical solutions, but also social and political solutions working together to construct an environment for controlled change.

It is inevitable that policies will evolve in the very near future to penalize excessively inefficient buildings. The motivation will be twofold, an attempt to increase incentives to improve efficiency and to recoup, at a national level, international carbon tax costs. But if these policies are to be made to work, they must be designed to be applied and evolved over a significant time interval, several decades. They will also have to be defined based on inputs from a wide variety of disciplines. We will try here to address the start of this problem in the context of the realities of what is available now and what may be done in the current climate.

There are many technical and political proposals that address some aspects of this problem. Here we will review these contributions and merge them into a novel adaptable energy efficiency rating system. We will focus on the interplay between the technology and policy for enacting energy efficiency. We propose a novel framework which provides incentives to building proprietors to improve energy efficiency. We do so by building upon the three main points of the United Nations Environment Programme (UNEP) sustainable building report (UNEP, 2007): (i) Determine energy efficiency objectives and energy performance standards; (ii) Formulate a methodological tool to verify the reductions on GHG emissions and implement fair comparisons between buildings; and (iii) Establish annual energy performance monitoring systems.

The key to achieving our challenging objective is constructing a unified and planned research line that is composed of a set of developments that address the apparently separate issues that underlie this problem. Thus the strands works that are described in this chapter should not be seen as separate independent activities, but as interacting elements, whose combined use can start to address this complex problem.

2. Context

Energy is obviously the true engine of industrial, social, and economic advancement that has benefited the developed world in the previous two centuries. This is a fact. However, energy consumption is a significant factor in the catastrophic destruction of the environment, health, and quality of life that we all have suffered during the same period. This is also a fact. This second fact will progressively worsen over the next few decades if nothing is done to combat this horrendous side effect. On the positive side, we are aware of these two facts, and actions have been triggered to address the issues. On the negative side, it is definitely not going to be easy.

The straightforward solution is removing the link between development and energy use. However, in the foreseeable future, this is simply not a possibility. Instead, all efforts focus not directly decreasing energy consumption, but rather on optimizing energy usage such that a similar level of development can be maintained at a smaller overall energy cost. This rather restrictive approach is what is generally call *energy efficiency* and will be the basis of the approach taken here.

There is no real starting point for the emergence of environmental concerns, but the Stockholm Environment Conference in 1972 can be regarded as the point where political attitude began to change. Momentum slowly built up and from the '80s onwards, international organizations and governmental institutions started creating initiatives to promote energy efficiency. Europe has been one of the pioneers, issuing its first Community policy in (European Union, 1986), and subsequently creating an extensive set of Directives and globally focused legislative policies. In the international scenario, the most significant action was the Kyoto Protocol (United Nations, 1998), leading to a limited number of countries agreeing to reduce their GHG emissions to achieve a 5 % reduction from 2008 to 2012 relative to those released in 1990.

Consequently, there is a clear worldwide interest in looking not for *joules* but for *negajoules*, i.e., the "non-consumed" energy. And we try and find those negajoules anywhere, starting from the main consumers: industry, transport, and others, each of which accounts for around a third of the world's energy consumption (IEA, 2010b), (IEA, 2010c). Buildings are responsible for the vast majority of the *others* category, contributing somewhere between 30 and 40 % of the consumption of primary energy and the corresponding greenhouse gas (GHG) emissions (UNEP, 2007).

Although buildings are a major energy consumer, we have to accept the fact we cannot simply reduce their number. However, they can be regarded as a massive source of potential energy savings (Uihlein & Edera, 2010). In the context of buildings then, the focus is on consumption reduction. The realization that buildings offer this potential is already reflected in policy documents. For instance, in the European Action Plan for Energy Efficiency (European Commission, 2006), 30 % of the priority measures are directed specifically towards buildings.

Buildings are one of the few markets where we have dramatically weak penetration of innovation. Although new buildings can easily incorporate new technologies, the long time scales and significant energy costs of building replacements, means that we will have a legacy environment for decades to come. This provides a remarkable opportunity for development, but it requires the cooperation of a great variety of disciplines: Science, Technology, Economy, and Policy. Together, they could provide a joint solution to achieve energy efficiency and propose novel approaches for succeeding in this ambitious enterprize.

In this work, we aim to propose novel approaches to achieving energy efficiency in buildings. We will follow the three main points indicated in the United Nations Environment Programme (UNEP) in order to succeed in having sustainable buildings (UNEP, 2007): (i) Determine energy efficiency objectives and energy performance standards; (ii) Formulate a methodological tool to verify the reductions on GHG emissions and implement fair comparisons between buildings; and (iii) Establish annual energy performance monitoring systems.

Firstly, we will summarize the legislative framework in which energy efficiency in buildings operates. Then we will describe the present achievements in energy efficiency, following the three previous main points. Finally, we will describe our own novel proposals to fulfill those defined objectives.

3. Legislative framework

3.1 Energy efficiency

Although nowadays energy efficiency is the worldwide accepted keyword to reach the objective of achieving a dramatic decrease in consumption and its consequent positive effect on environment, decades have been needed to come to this point. Among all the significant international efforts, Europe has become a leading actor in the energy field.

As far back in 1986, the European Council issued Resolution 86/C 241/01 (European Union, 1986), a seminal Community policy that fixed an objective of improving the efficiency of final energy demand by at least 20 % by 1995. The Community noticed that the availability of the energy required to fulfill the demand to maintain an optimum economic and social growth was a must. Nevertheless, given the prospects for supply and demand by that time, this availability was seriously uncertain.

In order to achieve such an ambitious objective, the Community adopted a set of innovative ideas that have led Europe in its titanic effort to combat energy consumption and environmental damage. The following list itemizes the most significant of these guidelines:

- It takes into account both the supply and the demand side of the problem. This was a disruptive approach since the majority of measures traditionally focused only on the former. Resolution 86/C 241/01 started considering that every kilowatt that was not consumed accounted for higher benefits that every extra kilowatt that was produced, thus empowering the definition of the term *energy efficiency*.
- It proposes the creation of a continental common framework to coordinate and harmonize efforts in committing to the overall objective. Following this principle, instead of defining rigid planning instruments it inclines for flexible and adaptable objectives. In addition,

it points out the need for the construction of monitoring systems capable of measuring the results of national policies. Finally, it suggests the construction of an internal energy market in order to integrate the different initial available resources to achieve the goal among countries.

- It requires cost effective implementations. Due to the existing direct connection between energy and economy, expensive actions appear as unreasonable to solve the global problem.
- It establishes a link between energy and environment. Despite the original concern regarding the availability of the required energy to keep the same level of social and economic development, the resolution paid also attention to the dramatic consequences of the energy consumption on the quality of life. Huge efforts have been invested in solving this issue during the following decades.
- It acknowledges technology as a key factor for achieving the objective. It points out the need to rely on technology to generate innovative solutions to give answer to the different issues that would arise in all fields related to energy.

In 1991 the European Community realized that the resolution 86/C 241/01 was doomed to fail if no *vigorous* measures were taken. Consequently, it created the so-called *Specific Actions for Vigorous Energy Efficiency* (SAVE programme). SAVE was firstly regulated as the Council Decision 91/565/EEC (91/565/EEC, 1991). 35 million euros were invested in 5 years in order to achieve the objectives fixed for 1995. This decision follows the same original ideas stated in its precedent, focusing efforts mainly on energy efficiency and coordination and harmonization. The former was promoted through the training of end users, creating pilot projects, and building a specific directive regarding efficiency in electricity (89/364/EEC, 1989). The latter was stimulated by the definition of technical standards and the construction of a network to share information among national, Community, and international parties regarding their activities in contributing to achieving the overall objective.

These decisions indicate that the European Community was inclined to set its energy policy on the basis of raising the awareness of the end user to the problem, relying on technology to support the needed innovations, and joining coordinated international efforts to enhance the previous measures and reach the challenging objective.

Decision 91/565/EEC apparently included all the fundamental aspects in its predecessor 86/C 241/01. However, it did not incorporate the relation between energy and environment as one of its pillars. This aspect was considered two years later in the Council Directive 93/76/EEC (93/76/EEC, 1993). It focused on limiting the CO_2 emissions by energy efficiency. It also takes this guideline a step further, noticing that buildings are a key objective for increasing energy efficiency due to: i) they represented a 40 % of the final energy consumption in the Community, and ii) it was an expanding sector. In consequence, this directive explicitly indicates the need for an energy efficiency certification scheme for buildings, as well as energy audits for companies with a high energy consumption. The specific legislation derived from directive 93/76/EEC will be extensively described in subsection 3.2.

External to the European environment, the international community was working in a similar direction. The most relevant milestone was the Kyoto Protocol (United Nations, 1998). The Kyoto Protocol aimed at giving solutions to the tremendous environmental concern through a dramatic reduction of the GHG emissions. The so called Parties of the Kyoto Protocol, a set of 38 countries around the world, came to an agreement on the objective of reaching from 2008 to 2012, a 5 % reduction on their GHG emissions referred to those in 1990. In order to provide the required flexibility for every member to achieve the overall goal, the Kyoto Protocol allowed the Parties to individually or jointly commit to this objective. In addition, the global objective was not fixed for each Party; instead, a particular percentage of limitation was established for each Party.

Consequently, the Kyoto Protocol included some of the features that had been already considered in the previous European legislation:

- The environmental concern.
- The need for a common framework based on objectives.
- The requirement regarding flexibility and coordination.

Let us go back to Europe and describe how the Kyoto Protocol was transposed into legislation. The first step to be taken for each Party is to legalize their own commitment. Europe performed this step in 2002, through a Council Decision (2002/358/CE, 2002) where the Kyoto Protocol was approved. The European Union agreed to select a joint achievement of its percentage of reduction. The subsequent percentages to be fulfilled for each member of the EU were also listed. Next, the EU issued a second decision (2006/944/EC, 2006), that specified the reference for these percentages, i.e., the levels of GHG emissions to which these percentages should be applied. Following the original guidelines that suggested flexibility and the creation of an internal market, the EU put in place an internal reallocation scheme. The set of the resulting reallocated percentages were defined in Decision (2006/944/EC, 2006). Finally, in this same Decision, the subsequent GHG emission levels allocated to each Member State in the EU for the first period of commitment (2008-2012) were also fixed.

The described first direct implementation of the Kyoto Protocol in Europe led to a decisive political framework to face the environmental and energetic issue in the EU. The final aim was to create a European high efficiency and low emissions economy. This challenging objective, namely the *20 20 by 2020*, was fixed in (European Commission, 2008) and its particularization was based on a set of four compromises:

- A 20 % reduction in GHG emissions with regard to those in 1990.
- A 20 % reduction in primary energy consumption, compared to the projected levels.
- A 20 % energy consumption coming from renewable energies.
- A 20 % increase in energy efficiency.

This overall objective was put into practice through the so called *climate and energy package*. This package was drafted as a Law in June 2009 through four complementary legislative pieces:

- The EU ETS, (EU Emissions Trading Scheme) (2009/29/EC, 2009), aimed at strengthen the European emissions allowances trade.
- Decision 406/2009/EC (406/2009/EC, 2009) in order to share the effort in reducing emissions among those sectors that are not covered by the EU ETS.

- Directive 2009/28/EC (2009/28/EC, 2009) that included the promotion of the use of renewable energy sources.
- Directive 2009/31/EC (2009/31/EC, 2009), to develop technologies for carbon collecting and storing.

In summary, the EU precisely followed the guidelines that were fixed back in the '80s. It built a coordinated continental framework based on a flexible regulation that defined an ambitious objective. This objective jointly considered both environment and energy, even extending the international agreements on the former. In addition, it achieved the committed GHG emissions reduction in a cost-effective way. This cost-effectiveness was heavily supported by mechanisms like the EU ETS with which the EU would fulfill the Kyoto Protocol compromise with an estimated cost below 0.1 % of its Gross Domestic Product. Furthermore, it would also achieve its own European aims for 2020. Finally, through the EU ETS, the EU is contributing to developing clean technologies that can support both its own internal progress and that of developing countries by channeling massive investments to promote energy efficiency.

Even in the realistic context of expecting local difficulties in achieving the objectives and massive international disagreement, the scene has been set. Energy and the associated emission reductions are now a legitimate political concern across the whole population. They are key issues in day-to-day political considerations and building processes that will enable energy reduction has become a topic of crucial importance.

3.2 Buildings

All the previous legislative pieces definitely construct a common framework to achieve the required reduction on energy consumption and its corresponding GHG emissions. On the other hand, specific regulation is needed to adapt the general guidelines to particular problems. Among the whole set of issues concerning energy efficiency, buildings are key as they represent a major energy consumer.

The European Energy Performance of Buildings Directive (EPBD) (European Council, 2002) define the basis of the promotion of energy efficiency in buildings. The EU recast this directive in 2010 (2010/31/EU, 2010) to include a set of amendments on the original legislation.

They take the seminal ideas set in the SAVE programme, which clearly indicated the importance of buildings as a key target for emission reduction through energy efficiency. Following the guidelines included in Resolution 86/C 241/01 the EPBD created a common framework that was particularized for buildings. This common framework specifies:

- A methodology to calculate the energy efficiency of a building.
- A certification scheme to provide meaningful and publicly available information regarding the energy efficiency of buildings.
- A set of minimum requirements to energy performance of existing and new buildings.

In addition, the EPBD fulfills the suggested requirement regarding flexibility through the definition of 30 energy efficiency standards. Each Member State is allowed to adopt a particular subset of these standards in order to construct its own regulatory framework. Despite this flexibility, the EU must guarantee the completion of the agreed commitments, thus it can make these standards mandatory if the national policies of the Member States are detected as non-compliant to those previously defined.

Finally, considering the individual European objective for 2020, the EPBD establishes an aggressive approach to promote energy efficiency in buildings. As an example, by the end of 2020, all new buildings should be *nearly-zero* buildings, which means that they have to have a very high performance and furthermore, their required energy must be provided by renewable energy sources produced on-site or in the vicinity of the building.

Although definitely promising, the EPBD does not seem enough to face the tremendous challenge of achieving a dramatic reduction of energy consumption and its subsequent GHG emissions. Several lines of action could be put in place to support the EPBD. Among others, due to the fact that building are not currently included into the EU ETS, we believe that doing so we would generate outstanding benefits in reducing their consumption. In Sections 4 and 5 we will analyze the current situation and future proposals, following the three principal guidelines stated in the UNEP to achieve sustainable buildings.

4. Current situation

4.1 Changing standards

The first guideline indicated in the UNEP states the need for determining energy efficiency objectives and energy performance standards. This requirement is essential if we intend to reach a global energy efficiency framework. Some standardization initiatives have arisen in the recent years, which create a common framework for defining energy efficiency in buildings.

The American Society of Heating, Refrigerating, and Air-Conditioning Engineers (ASHRAE) has been working for decades on constructing standards, including ones to promote energy efficiency. It published the standard 189.1, that provided a set of minimum requirements for the design, construction, and operation and maintenance of high-performance *green buildings*. The last update of this standard was issued in 2009 (American Society of Heating, Refrigerating, and Air-Conditioning Engineers, 2009a). The standard includes mandatory prescriptions with respect to the building's envelope and HVAC and Service Water Heating (SWH) systems, in order to achieve energy efficiency, together with site sustainability, water use efficiency, and indoor environment quality. Regarding energy efficiency, it is important to notice that the standard 189.1 states as mandatory that the energy consumption measurements should be captured and stored during at least 36 months. Based on these energy measurements, reports on consumption should be hourly, daily, monthly, and annually generated.

Furthermore, the ASHRAE detailed the energy prescriptions for buildings in a second standard that was first issued in 1975, having a recent update in 2010 (American Society of Heating, Refrigerating, and Air-Conditioning Engineers, 2009b). Once again it provides the minimum requirements for the design, construction, operation and maintenance, and the use of on-site renewable energy sources to achieve energy efficiency. It includes detailed rules with respect to the envelope, HVAC and SWH systems, power, lighting, and electric motors. In addition, it described the criteria for monitoring the compliance of actual implementations of the standard with the fixed requirements.

Returning to Europe, buildings are included in the European labeling scheme that was set down in law in Directive 2010/31/EU (2010/31/EU, 2010). This directive makes it mandatory for all energy consuming products to be certified in respect to their energy performance. Buildings, as one of these energy consuming products, must commit to this requirement. Consequently, at present all new projected buildings must be certified. A set of software tools, the majority of which are based on DOE-2 (DOE-2, 2010), take as input variables the building's construction materials, geographic location, its operation, its equipment for lighting, HVAC, etc., and the weather conditions at their site. Based on these variables, they predict the theoretical energy consumption of the building. This predicted consumption is then compared to a certain ad hoc reference to produce a ratio that expresses the energy efficiency of the building under study. Classifying this ratio into a set of intervals defined by some predefined thresholds finally produces a label that illustrates the performance of the building.

Including buildings into a labeling scheme provides evident benefits to the promotion of energy efficiency because:

- It shows meaningful information to end users regarding the energy efficiency of the building.
- It adds energy efficiency as a new variable to be considered in the decision on buying a house. Thus buildings that were constructed taking into account efficiency principles are best candidates to be more easily sold and at higher prices.
- It promotes innovation and technology in order achieve sustainable buildings.
- It supports the diffusion of the energy efficiency guidelines to generate public consciousness about the energy and environmental problem.

Nevertheless, on the negative side, we can remark that, to the best of our knowledge, there is no global standard for energy efficiency labeling of buildings. For instance, although Directive 2010/31/EU compels every consuming product to be labeled according to its energy efficiency, there is no fixed procedure on how this labeling process must be implemented. In fact, there is not even a unique number of bands, going from seven in countries like France to fifteen in Ireland. In addition, the required classification of buildings into different types is not standardized, nor does the definition incorporate a set of climatic areas, both key to implement a fair comparison between the building under study and its reference. In consequence, a single and global procedure to calculate an energy efficiency index for buildings and its subsequent label are required.

4.2 Methodological verification framework

Suitable methodological frameworks for the verification of reductions in energy consumption and GHG emissions are key to the construction of national, regional, and worldwide inventories. At present, there are examples of these inventories collected by international agencies, such as the International Energy Agency (IEA) (IEA, 2010a), (IEA, 2010b), or the United Nations Statistics Division, based on the work of the Carbon Dioxide Information Analysis Center (CDIAC) of the United States Department of Energy (United Nations Statistical Division, 2009). Also available are national inventories in many countries around the world. For example, the UK (AEA, 2009) or Spain (Ministerio de Medio Ambiente, y Medio Rural y Marino, 2010) among others. Focusing on buildings, the most detailed reports are issued by the U.S. Energy Information Administration (EIA) (EIA, 2011). Both the Commercial Buildings Energy Consumption Survey (CBECS) (CBECS, 2011) and the Residential Energy Consumption Survey (RECS) (RECS, 2011) provide data regarding building energy consumption across the US.

These inventories are key in fulfilling the harmonization and coordination requirements. Furthermore, they are also needed to construct a solid reference for the calculation of energy performance. Note, as we will see in Section 5, our proposed certification scheme and its corresponding labeling system are based on a ratio between the consumption of the building under study and that of a reference building. The more information we have, the more accurate and fair the resulting energy efficiency measurement will be.

All the previous initiatives form an interesting basis from which we can start to construct a complete worldwide inventory. Nevertheless, further efforts are needed in order to reach the level of completeness required for such an ambitious enterprize as the one regarding energy efficiency.

The most important pending issue is the construction of a standard and global model to be followed in every energy consumption or emissions inventory. At present it is difficult to construct a fair comparison between building's performances as data could be collected or aggregated following different rules that surely impact on the final energy efficiency measurement. The required model must be flexible in order to be suited for any type of international, national, regional, or local policy. In consequence, it must provide a common framework through which the different entities responsible for energy and environment will be able to construct their particular data structures.

There is also a set of detailed needs regarding different aspects of the common and global energy consumption and emissions inventory. Firstly, inventories must be constructed separating buildings depending on their type. With the exception of CBECS and RECS, inventories do not separate data from different types of buildings. Once again, this fact interferes with a fair comparison procedure because higher levels of consumption may be due to the inherent operation of the building (consider hospitals compared to residential buildings for example) and not to a deviation from the good practices in its use or construction. Secondly, annual inventories could be adequate to identify trends over long periods of time, but they are likely to be insufficient to detect energy efficiency issues depending on season, or use (consider for instance the working days against holidays in an office building). Hence temporal resolution (months, days...) must be added to the collected data in inventories. This temporal separation of consumptions could be key to perform a detailed energy analysis and provide an excellent means of improving energy efficiency. Following the same inspiration, a global and common energy consumption and emissions inventory should include detailed spatial information. Consider for example a residential multi-floor building; its energy efficiency is fully dependent on each householder's use or maintenance. This way, consumptions should refer to flats, areas, floors, or rooms in the building. Consequently, this would allow a better understanding of the building's performance and the underlying problems, making the entities responsible for its operation and maintenance capable of providing solutions and increasing their energy efficiency.

There is never going to be an ideal situation for the data acquisition and waiting for adequate levels of information is not an option. Hence despite the insufficient data to support the obvious and clear requirements, we need now a procedure that can evolve from the current state of affairs. Simply starting the process and including a mechanism to evolve the structure of the framework against an improved data models has clear and immediate benefits. In fact it will obviously provide the momentum for its own improvement, simply by initiating the process of collecting the data.

4.3 Monitoring system

The development of a monitoring system able to collect data regarding consumption and related emissions, CO_2 levels or any other environmental agent, has been a mature research area since the previous century. Industry has already developed a great variety of devices and they are readily commercially available.

The last decade has seen the introduction of Wireless Sensor Networks (WSN) (Elson & Estrin, 2004), which have greatly extended the applicability of these sensors. The standardization process generated both ZigBee (Alliance, 2002) and IEEE 802.15.4 (IEEE, 2006) standards in 2002 and 2006 respectively. Today, WSN are widely used to monitor the environment and interact with it.

Energy usage has become one of the many applications of this technology and has received increasing levels of attention in recent years. Even so, although most large buildings contain management and monitoring systems, to the best of our knowledge, no monitoring system has been specifically tailored to produce energy efficiency measurements. Hence, to fulfill this objective, a monitoring application framework, that can be adapted easily and cheaply to specific building problems, is needed.

In addition, we believe that a higher temporal resolution is needed in order to detect the source of the eventual deviation of the building from the fixed objectives. The UNEP required annual monitoring reports, but those would not include the precise details needed to perform an energy diagnosis that could point out seasonal or operational issues. Consider for example a highly efficient building in summer with a poorly designed heating system or an optimum performance during working days but not on weekends. Furthermore, this same philosophy should be extended to space allowing to detect failures in specific floors or rooms in the building.

The power of an energy efficiency monitoring system goes beyond its original conception; it obviously can be used to enhance the analysis of a building's performance for maintenance and operation purposes. Additionally different studies, for example (Darby, 2000), have demonstrated how consumption in buildings could be reduced simply by providing understandable feedback to end users. Consequently, the monitoring system appears to be even more significant than initially thought. Also, these monitoring systems could be connected to local, regional, or national governments, so that the collected information regarding energy efficiency could be uploaded to their respective inventories.

5. Novel proposals

This last section is devoted to the description of a set of proposals developed by our team in order to give answer to the previously listed requirements. Some of them are already available for the scientific community and others are being subject of our current work. This line of research in our group presents two key features. First, the work carefully follows the requirements set out by the associated international organization, such as the United Nations, in order to ensure the proposal is as compliant as possible to current objectives. Second, it is an interdisciplinary piece of work. It incorporates components from a diverse set of disciplines, each of which is designed to provide an answer to specific problems, but always including the requirement of cooperation with other elements of the solution. In this way we believe we can provide a framework for achieving energy efficiency in buildings.

5.1 New standards

The first requirement to accomplish worldwide objectives is to have a homogeneous scheme. This issue includes defining a common measure regarding energy efficiency. In (Rodríguez González et al., 2011) we proposed a universal energy efficiency index for buildings ($\mathrm{EEI_B}$) giving a measure of the energy efficiency to a building as the ratio between its actual consumed energy and that of a reference building. The $\mathrm{EEI_B}$ is defined as:

$$\mathrm{EEI_B} = \frac{C_{\mathrm{AB}}}{C_{\mathrm{RB}}}, \tag{1}$$

where C_{xB} is the energy consumption in the actual building (AB) under study and the reference building (RB) respectively.

In the initial stages of the framework we consider only the total energy consumption of the building; these data are readily available from the energy providers. It can also be easily evolved to more sophisticated forms, as more data become available. In addition, if the building under study has not yet been constructed, its energy consumption can also be estimated using any of the available software tools using DOE-2 (DOE-2, 2010) as their simulation engine. The method of obtaining the actual energy consumption of a building (in operation) to be certified based on the energy suppliers reports, avoids the use of complex simulations tools that require a huge amount of input data, which is seldom available. Furthermore, the $\mathrm{EEI_B}$ will be calculated from actual values of the consumed energy in the building rather than from estimates.

In order to perform the required ratio that defines $\mathrm{EEI_B}$, we need a reference consumption C_{RB}. We will classify buildings by their type, thus separating hospitals from residences or office buildings as the consumption profiles depend directly on their particular operation. The reference value of the consumptions of all buildings of a particular type is defined as a function of the total constructed area, which again can be built now, in terms of readily available data and can easily be evolved. Using the defined reference, we are accepting an inherent premise: "any square meter of any building of a specific type consumes the same energy". In consequence, C_{RB} is calculated as the energy consumption per square meter of all the buildings of the same type i, times the floorspace of the AB, S_{AB}

$$C_{\mathrm{RB}} = \frac{C_i}{S_i} \cdot S_{\mathrm{AB}}. \tag{2}$$

The proposed index is perfectly suitable for a certification scheme based on a labeling system. This labeling system also needs to be defined in a common and accepted format, such as the one we suggest in (Rodríguez González et al., 2011), which is summarized in Figure 1.

$$
\begin{aligned}
&A, && EEI_B < 0.40;\\
&B, && 0.40 \le EEI_B < 0.65;\\
&C, && 0.65 \le EEI_B < 1.00;\\
&D, && 1.00 \le EEI_B < 1.30;\\
&E, && 1.30 \le EEI_B < 1.60;\\
&F, && 1.60 \le EEI_B < 2.00;\\
&G, && 2.00 \le EEI_B.
\end{aligned}
$$

Fig. 1. Example of label and thresholds based on the EEI_B

In consequence, EEI_B provides a flexible means to monitor energy efficiency and update the resulting index whenever a new set of data is collected. Hence it can be used to verify objectives in energy policies. In addition, it promotes energy efficiency in buildings by upgrading the reference values as innovation is incorporated into buildings to increase their energy performance.

The limitation of the scheme is that the index provides an indicator of efficiency, but it does not identify the source of the problem. However it is relatively simple to extend the indexing process into a diagnostic tool.

5.2 Methodological verification framework

All of the previous sections implicitly assume that there is a common means of describing a building. A standardized and universal way of measuring energy efficiency in buildings will not be able to reach the fixed objectives if there is no common definition of those buildings to apply it to. The question is how could we implement fair comparisons of energy efficiency achievements amongst different regions or nations, thus verifying their policies, if we do not have an agreed base reference? Furthermore, how could we generate a worldwide database if the information is not provided in an accepted format?

If we could build a common description, it also could then be used to support information mining from a specific set of measurements that have been applied to it. It is relatively straightforward to define, if the description is limited to type and active surface area. However, if the description is to support a more detailed and extensible description, much more care has to be taken.

This problem of definition is compounded by a need to have a description that can represent an arbitrary set of resolutions. By this we mean that the building may be described simply by its type and its workable surface area, or a more detailed description at the level of floors and rooms. At some later date we may require a more complex description in terms of the materials that are used to construct the walls, or the locations of the heating pipes within the building. This problem of data-representation is twofold; we need a common, but flexible, descriptive language; and we need to actually build the data. The last point is obviously the most costly. Building owners will obviously require something intuitively easy to use. To do this we must define an approach that is extensible. The description model must be: intuitive, expandable in terms of details of the description, and extensible in terms of being able to incorporate new concepts and new elements within the model.

Fortunately, this type of data description problem is very generic and has been considered in detail by the information technology community. Such an extensible data description can be achieved by the use of an *ontological* model, which can be used to construct a common framework in which to define buildings through their subcomponents and interrelationships between them. An ontology is a semantic specification of a certain domain or context (Gruber, 1993). It includes:

- Classes: concepts or abstractions. For example, person, building, floor, room.
- Individuals: class instantiation. For example, John, NY Hotel.
- Attributes: features of a class and all its individuals. For example, age, name, height, weight.
- Relations: interactions and links between the classes. For example: part-of, subclass-of; Hospital or Residential are subclasses of Building.
- Functions: relations defined by the user between different classes in the ontology. For example, `isFather(Person, Person)`, `compoundOf(Room, Object)`.
- Axioms: theorems that are declared over relations that the elements of an ontology must fulfill. For example: every room must be composed by a Door, a set of Walls, and Corners.

An ontological model of the building provides a perfect framework to define its singular features, create the appropriate relations between its composing elements, and finally connect those to the actual collected data in a semantic form. The very nature of an ontological description also supports evolution of the definition and the standardization of information exchange.

The ontology of a building mainly consists of a structural layer and a functional layer. The former describes the geometric structure of the building (floors, rooms, etc.), the construction materials, its topological representation, and the metrics characterizing the previous elements (length, height, area, etc.). Additionally, it also used to describe the most significant devices within the building and the features they provide. The latter describes the knowledge that has been generated regarding a certain field. For the purpose of this work, this knowledge revolves around energy efficiency. The knowledge can be introduced in the system by an expert or acquired during the operation. These two layers must interact and feedback each other.

The ontological description of a building is currently under the scrutiny of today's research community. One of the earliest examples of a building description ontology can be found in (Bonino & Corno, 2008). Here the ontological model called *DogOnt* is used to construct models of domotic system for home automation from the description of the building itself and the different devices deployed throughout it. Specifically, it includes 7 classes that define the building and its environment, the functionalities of the devices, their state and the network linking them, and commands and notifications for the operation of the domotic system. This work was extended in (Bonino et al., 2011) to add a new *Energy Profile ontology* to DogOnt indicating the energy consumption of the devices. This clearly demonstrates how an ontological specification can be extended and enhanced to incorporate new concepts and functionalities. In this extended model devices are split into different categories depending on their energy consumption profile. In addition, the energy consumption is also divided into different levels depending on the state of the device. Finally, consumption is modeled through the state of the device and its either nominal or actual value, depending on whether real data is available or not.

This particular ontology does provide an excellent starting point for a building description and a starting framework for monitoring processes. However, it does not include the key factors regarding the energetic performance of the building. To do so, we are currently developing an extension of this ontology through a set of modules that incorporate the data components required to describe energy efficiency of the building, as well as the interrelations between the different indices. This process is producing a useful tool allowing a systematic description of a building in terms of its spatial structure, its temporal evolution, and its energy efficiency. Once constructed the ontology is designed to receive data from all the monitoring devices placed around the building. These devices collect information regarding both the environment and the actual consumptions within the building, which can in part be processed in real time, or in its entirety be post processed.

Through the functional layer of the ontology, we are able to obtain information regarding:

- Consumption patterns depending on the user's activity.
- Detection of the user's activity from the data provided by the devices monitoring the environment and the consumptions, coupled with the structure of the building.
- Deviations from usual energy efficiency values in space and time.
- Current status of a specific device.

5.3 Monitoring system

The third and last guideline stated in the UNEP in order to succeed in obtaining energy efficient buildings is the development of monitoring systems capable of providing annual energy performance reports.

The immediate issue regarding a monitoring system is how to collect the required significant data. As we presented in Subsection 4.3, a complete family of devices are commercially available at present. In addition, WSN have paid special attention to this matter in the last decade, thus providing solid solutions for developing monitoring systems able to measure the set of variables needed to study the energy performance of a building (mainly temperature, light, humidity, and consumption). There is still a minor issue regarding the optimum design and deployment of such sensors throughout the building in order to produce the data with the required temporal and spatial resolution.

Nevertheless, under this perspective, monitoring is just a means to collect data, but its true significance resides in the information that can be extracted from them. Consequently, the main reason why this monitoring system is definitely needed is that it is required to construct a complete worldwide inventory of energy consumption that will allow the construction of a fair comparisons between different initiatives and achievements, thus verifying the national and regional policies put in place.

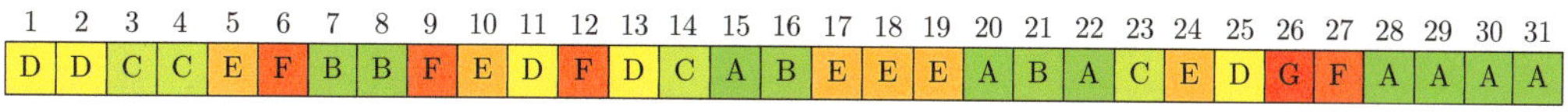

Fig. 2. Daily energy efficiency landscape.

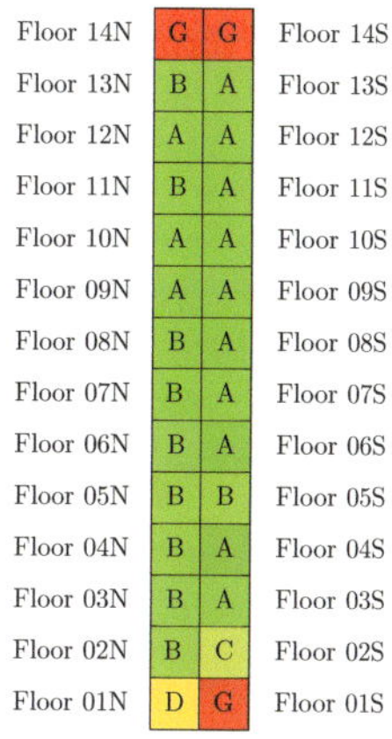

Fig. 3. Spatial floor energy efficiency landscape.

Construction of a basic energy indexing scheme is a must. However, we could significantly extend the benefits of this monitoring system by designing it with a higher temporal and spatial resolution. Doing so, we would be able to provide a powerful tool to implement an accurate diagnosis of the possible sources of failures in the energy performance of the building.

Fortunately this "extension" of the indexing mechanism is perfectly in line with the design and operation of an ontological descriptive language, described in the previous Subsection 5.2. This building's ontology quote naturally and transparently allows detailed inspection of its energy efficiency to whichever temporal and spatial resolution is required. At this point it is worth remembering that the basic description is simply the building location, its type and its workable surface area. However, an ontological description quite naturally allows us to expand this description to whatever level of detail is required. Additionally, this expansion can be selective, expanding specific parts of the building in detail, whilst having poor resolution of details in the remaining areas.

When this type of decomposition is applied to an index, EEI_B, we call it an *energy efficiency landscape* (EEL_B) of EEI_B. We define an EEL_B as the temporal (days, months...) and/or spatial (rooms, floors,...) disaggregation of the global consumption of every energy source (electricity, gas, etc..), and its corresponding emissions for the building. Some examples of EEL_B are shown in Figures 2 and 3. As it can be observed, they perfectly serve the purpose of detecting temporal and spatial deviations by simple visual inspection.

The level of decomposition is dependent upon need. An energy landscape provides immediately the location and time of anomalies. It is these anomalies that need to be studied, allowing building managers to focus their attention on the relevant areas. Importantly, as the basis for the decomposition is the relative contribution to the overall energy index, comparisons between energy sources are possible and meaningful. The landscaping process takes us from global policy directly into the local decision making process of building management. It also provides information that could inform future policy, hence we have a system that takes us full circle. In the context of the ontological description, only data that captures details of the area of the anomaly need be added to the model. Hence, workload is limited to only necessary changes, further enhancing the efficiency of the system.

6. Conclusions

Implementing energy efficiency in buildings is a complex problem. It involves different and heterogeneous disciplines such as Policy, Technology, and Science among many others. In addition, developments in any of those must be coordinated with the others and moreover put in place in a real scenario and in a cost-efficient way. Consequently, a well defined and planned strategy is key to achieving this challenging objective.

We accept this premise in the field of research. In order to follow a sensible planning for our research objectives, we chose the three main points of the UNEP to act as guidelines of our work. Hence, we have started analyzing the state of the art that has been reached in defining energy efficiency objectives and standards, developing a methodology to verify accomplishments, and designing monitoring systems to measure energy performance.

Then we have extended the results obtained in each of these three points through a set of novel proposals developed by our team. This on-going research must be observed as a cooperative framework developed for the overall objective rather than a collection of separate pieces of work. Thus, the definition of a universal energy efficiency index and the modeling of buildings through an ontology will be then used to deploy monitoring systems capable of both constructing a worldwide energy efficiency inventory, and serving as a diagnosis tool for operation and maintenance companies. On the whole, we believe that the proposed framework will significantly contribute to succeeding in implementing energy efficiency in buildings.

7. References

2002/358/CE (2002). Council Decision of 25 April 2002 concerning the approval, on behalf of the European Community, of the Kyoto Protocol to the United Nations Framework Convention on Climate Change and the joint fulfilment of commitments thereunder, *Official Journal of the European Communities* , 15 May 2002, pp. 1-20.

2006/944/EC (2006). Commission Decision of 14 December 2006 determining the respective emission levels allocated to the Community and each of its Member States under the Kyoto Protocol pursuant to Council Decision 2002/358/EC, *Official Journal of the European Communities* , 16 Dec. 2006, pp. 87-89.

2009/28/EC (2009). Directive 2009/28/EC of the European Parliament and of the Council of 23 April 2009 on the promotion of the use of energy from renewable sources and amending and subsequently repealing Directives 2001/77/EC and 2003/30/EC, *Official Journal of the European Communities* , 5 Jun. 2009, pp. 16-62.

2009/29/EC (2009). Directive 2009/29/EC of the European Parliament and of the Council of 23 April 2009 amending Directive 2003/87/EC so as to improve and extend the greenhouse gas emission allowance trading scheme of the Community, *Official Journal of the European Communities* , 5 Jun. 2009, pp. 63-87.

2009/31/EC (2009). Directive 2009/31/EC of the European Parliament and of the Council of 23 April 2009 on the geological storage of carbon dioxide and amending Council Directive 85/337/EEC, European Parliament and Council Directives 2000/60/EC, 2001/80/EC, 2004/35/EC, 2006/12/EC, 2008/1/EC and Regulation (EC) No 1013/2006, *Official Journal of the European Communities* , 5 Jun. 2009, pp. 114-135.

2010/31/EU (2010). Directive 2010/31/EU of the European Parliament and of the Council of 19 May 2010 on the energy performance of buildings, *Official Journal of the European Communities* , 18 Jun. 2010, pp. 13-35.

406/2009/EC (2009). Decision No 406/2009/EC of the European Parliament and of the Council of 23 April 2009 on the effort of Member States to reduce their greenhouse gas emissions to meet the Community's greenhouse gas emission reduction commitments up to 2020, *Official Journal of the European Communities* , 5 Jun. 2009, pp. 136-148.

89/364/EEC (1989). 89/364/EEC: Council Decision of 5 June 1989 on a Community action programme for improving the efficiency of electricity use, *Official Journal of the European Communities* , 9 Sept. 1989, pp. 32-34.

91/565/EEC (1991). 91/565/EEC: Council Decision of 29 October 1991 concerning the promotion of energy efficiency in the Community (SAVE programme), *Official Journal of the European Communities* , 8 Nov. 1991, pp. 34-36.

93/76/EEC (1993). Council Directive 93/76/EEC of 13 September 1993 to limit carbon dioxide emissions by improving energy efficiency (SAVE), *Official Journal of the European Communities* , 22 Sep. 1993, pp. 28-30.

AEA (2009). End User GHG Inventories for England, Scotland, Wales and Northern Ireland: 1990, 2003 to 2007.
URL: *http://www.airquality.co.uk/reports/cat07/0911120930_DA_End_Users_Report_2007_Issue_1.pdf*

Alliance, Z. (2002). Zigbee.
URL: *http://www.zigbee.org*

American Society of Heating, Refrigerating, and Air-Conditioning Engineers (2009a). Standard 189.1-2009: Standard for the Design of High-Performance Green Buildings Except Low-Rise Residential Buildings, ANSI/ASHRAE/USGBC/IES.

American Society of Heating, Refrigerating, and Air-Conditioning Engineers (2009b). Standard 90.1-2010: Energy Standard for Buildings Except Low-Rise Residential Buildings, ANSI/ASHRAE/IESNA.

Bonino, D. & Corno, F. (2008). Dogont-ontolgoy modeling for intelligent domotic enviornments, *International Semantic Web Conference*, number 53, pp. 790–803.

Bonino, D., Corno, F. & Razzak, F. (2011). Enabling machine understandable exchange of energy consumption information in intelligent domotic environments, *Energy and Buildings* 43: 1392–1402.

CBECS (2011). The cbecs website.
URL: *http://www.eia.doe.gov/emeu/cbecs/*

Darby, S. (2000). Making it obvious: Designing feedback into energy consumption, *Proceedings of the 2nd International Conference on Energy Efficiency in Household Appliances and Lighting*.

DOE-2 (2010). DOE-2, Building Energy Use and Cost Analysis Software, developed by Lawrence Berkeley National Laboratory (LBNL), funding from the United States

Department of Energy (USDOE).
URL: *http://www.doe2.com/*

EIA (2011). The eia website.
URL: *http://www.eia.gov/*

Elson, J. & Estrin, D. (2004). Sensor networks: a bridge to the physical world, pp. 3–20.

European Commission (2006). Communication COM(2006) 545 final, Action Plan for Energy Efficiency: Realising the Potential.

European Commission (2008). Communication COM(2008) 30 final, 20 20 by 2020: Europe's climate change opportunity.

European Council (2002). Directive 2002/91/EC of the European Parliament and of the Council of 16 December 2002 on the energy performance of buildings, *Official Journal of the European Communities* , 4 Jan. 2003, pp. 65-71.

European Union (1986). Council Resolution of 16 September 1986 concerning new Community energy policy objectives for 1995 and the convergence of the policies of Member States, *Official Journal of the European Communities* C 241: 1–3.

Gruber, T. R. (1993). A translation approach to portable ontologies, *Knowledge Acquisition* 5(2): 199–220.

IEA (2010a). CO2 Emissions from Fuel Combustion, 2010 edition, IEA Publications.

IEA (2010b). Energy Balances of non-OECD Countries, 2010 Edition, IEA Publications.

IEA (2010c). Energy Balances of OECD Countries, 2010 Edition, IEA Publications.

IEEE (2006). Standard 802.15.4.
URL: *http://standards.ieee.org/getieee802/download/802.15.4-2003.pdf*

Ministerio de Medio Ambiente, y Medio Rural y Marino (2010). Inventario de Gases de Efecto Invernadero de España.
URL: *http://www.mma.es/secciones/calidad_contaminacion/atmosfera/emisiones/pdf/Sumario_de_Inventario_Nacional_Emisiones_GEI_serie_1990-2008.pdf*

RECS (2011). The recs website.
URL: *http://www.eia.doe.gov/emeu/recs/*

Rodríguez González, A. B., Vinagre Díaz, J. J., Caamaño, A. J. & Wilby, M. R. (2011). Towards a universal energy efficiency index for buildings, *Energy and Buildings* 43: 980–987.

Uihlein, A. & Edera, P. (2010). Policy options towards an energy efficient residential building stock in the EU-27, *Energy and Buildings* 42: 791–798.

UNEP (2007). Buildings and Climate Change: Status, Challenges and Opportunities, UNEP Publications.

United Nations (1998). Kyoto Protocol to the United Nations Framework Convention on Climate Change, United Nations Publications.

United Nations Statistical Division (2009). Greenhouse Gas Emissions.
URL: *http://unstats.un.org/unsd/environment/air_co2_emissions.htm*

10

Improving Air-Conditioners' Energy Efficiency Using Green Roof Plants

Fulin Wang[1] and Harunori Yoshida[2]
[1]Tsinghua University, Beijing,
[2]Okayama University of Science, Okayama,
[1]China
[2]Japan

1. Introduction

Environment and energy issues are considered to be most urgent things nowadays even in future. A lot of researches have been conducted to study how to prevent global warming and reduce energy consumption. In the field of building and urban environment, green roof attracts a lot of researchers' attention because it is considered to be a good solution for improving urban thermal environment by mitigating heat island and to reduce building cooling energy consumption by reducing cooling load. Alexandria et al. (2008) analyzed how much the urban canyon temperature can be decreased due to green walls and green roofs. Takebayashi et al. (2007) compared the building surface heat transfer of green roofs with common roofs and high reflection roofs. Kumar et al. (2005) developed a mathematical model to evaluate the cooling potential and solar shading effect of green roofs. Wong et al. (2003) analyzed the thermal benefits of green roofs in tropical area. Di et al. (1999) measured an actual green wall to analyze how much cooling effect is achieved. Elena (1998) analyzed the cooling potential of green roofs. Besides studying the green roofs' benefits of heat island mitigation and thermal isolation, the cost vs. benefit is also analyzed (Clerk et al., 2008) and green roof plants selection is analyzed as well (Spala et al., 2008).

However, researchers seldom focus on how to improve air-conditioners' energy efficiency utilizing the cooling effect and solar shading of green roof plants. Therefore this chapter describes a new system combining the green roof plants with air-conditioners outdoor units for the purpose of utilizing the cooling effect and solar shading of green roof plants (Wang et al, 2008, 2009). Figure 1 shows the structure of the combination system. The outdoor units of air conditioner are set under the plants and let air flow through plants and cooled down by plants. Also the plants shade solar radiation to prevent the outdoor unit from absorbing solar energy and raising surface temperature.

2. System design

Different type of green roof plants, including tree, grass, moss, vine, etc. can be used to construct the system. Different type of plant needs different system structure.

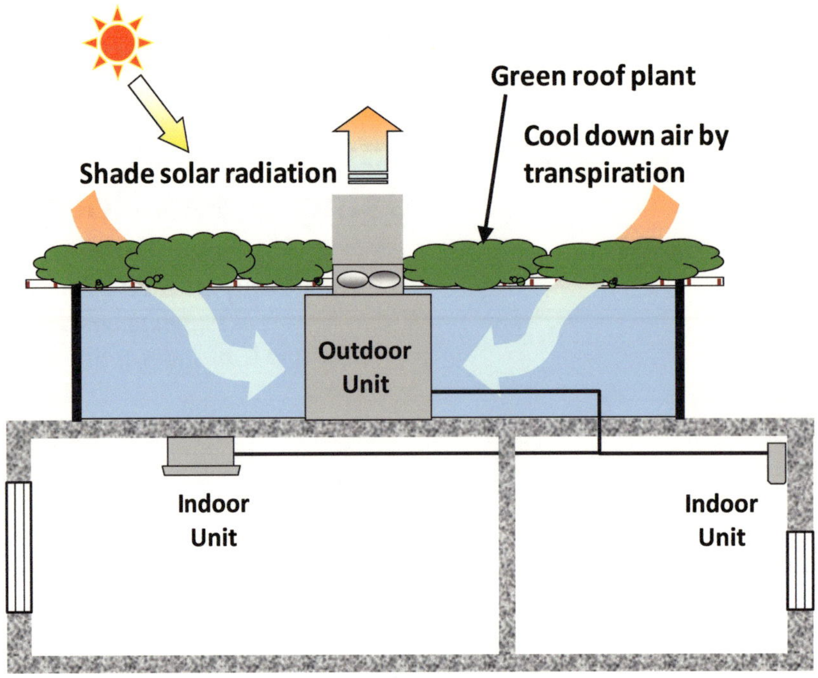

Fig. 1. The system combining the green roof plants with air-conditioner outdoor units.

2.1 Trees

Outdoor units of air-conditioner can be put under threes (Figure 2). Trees can shade solar radiation to prevent outdoor unit from absorbing solar radiation. Trees can cool down air as well by transpiration and the cooled air flows down and is sucked into air-conditioner outdoor units.

Fig. 2. Combination of air-conditioner outdoor units with trees.

2.2 Grass and moss

Grass or moss type green roof plants can be lifted up over the air-conditioner outdoor units, as shown in Figure 3. Gaps between plant blocks are needed to let cooled air flow through. The hung up plants can shade the solar radiation as well.

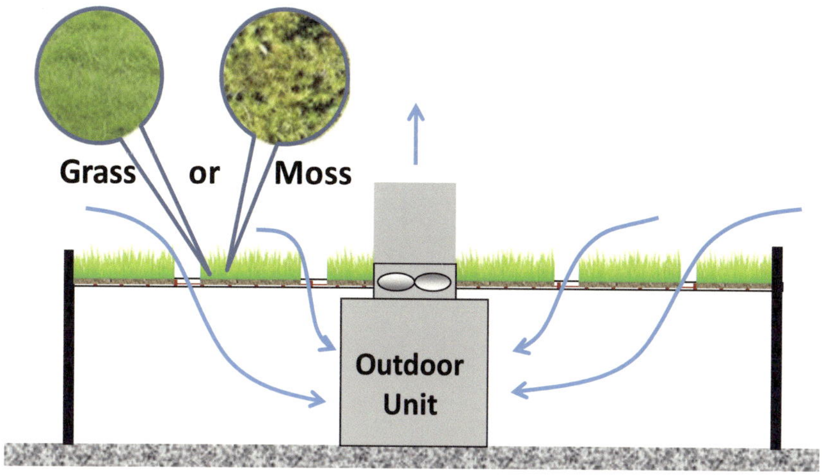

Fig. 3. Combination of air-conditioner outdoor units with hung up grass or moss.

2.3 Vine

Vine type green roof plants are most suitable for combining with air-conditioner outdoor units because of the higher transpiration rate. Further, vine type plants suits for being cultivated using hydroponic technology. Figure 4 shows a typical hydroponic cultivation system, which consists of fertilizer tanks, fertilizing controller, fertilizer adjustment tank, circulation pump, and cultivation unit. Hydroponic technology can use a small area cultivation unit to grow a larger area of plants. So in the area covered by plants, only a small part needs to support the relatively heavy cultivation system, while most part only supports the plant vine, which is so light that its weight can be neglected from the viewpoint of roof supporting ability. Figure 5 shows an example of one hydroponic-cultivated tomato tree. From the figure, it can be seen that the cultivation unit only occupies a small part of the area covered by the plant.

Compared with soil-cultivated green roof, hydroponic-cultivated green roofs are light enough to set on existing buildings, which did not consider the weight of cultivation soil during design phase so it cannot burden the weight. Further, the hydroponic-cultivated green roof plants are light enough to be lifted up over the outdoor unit of air-conditioners, which makes the combination system feasible.

3. Experimental study on the cooling and shading effect of hydroponic-cultivated plant

For the purpose of check the energy saving potential of the combination system, experimental devices are set up on the roof of a five-story office building in Osaka Japan.

The air temperature cooled by the green roof plants, plant transpiration rate, solar radiation shading rate, and inlet and outlet air temperatures at the outdoor unit of an air-conditioner, etc. were measured.

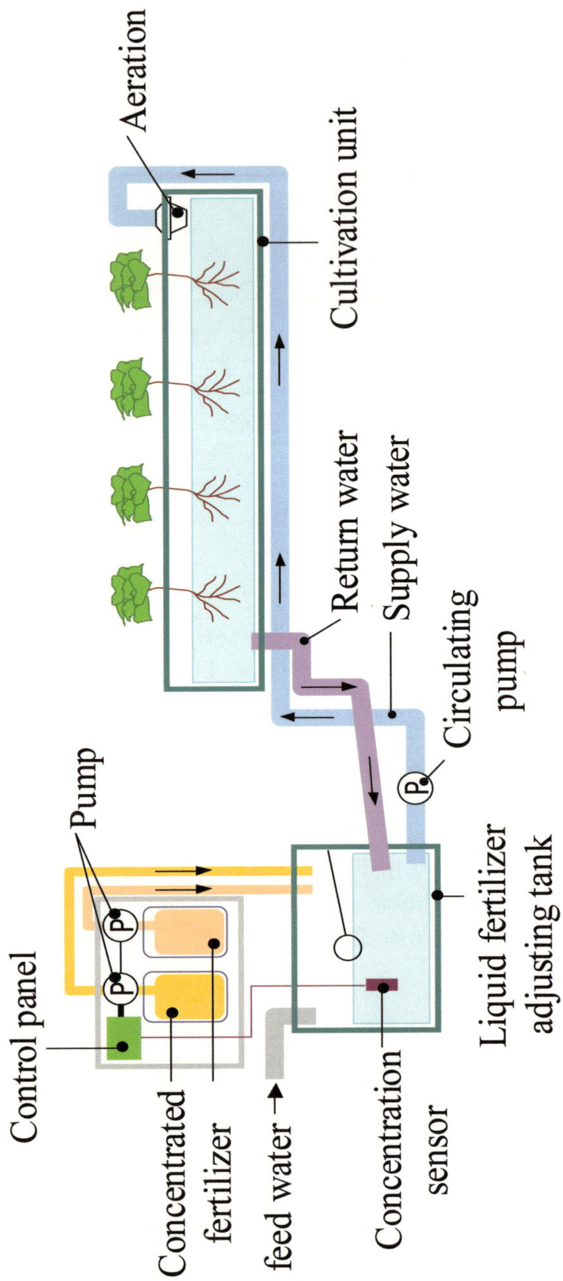

Fig. 4. An example hydroponic cultivation system.

Fig. 5. A hydroponic-cultivated tomato tree (Small floor area occupied by the cultivation-unit compared with the large floor area covered by the plant).

3.1 Experiment setup

Experimental device and measurement points are shown in Figure 6. Measured items and instruments are shown in Table 1. Data are recorded with five minutes interval using a data logger. The hydroponic cultivation device is set on a stage with 1.4 meters height on the roof. The vertical walls of the stage are closed using isolation material to let air can only flow from upside and go through the plants. The upper side of the stage is covered by metal mesh, where the plants spread vines and air can flow through as well. Three ducts equipped with a fan respectively are connected to the south wall of the stage to suck air under the stage, acting as the same function of the outdoor unit fans of air-conditioner.

Sweet potato is selected as the plant because it has high transpiration rate, grows fast, and is strong against wind. The sweet potato was planted on 22nd May 2007. One month later, it began to grow fast with a speed of 0.8 m^2 of horizontal projection area per day until the end of September. Figure 7 shows the plant growing situation at the end of September. In October, its growth slowed down and withered at the beginning of November. So generally speaking, the air-cooling and solar shading effect can be utilized for two months of August and September, which are period with large cooling load. This indicates that the combination system is meaningful for actual application.

For the purpose of check the cooling potential of the plants, the following experiments are conducted.

- Set the flow rate of air flowing through the plants at 2000 m^3/h, 4000 m^3/h, and 6000 m^3/h, to check the relations between temperature decrease and air flow rate.
- Sprinkle water over and under the plants to check how much the air-cooling effect can be improved, focusing on cooling air temperature down as low as to wet-bulb

temperature. Considering it might damage the plants growth, the water sprinkling experiments are conducted twice a day on 11:00 - 12:00 and 14:00 - 15:00.

Mark	Item	Point	Instrument
WD	Wind direction	A	vane-type
WS	Wind speed	A	3-cup anemo meter
RU	Solar radiation	B,D	pyranometer
RD	Solar radiation (under green)	2,3	pyranometer
PAR	Photosynthetically active radiation	D	photon sensor
NRF	Net radiation (roof)	C	net radiometer
NRG	Net radiation (green)	4	net radiometer
IR	Infrared radiation	B	infrared radiometer
WF	Water flow	F	flowmeter
TR	Outside air temperature	A	thermohygrometer
HR	Outside air relative humidity	A	thermohygrometer
TU	Temperature (over green)	1~8	thermocouple thermo recorder
HU	Ralative humidity (over green)	1~8	thermo recorder
TD	Temperature (under green)	1~8	thermocouple thermo recorder
HD	Ralative humidity (under green)	1~8	thermo recorder
TDU	Temperature (in duct)	▲	thermocouple
TL	Leaf temperature	1~4	thermocouple
TW	Wall temperature	×	thermocouple
TF	Floor temperature	⨯	thermocouple
TH	Temperature of suction opening	○	thermocouple
THO	Temperature of supply opening	●	thermocouple

Table 1. Measured items and instruments

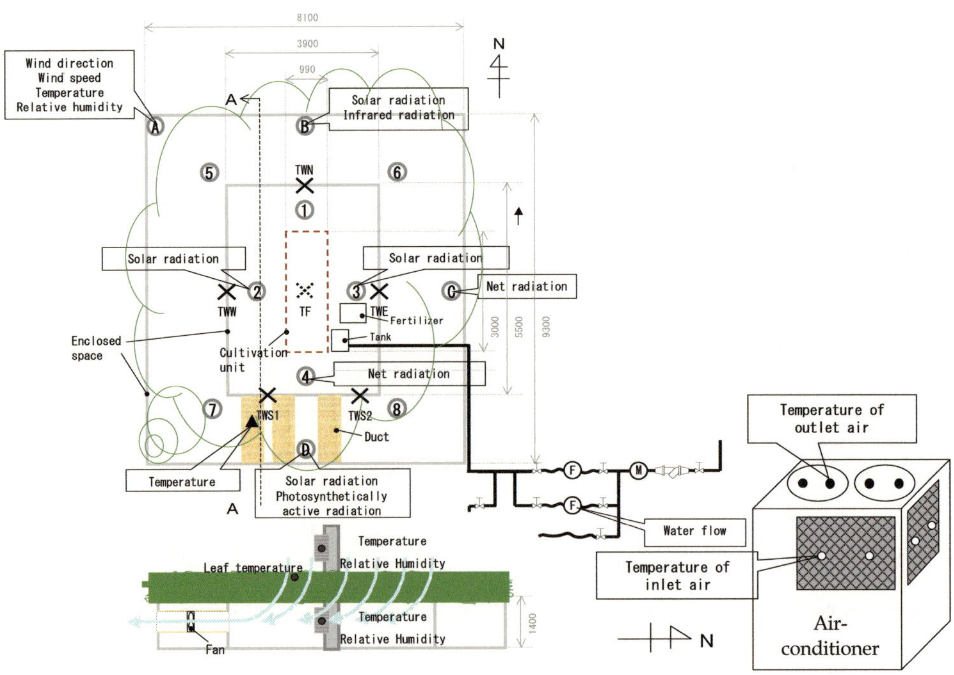

Fig. 6. Experimental device and measurement items.

Fig. 7. Experimental system and plants growing situations at the end of September.

3.2 Experimental results

The experimental period is from August 15th to September 23rd. Although data are recorded 24 hours a day, the cooling affect can only be observed during the period when photosynthetic radiation is active enough to trigger plant transpiration. The observed transpiration and cooling effect are mainly in the period of 10:00 - 16:00, therefore the data during this time are used for analysis.

1) Air temperature decrease

For the experimental period, the daily temperature decreases of maximum, minimum, average of 10:00 - 12:00, and average of 13:00 - 16:00 are as shown in the upper part of Figure 8. The daily average temperature decrease for clear and no-water-sprinkling days is 1.3 ^{o}C. While the average temperature decrease for rainy or water-sprinkle days is 3.0 ^{o}C, which is 2.3 times of that in clear and no-water-sprinkling day. However, the temperature decrease does not differ much for the air flow rate of 2000, 4000, and 6000 m^3/h.

2) Plants transpiration

The daily sum and 10:00 - 16:00 sum of transpired water are shown in the lower part of the Figure 8. The maximum and average daily summed transpirations are 8.3 and 6.3 kg/m^2 horizontal area respectively for clear and no-water-sprinkle days. For the days when water was sprinkled upon the plants, the transpiration is relatively small because leaves were wet and the transpiration temporally stopped.

The comparison of different plant transpiration rate is shown in Figure 9. The maximum daily transpiration rate (kilogram water per square meter of horizontal projection area) of hydroponoic-cultivate sweet potato is 13.8 times of sedum without water-sprinkle and 1.8 times of sedum with water-sprinkle. The maximum transpiration of hydroponoic-cultivate sweet potato is similar to a single tree, which indicates the hydroponoic-cultivate sweet potato can transpire as much water as a tree, while a tree has 20 times more leaf volume per unit horizontal projection area than sweet potato.

3) Solar radiation shading

The solar radiations over and under the plants on a typical day are shown in Figure 10. Even the solar radiation is as high as 1000 w/m2, the measured solar radiation under the plants is no more than 10 W/m2. That is to say more than 99% solar radiation is shaded by the plants. Therefore hydroponic-cultivated green roof plant can shade solar radiation enough to ignore the influence of solar radiation to the outdoor unit of an air-conditioner.

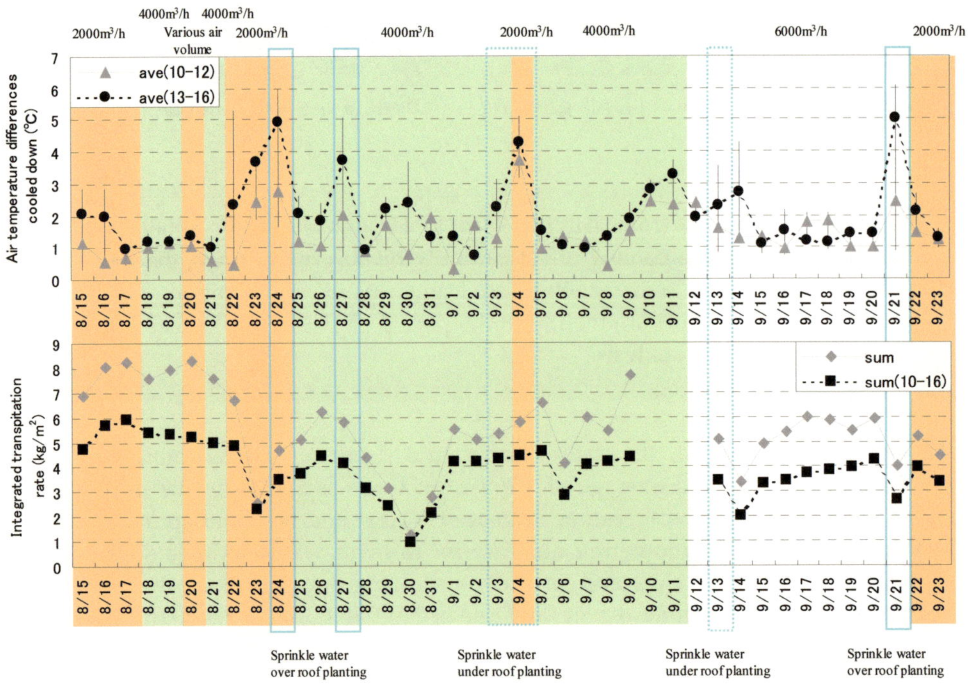

Fig. 8. Experimental results of air temperature decrease (upper graph) and transpiration rate (lower graph).

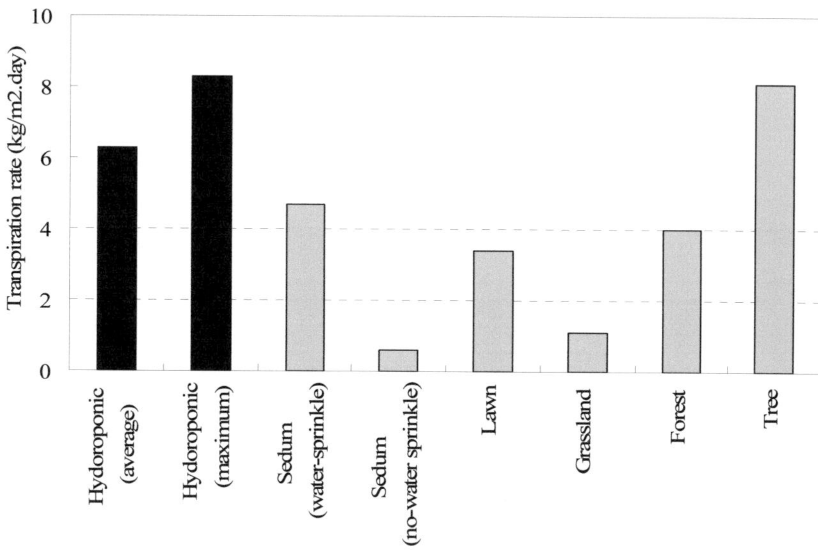

Fig. 9. Transpiration comparison.

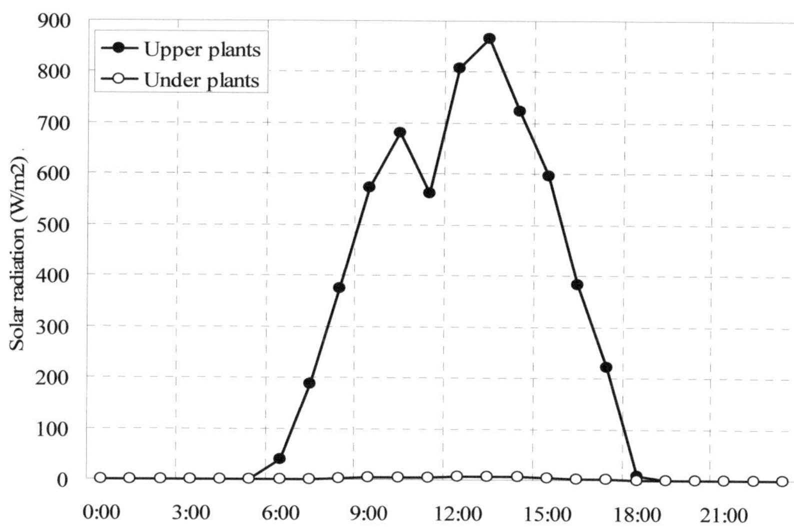

Fig. 10. Solar radiation up and under plants.

4. Estimation of energy saving effect based on experimental data

Based on experimental data, energy saving potential of the combination systems can be estimated using the measured air temperatures before and after being cooled down by the plants and the solar radiation after being shaded.

The calculation flow for estimating the energy saving is shown in Figure 11. Firstly calculate the air cooling effect ΔT caused by plants transpiration and solar shading. Then calculate air-conditioners' energy consumption using an air-conditioner energy consumption model.

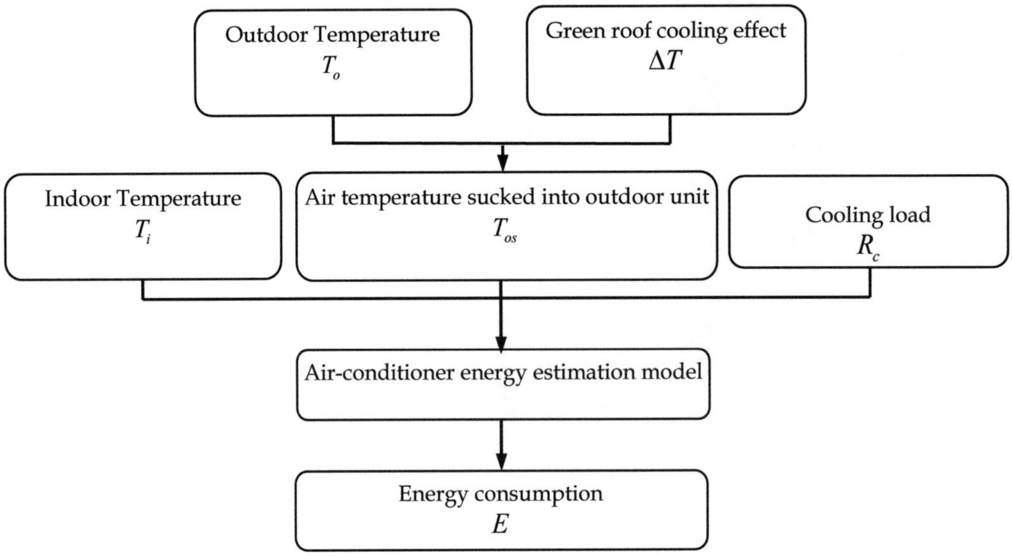

Fig. 11. Calculation flow for energy consumption of air-conditioners.

4.1 Air cooling effect

The air temperature decreasing is caused by two reasons. The first is the plants transpiration. The measured temperature differences between the air over and under the plants are used for the calculation. The second is the equivalent air temperature decrease caused by solar shading. If there is no plant shading, the outdoor unit will absorb solar radiation and its surface temperature will rise. This part of heat will raise the temperature of the air sucked into the outdoor unit. The air-conditioners' energy efficiency will be decreased by the temperature increasing of sucked in air. The air temperature increase is calculated using the following equations.

$$\Delta\theta_s = \frac{q_L + \alpha_S q_S}{mC_{Pa}} \tag{1}$$

4.2 Air-conditioner energy consumption model

A regression model fitted using air-conditioner manufacturer's specification data is used to calculate the air-conditioner's energy consumption (Wang et al., 2005).

$$RE = (a_1\theta_{of}^{\ 2} + b_1\theta_{of} + c_1)(a_2\theta_i^2 + b_2\theta_i + c_2)(a_3CA^2 + b_3CA + c_3) + d \tag{2}$$

$$\theta_{of} = \theta_o - \Delta\theta_T + \Delta\theta_s \tag{3}$$

The energy saving is calculated for four types of typical air-conditioner made by four different manufactures, including Gas-engine Heat Pump (GHP) and Electricity-driven Heat Pump (EHP). The average nominal primary energy Coefficient of Performance (COP) is 1.4 and manufacture year is 2005. Different manufacturer's air-conditioners have different efficiency improvement ratio accompanying to outdoor temperature decreasing. So the energy saving is different. The daily average energy savings are shown in Figure 12. The maximum energy saving rate is 12% for the air-conditioners with high efficiency improvement ratio, and 3% for the air-conditioners whose efficiencies improve little accompanying to outdoor air temperature decreasing.

The summed energy savings for the experimental period are shown in Table 2. If water-sprinkle is not conducted, the energy saving ratio is 4% for the air-conditioners with high efficiency improvement ratio, and 1% for the air-conditioners with low efficiency improvement ratio. If water-sprinkle is conducted for two hours a day, the energy saving ratios are 9% and 2% for the air-conditioners with high and low efficiency improvement ratio respectively. Among the total energy saving, about 10% is from solar shading and 90% is from transpiration.

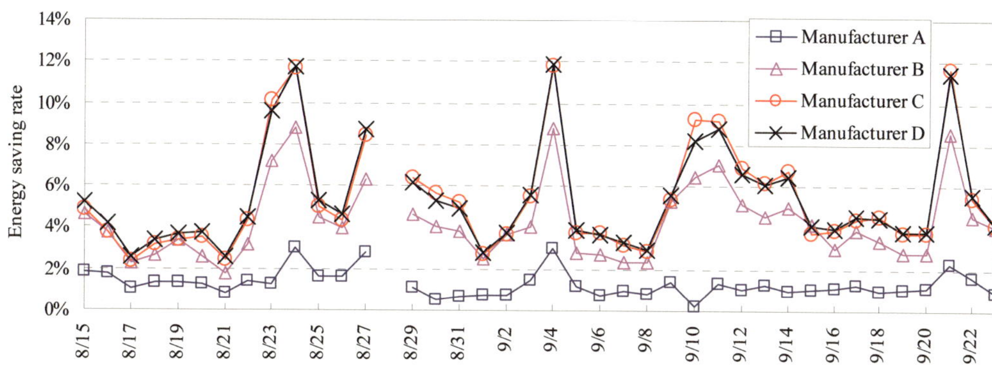

Fig. 12. Estimated daily average energy saving rate achieved by using hydroponic-cultivated sweet potato.

The reasons for why the energy saving ratios at the air-conditioners of the Manufacturer A are roughly 4 times less than these of the others are not further studied because the detailed information is not available about how the air-conditioner operation is tuned corresponding to the different outdoor temperature. The possible reasons might be that different manufactures use different actions or components to tune the running of air-conditioner accompanying to the change of outdoor air temperature. For example if outdoor air temperature decreases, for the air-conditioner with variable speed drive compressor can decrease the compressor's rotational speed to meet the requirement of low compression rate and maintain high efficiency as well. Low compressor rotational speed directly relates to low energy consumption because the energy consumption of air-conditioners is roughly linear to compressor's rotational speed. While for the air-conditioner with constant speed drive compressor can only decrease compression rate through increasing refrigerant flow rate and bypassing the redundant refrigerant flow. The larger flow rate needs larger energy input, which will counteract the energy saving benefitted from the decrease of compression rate. Further the flow rate larger than rated value will cause compressor efficiency decrease. So the energy saving rate might be small when outdoor air temperature decreases.

	Group 1			Group 2			Group 3					
	Manufacturer A (GHP)			Manufacturer B (EHP)			Manufacturer C (GHP)			Manufacturer D (EHP)		
	Energy consumption		Energy saving rate	Energy consumption		Energy saving rate	Energy consumption		Energy saving rate	Energy consumption		Energy saving rate
With or without green roo	With	Without		With	Without		With	Without		With	Without	
sum of Clear and no-water-sprinkle days	2830.5	2859.7	1.0%	2376.4	2449.9	3.0%	2673.1	2767.7	3.4%	2227.6	2317.9	3.9%
sum ofwater-sprinkle days	1024.9	1046.7	2.1%	905.4	970.9	6.7%	965.5	1050.3	8.1%	875.2	963.8	9.2%
sum of rainy days	1011.5	1019.6	0.8%	766.0	810.7	5.5%	857.5	917.6	6.5%	703.4	758.4	7.2%

Table 2. Summed annual energy saving achieved by hydroponic-cultivated sweet potato.

5. Transpiration and cooling effect modeling

Because the energy saving potential estimation results based on experimental result might depend on the local climate conditions, for the purpose of developing a universal method to estimate the energy saving potential, modeling of the transpiration and cooling effect are needed. Therefore it is necessary to develop plant transpiration model, which is used to estimate the plant transpiration rate, and boundary layer model, which is used to calculate the air temperature decreased by plant transpiration.

5.1 Transpiration model

Plants control their transpiration rate by adjusting the opening of stomata, as shown in Figure 13. So the transpiration rate can be calculated through dividing the water vapour partial pressure deficit $(W_i - W_a)$ by the stomatal resistance r_s and boundary layer air resistance r_a, as shown in Equation 4. The resistances can be calculated from their reciprocals, stomatal conductance g_s and boundary layer air conductance g_a, respectively (Equation 5). Boundary layer air conductance g_a is usually used a constant value of 1.13 mol/(m^{2}s) (Kadaira et al., 2005). Here Equation 6 (Campbell and Norman, 1998) is introduced to improve the model accuracy by considering the outdoor wind speed. Regarding the stomatal conductance g_s Jarvis (1976) proposed a model to estimate it using the leave surface temperature T, photosynthetically active radiation (PAR) Q and water vapour saturation pressure deficit D (Equation 7). Based on Jarvis model, Kosugi et al. (1995) improved the models of $f(Q)$, $f(T)$, and $f(D)$ and found that the models shown in Equation 8, 9 and 10 can suit for more plant so these models are used in this research. Further Kadaira et al. (2005) proved through experiment that in Equation 9 ambient air temperature θ can substitute for the leaf surface temperature T with acceptable accuracy. Kadaira et al. (2005) also found that the PAR can be accurately estimated by multiplying global sky radiation J by a correlation coefficient of C_Q with the value of 2.1 (Equation 11). To use these models to estimate the transpiration rate, seven coefficients need to be fitted using the data of the plant that the system uses, i.e. $g_{s\max}$ and a in Equation 5, T_o, T_h, and T_l in Equation 6, b_1 and b_2 in Equation 10.

$$M = \frac{M_w(W_i - W_a)}{r_s + r_a} \tag{4}$$

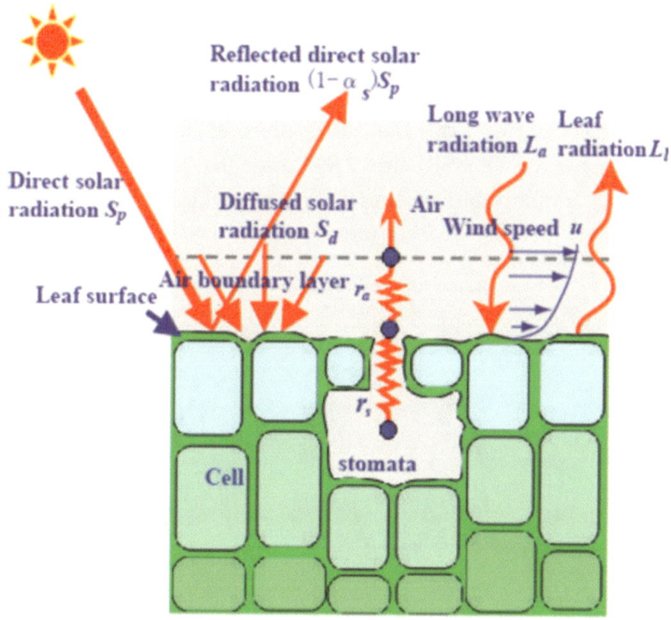

Fig. 13. Plant transpiration mechanisms.

$$r_s = \frac{1}{g_s},\ r_a = \frac{1}{g_a} \tag{5}$$

$$g_a = 0.147C_t\sqrt{u/l} \tag{6}$$

$$g_s = g_{s\max} \cdot f(Q) \cdot f(T) \cdot f(D) \tag{7}$$

$$f(Q) = \frac{g_{s\max} \cdot Q}{Q + g_{s\max}/a} \tag{8}$$

$$f(T) = \left(\frac{T - T_l}{T_o - T_l}\right)\left\{\left(\frac{T_h - T}{T_h - T_o}\right)^{\left(\frac{T_h - T_o}{T_o - T_l}\right)}\right\} \tag{9}$$

$$f(D) = \frac{1}{1 + (D/n_1)^{n_2}} \tag{10}$$

$$Q = C_Q J \tag{11}$$

The transpiration model is validated using the data measured in the former mentioned experiment.

The measured solar radiation, wind speed, ambient air temperature, plant cooled air temperature, leaf surface temperature, stomatal conductance, transpiration rate, etc. are used to fit the seven coefficients mentioned above. The fitted coefficients and fitting accuracy (Root Mean Square Error, %RMSE) are shown in Table 3. Then the fitted equations are used to calculate the leaf stomatal conductance and transpiration rate. The calculate transpiration rates are compared with measured ones, as shown in Figure 14. The average and root mean square error (RMSE) are 2.9% and 18.8%, which show that the model accuracy is acceptable for simulation study the energy performance of the combination system.

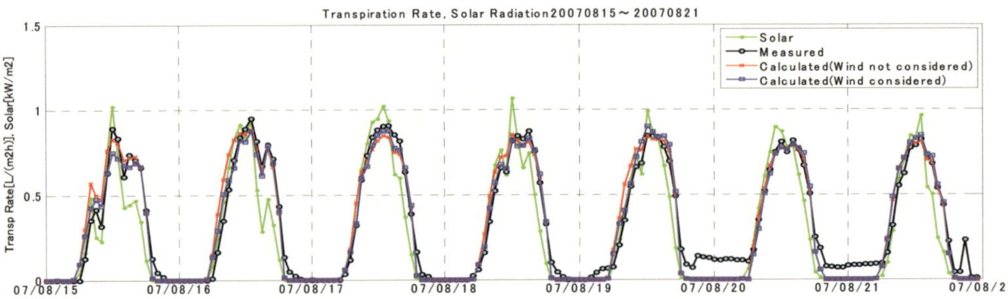

Fig. 14. Comparison of calculated and measured transpiration rate.

$g_{s\max}$	a	b_1	b_2	T_o	T_h	T_l	%RMSE
2.464E+00	5.174E-03	4.898E-01	1.979E+00	3.748E+01	4.350E+01	2.591E+01	8.70%

Table 3. Coefficients fitted for calculating stomatal conductance.

5.2 Boundary layer model

After the transpiration rate is obtained, next step is to use it to calculate how much the air temperature can be decreased by the transpiration. A physical model is considered, as shown in Figure 15.

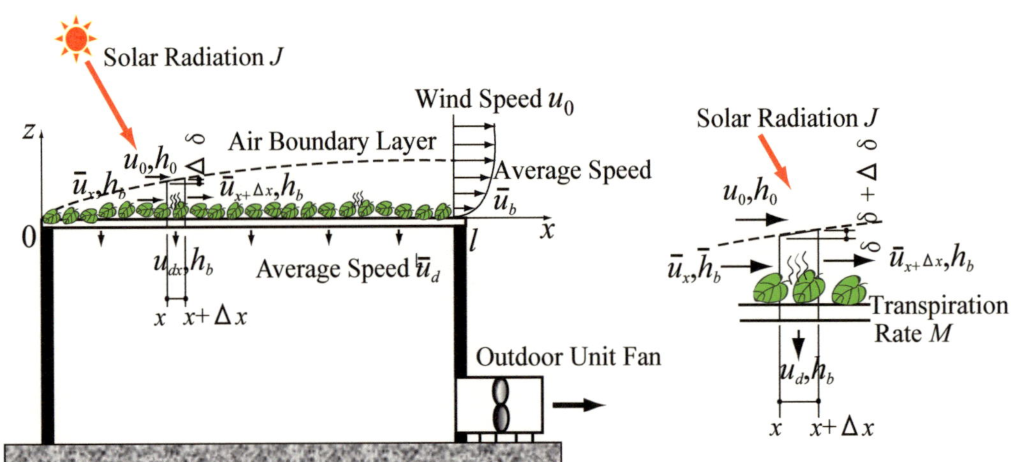

Fig. 15. Air boundary layer model.

When air flows along the plant leaf surface, an air boundary layer forms. If we take a micro air volume with the x-length of Δx, y-length of one meter, and z-length of boundary thickness as the analysis object, the mass conservation is shown in Equation 12 and energy conservation is shown in Equation 13. If the thermal stored in leaf is ignored, the leaf energy conservation is shown in Equation 14. If substituting Equation 14 into Equation 13, the energy conservation will be as shown in Equation 15. Because it is difficult to solve Equation 12 and 15, and what we want to know is not the temperature distribution but the average temperature, we can take the total boundary lay as the analysis object. Thus the mass conservation becomes as shown in Equation 16 and energy conservation becomes as shown in Equation 17. If substituting Equation 14 and Equation 16 into Equation 17, the energy conservation becomes as shown in Equation 18. Further, because the volume of boundary layer V_b is so small (it is 88.7 m^3 for the green roof with a length of 10 m when calculating using boundary layer thickness Equation 20 (Kato, 1964) and boundary layer volume Equation 21) comparing to the air volume entering the boundary layer (usually several thousands m^3/h) that it is safe to ignore the energy change of boundary layer air, i.e. $\Delta H \approx 0$. Therefore energy conservation becomes as shown in Equation 19. The physical meaning of Equation 19 is that the solar radiation energy absorbed by plant leaves becomes the energy difference between the air flowing into and flowing out of the boundary layer if the leaves temperature change and boundary layer air energy change are ignored.

$$u_0 \Delta\delta + \bar{u}_x \delta - \bar{u}_{x+\Delta x}(\delta + \Delta\delta) - u_{dx}\Delta x = 0 \quad (12)$$

$$[\rho_0 u_0 \Delta\delta h_0 + \rho_b \bar{u}_{b,x} \delta h_x - \rho_b \bar{u}_{b,x+\Delta x}(\delta + \Delta\delta) h_{x+\Delta x} - \rho_b u_{d,x} \Delta x h_x - q\Delta x + M \Delta x r_w]\Delta t = \Delta H \quad (13)$$

$$\alpha J + q - r_w M = 0 \quad (14)$$

$$[u_0 \rho_0 \Delta\delta h_0 + \rho_b \bar{u}_{b,x} \delta h_x - \rho_b \bar{u}_{b,x+\Delta x}(\delta + \Delta\delta) h_{x+\Delta x} - \rho_b u_{d,x} \Delta x h_x + \alpha J \Delta x]\Delta t = \Delta H \quad (15)$$

$$u_0 \delta_l - \bar{u}_b \delta_l - \bar{u}_d l = 0 \quad (16)$$

$$\left(h_0 \rho_0 u_0 \delta_l - \bar{h}_b \rho_b (\bar{u}_b \delta_l + \bar{u}_d l) - ql + Mlr_w\right)\Delta t = \Delta H \quad (17)$$

$$\left(h_0 \rho_0 u_0 \delta_l - \bar{h}_b \rho_b u_0 \delta_l + \alpha J l\right)\Delta t = \Delta H \quad (18)$$

$$h_0 \rho_0 u_0 \delta_l - \bar{h}_b \rho_b \bar{u}_0 \delta_l + \alpha J l = 0 \quad (19)$$

$$\frac{\delta}{x} = 0.380 \Big/ \left(\frac{u_0 x}{\nu}\right)^{1/5} \quad (20)$$

$$V_b = \int_0^l \delta dx = 0.211 \left(\frac{u_0}{v}\right)^{-1/5} l^{9/5} \quad (21)$$

Further, from Equation 16, we can deduce the speed $\bar{u}_b$ of the air flowing out of boundary layer at the end of the plant stage, as shown in Equation 22. If wind speed u_0 is small, the

$\bar{u}_b$ might be minus. This means the air volume sucked by the air-conditioner fan is large enough to suck all the boundary layer air to the underside of the plant. When $\bar{u}_b$ <0, the energy conservation will become as shown in Equation 23. From Equation 19 and 23, we can deduce the equation for calculating the average enthalpy of boundary layer air $\bar{h}_b$, as shown in Equation 24. Same as the enthalpy, the equations for calculating humidity ratio $\bar{x}_b$ can be deduced, as shown in Equation 25. Where, the average speed of air flowing into the underside of plant is calculated using Equation 26. Further from the enthalpy definition Equation 27, we can deduce the equation for calculating air temperature, as shown in Equation 28. Thus, the air temperature changed by the plant transpiration $\Delta\theta_T$ can be calculated, as shown in Equation 29.

$$\bar{u}_b = u_0 - \bar{u}_d \frac{l}{\delta_l} \tag{22}$$

$$h_0 \rho_0 u_d l - \bar{h}_b \rho_b \bar{u}_d l + \alpha J l = 0 \tag{23}$$

$$\begin{aligned} \bar{u}_b > 0 : \bar{h}_b &= \frac{h_0 \rho_0 u_0 \delta_l + \alpha J l}{\rho_b u_0 \delta_l} \\ \bar{u}_b < 0 : \bar{h}_b &= \frac{h_0 \rho_0 \bar{u}_d + \alpha J}{\rho_b \bar{u}_d} \end{aligned} \tag{24}$$

$$\begin{aligned} u_b > 0 : \bar{x}_b &= \frac{x_0 \rho_0 u_0 \delta_l + Ml}{\rho_b u_0 \delta_l} \\ u_b < 0 : \bar{x}_b &= \frac{x_0 \rho_0 \bar{u}_d + M}{\rho_b \bar{u}_d} \end{aligned} \tag{25}$$

$$\bar{u}_d = \frac{V}{l} \tag{26}$$

$$\bar{h}_b = C_{pa} \bar{\theta}_b + \bar{x}_b \left(C_{pw} \bar{\theta}_b + r_{w0} \right) \tag{27}$$

$$\bar{\theta}_b = \frac{\bar{h}_b - \bar{x}_b r_{w0}}{C_{pa} + \bar{x}_b C_{pw}} \tag{28}$$

$$\Delta\theta_T = \theta_0 - \bar{\theta}_b \tag{29}$$

To check the model accuracy, the measured air temperature changes are compared with the model calculated ones from August 15 to September 23. The comparison of the first one week is shown in Figure 16. The average error of the whole comparison period is 3.32% and the %RMSE is 94.02%. The model accuracy is not so high. But considering that the model is a totally physical model so it can be easily used for all situations, the model is acceptable for the further simulation study.

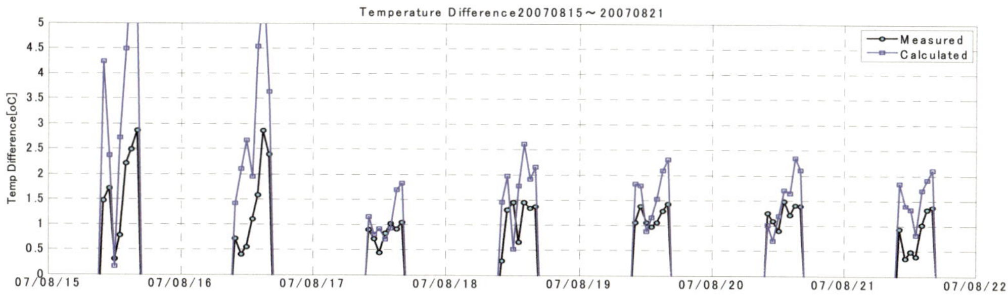

Fig. 16. Comparison of calculated and measured air temperature difference.

6. Methodology for predicting energy saving effect using simulation

6.1 Simulation flow

The flow of predicting the energy saving of the combination system is shown in Figure 17. The following five modules are used to calculate the energy savings.

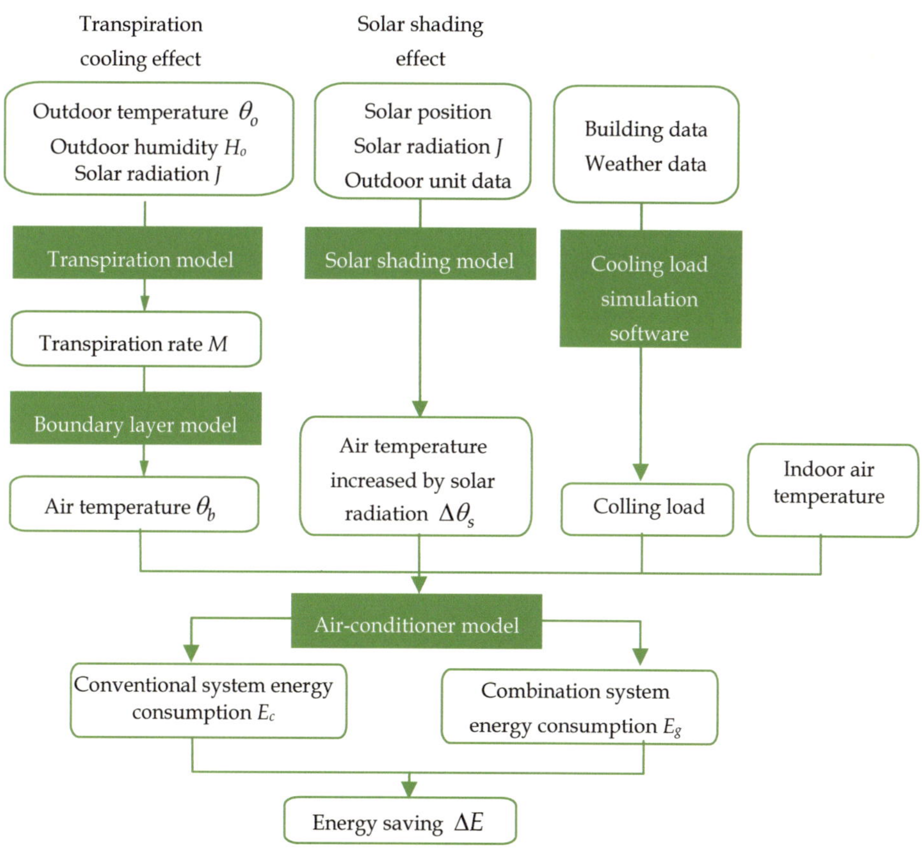

Fig. 17. The flowchart of predicting energy consumption of the system combining the green roof plants with air-conditioner outdoor units.

1. The plant transpiration model, which is used to estimate the plant transpiration rate M.
2. Boundary layer model, which is used to calculate the air temperature θ_b , which is the temperature after being decreased by plant transpiration.
3. Solar shading model (Equation 1), which is used to consider the solar shading effect.
4. Cooling load simulation software, which is used to simulated the cooling load of a given building.
5. Air-conditioner model, which is used to calculate an air-conditioner's energy consumption give the outdoor air temperature, indoor air temperature, and cooling load. The regression model fitted using manufacturers' specification data is used (Equation 2).

6.2 Simulation results

The former explained methodology is used to check the energy saving potential of the air-conditioners made by six main air-conditioner manufactures. They are divided into to three groups: high, middle and low improvement of energy efficiency accompanying to the outdoor air temperature decreasing. The cooling loads of a standard office building at four typical climate areas of Sapporo, Tokyo, Osaka, and Naha in Japan are simulated and used to estimate the energy consumed by the three groups of air-conditioner at the conditions of with and without combining outdoor unit with hydroponic-cultivated sweet potato. The results are shown in Table 4. The energy saving rate is similar at four different areas. It is about 8%, 6% and 1% for the high, middle and low efficiency-improvement group respectively.

Manufactuer	Group 1			Group 2			Group 3					
	A			E			C			D		
Energy related to use or not use roof plant	Energy consumption [kWh]		Energy saving rate	Energy consumption [kWh]		Energy saving rate	Energy consumption [kWh]		Energy saving rate	Energy consumption [kWh]		Energy saving rate
	Use	Not Use		Use	Not Use		Use	Not Use		Use	Not Use	
Sapporo	32396.30	32449.91	0.2%	26876.48	28371.19	5.3%	24250.59	26250.85	7.6%	24733.34	26492.37	6.6%
Tokyo	6353.16	6430.23	1.2%	5396.26	5726.88	5.8%	5252.81	5690.67	7.7%	4970.74	5355.23	7.2%
Osaka	10314.08	10438.78	1.2%	9595.27	10211.44	6.0%	8669.72	9465.12	8.4%	9089.36	9880.86	8.0%
Naha	12675.15	12801.64	1.0%	12050.77	12845.89	6.2%	10430.95	11445.15	8.9%	11388.72	12430.08	8.4%

Table 4. Energy saving potentials of the combination system at four typical climate areas.

7. Summary

This chapter describes how to improve energy efficiency of air-conditioners by combining the air-conditioner outdoor units with green roof plants for the purpose of utilizing the plant transpiration cooling effect and solar shading effect.

Experimental system was set up to measure the actual plant cooling and solar shading effect. The measurement results show that: 1) the air temperature can be cooled down by the hydroponic-cultivated sweet potato by 1.3°C in average for clear day and 3°C in average when water was sprinkled; 2) more than 99% solar radiation can be shaded by the plants.

Based on the experimental results, energy saving potential of the combination system was estimated for typical air-conditioners from different manufacturers. The results show that for clear days, the energy saving ratio is about 4% for air-conditioners with high efficiency improvement ratio and 1% for air-conditioners with low efficiency improvement ratio. If water-sprinkle is conducted two hours per day, the energy savings are 9% and 2% respectively.

Further, plant transpiration model was described, which is used to calculate the transpiration rate. And boundary layer model was described as well, which is used to calculate the temperature decrease caused by the plant transpiration. The models were validated using the experimental data. The validation results show that: 1) the average and %RMSE of transpiration model are 2.9% and 18.8% respectively. The average error of boundary layer model is 3.32% and the %RMSE is 94.02%. The model accuracies are not very high, which imply that the models need further improvement. Considering that the models are physical model and it can be easily used for all situations, the models are acceptable for simulation study.

Finally the methodology of predicting the energy saving potential using the models is explained. Four typical climate areas in Japan are selected to simulate the cooling loads and energy saving potentials of combining green roof plant with air-conditioner outdoor units. The simulation results show that the energy saving rates are similar for the four typical climate regions, which are about 8%, 6% and 1% respectively for the air-conditioners with high, middle and low efficiency-improvement ratio accompanying to air temperature decreasing.

8. Acknowledgement

The research described in this chapter is financially supported by Nissan Science Foundation, the Japan Society for the Promotion of Science (Project No. 17760468, 2004) and the Kansai Electric Power Co. Inc. The hydroponic cultivation system are provided and technically supported by Kyowa Corporation Ltd. Here the grateful acknowledgements are expressed to all the supporters.

9. Nomenclature

a : Reaction efficiency of stoma opening corresponding to light
a_i, b_i, c_i, d, $i=1,2,3$: Coefficients fitted using manufacturer specification data
CA: Cooling amount produced by an air-conditioner (kW)
C_{pa}: Specific heat of dry air, 1.005 kJ/(kgDA°C)
C_{pw}: Specific heat of water vapor, 1.846 kJ/(kg°C)
C_t: Outdoor wind turbulent coefficient, 1.4
C_Q: Correlation coefficient between PAR and sky global radiation, 2.1
g_a : Conductance of leaf surface boundary layer, mol/m^{2}s
$g_{s\max}$: The maximum stomatal conductance, mol/m^{2}s
g_s : Stomatal conductance, mol/m^{2}s
D : Saturation pressure deficit, kPa
E: Power consumption of air-conditioner, kW
h: Air enthalpy, kJ/kgDA
J: Global sky radiation, kW/m^2
l: Length of green roof, m

m : Outdoor unit fan air mass flow rate (kg/s)
M: Transpiration rate, kg/m^{2}s
M_w : Molar mass of water, 0.01802 kg/mol
n_1 : The saturation pressure deficit when stomatal conductance becomes half
n_2 : Curvature of the saturation pressure deficit function
q: Sensible heat exchange between leaves and air, kW/m^2
q_L : Long wave radiation between outdoor unit and its surroundings (W)
q_s : Short wave radiation (i.e. global solar radiation) (W)
Q : photosynthetically active radiation (PAR), μ mol/m^{2}s
r_a : Boundary layer resistance, m^{2}s/mol
r_s : Stomatal resistance, m^{2}s/mol
r_{w0} : Latent heat of water evaporation at 0°C, 2501 kJ/kg
T : Leaf surface temperature, °C
T_o : The most proper temperature, °C
T_h : Upper applicable temperature, °C
T_l : Lower applicable temperature, °C
u: Air flow speed, m/s
V: Air volume flow rate, m^3/s
W_i : Saturation partial pressure of water vapor at leaf surface, mb/mb
W_a : Water vapor partial pressure, mb/mb
x : Air humidity ratio, kg/kgDA
α : Absorption ratio to short wave radiation, leaves of sweet potato: 0.56; outdoor unit of air-conditioner: 0.76
α_s : Short wave radiation absorption ratio of outdoor unit surface
ρ : Air density, kg/m^3
δ : Thickness of boundary layer, m
Δt : Time interval, s
ΔH : Energy change of the air in the boundary layer, kJ
ν : Kinematic viscosity, m^2/s
θ : Air dry bulb temperature, °C
θ_i : Indoor air wet bulb temperature, °C
θ_o : Outdoor air dry bulb temperature, °C
θ_{of} : Final outdoor air dry bulb temperature after transpiration cooling and solar radiation absorption, °C

$\Delta\theta$: Change of air dry bulb temperature, °C

Subscription

o: Air out of the boundary layer
b: Air in the boundary layer
d: Air flowing into the underside of green roof plant
g: Use hydroponic-caltivated green roof plant

l: Length of the green roof
L: Long wave
S: Solar radiation
T: Transpiration

Superscript

t : Time step t
$^{-}$: Average

10. References

Alexandria Eleftheria, Jones Phil (2008), Temperature decreases in an urban canyon due to green walls and green roofs in diverse climates, Building and Environment, Vol. 43, pp. 480–493

Clerk Corrie, Adriaens Peter, Nriantalbot Albot, Talbot F. Brian (2008), Green Roof Valuation: A Probabilistic Economic Analysis of Environmental Benefits, Environmental Science Technology, Vol. 42, pp. 2155–2161

Campbell, G. S. & Norman, J. M. (1998) An Introduction to Environmental Biophysics Second edition. p101, p108, Springer, New York, US

Di H. F. and Wang D. N. (1999), Cooling Effect of Ivy on a Wall, Experimental Heat Transfer, Vol. 12, pp.235-245

Elena Palomo Del Barrio (1998), Analysis of the green roofs cooling potential in buildings, Energy and Buildings, Vol. 27, pp.179-193

Jarvis, P.G. (1976) The interpretation of the variations in leaf water potential and stomatal conductance found in canopies in the field. Phil.Trans.R.Soc.Lond.B, 273, pp.593-610.

Kadaira, A., Yoshida, H., Ito, M. et al. (2005) A study on the climatic mitigation effect of trees in an area around a residential complex. Journal of Environmental Engineering, Architecture Institute of Japan, No. 598, pp. 71-77.

Kato, K. (1964), Heat transfer. Yokendo Co., ltd., p105.

Kumar Rakesh, Kaushik S.C. (2005), Performance evaluation of green roof and shading for thermal protection of buildings Building and Environment, Vol. 40, pp. 1505–1511

Spala A., Bagiorgas H.S., Assimakopoulos M.N., Kalavrouziotis J., Matthopoulos D., Mihalakakou G. (2008), On the green roof system. Selection, state of the art and energy potential investigation of a system installed in an office building in Athens, Greece, Renewable Energy, Vol. 33, pp.173-177

Takebayashi Hideki, Moriyama Masakazu (2007), Surface heat budget on green roof and high reflection roof for mitigation of urban heat island, Building and Environment, Vol. 42, PP. 2971–2979

Wong Nyuk Hien, Chen Yu, Ong Chui Leng, Sia Angelia (2003) Investigation of thermal benefits of rooftop garden in the tropical environment, Building and Environment, Vol. 38, pp.261-270

Wang Fulin, Yoshida Harunori, Masuhara Satoshi, Kitagawa Hiroaki, and Goto Kyoko (2005), Simulation-Based Automated Commissioning Method for Air-Conditioning Systems and Its Application Case Study, Proceedings of the 9th International Building Performance Simulation Association Conference, Montreal, Canada, pp. 1307-1314

Wang Fulin, Yoshida Harunori, Yamashita Michiko, Improving Air-conditioners' Energy Efficiency Using Hydroponic Roof Plants, Proceedings of the 29th AIVC Conference in 2008, pp. 131-136, 2008.10

Wang Fulin, Yoshida Harunori, Yamashita Michiko, Prediction of the Energy Savings of a System Combining Air-conditioner's Outdoor Unit with Hydroponic-cultivated Roof Plant, Proceedings of the 6th International Symposium on Heating, Ventilating and Air Conditioning - ISHVAC09, Nanjing, China, pp. 419-427, 2009.11

Part 4

Energy Efficiency on Supply Side

11

Criteria Assessment of Energy Carrier Systems Sustainability

Pedro Dinis Gaspar, Rui Pedro Mendes and Luís Carrilho Gonçalves
University of Beira Interior - Faculty of Engineering - Electromechanical Eng. Dept.
Portugal

1. Introduction

Energy carrier systems of renewable sources have become widely used due to world's need to reduce the fossil fuel consumption and consequently greenhouse effect. However, the energy density of these systems is much lower than fossil fuels or nuclear fission. Besides, energy outlooks (IEA, 2011) show that energy demand around the world will continue its increasing trend. In turn, the wide scale construction of power plants based in fossil fuel cannot continue due to negative environmental effects. Also, the latest accident in reactor n.^{er} 3 of the Fukushima Daiichi Nuclear Power Station in Japan in consequence of the March 11 earthquake and tsunami increased the fear of radiation effects and the discussion about nuclear safety. Thus, it is relevant to provide an up-to-date assessment of the global sustainability of current and future energy carrier systems for electricity supply based on fossil fuel and renewable energy sources. It includes the analysis of energy carrier systems based on fossil fuels: coal, natural gas and oil and on renewable ones: wind, solar photovoltaic, geothermal, hydro, hydrogen, ocean (wave and tidal power), and nuclear.

The sustainability assessment of an energy conversion process into electrical energy is carried out in technological, economical, environmental and social dimensions. A solid basis for a state-of-the-art interdisciplinary assessment using data obtained from the literature supports the sustainability comparison. Thus, indicators that best describe the technologies and that are related to each of the abovementioned dimensions are defined to quantify the sustainability of energy carrier systems. These indicators are: efficiency of electricity generation, lifetime, energy payback time, capital cost, electricity generation cost, greenhouse gases emissions during full life cycle of the technology, land requirements, job creation and social acceptance. A criteria based on membership functions is exposed in order to determine a global sustainability index that quantifies how sustainable each energy carrier system is. The multi criteria analysis is performed considering different weighting functions applied to sustainability indexes in order to assess, today and in the near future, energy carrier systems that should be used in the mix of energy conversion systems to electricity. This work extends the research developed by Mendes et al. (2011a) and Mendes et al. (2011b).

2. Energy carrier systems

This section is devoted to describe the different types of energy carrier systems for conversion into electricity. The world energy source share of electricity generation in 2009 is shown in Fig. 1 (IEA, 2011a). Energy carrier systems are subdivided in renewable energy sources (wind, solar photovoltaic, geothermal, hydro, hydrogen and ocean) and fossil fuel (coal, natural gas and oil). Among these renewable energy sources, the wind and solar photovoltaic energies carrier systems are those with higher growth. Nowadays, nuclear energy can be considered as "an almost" renewable energy because new generation's nuclear plants can reuse uranium and its derivates. Thus, this energy carrier system is included into renewable section. Their advantages, disadvantages and capture technology are presented. The installed power, worldwide production and perspective of future increase are quantified for each energy carrier system.

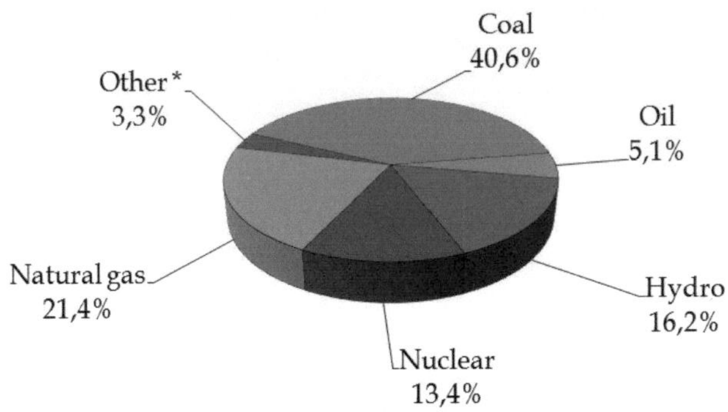

*Other: wind, solar photovoltaic, geothermal,, biofuels and waste, and heat.

Fig. 1. 2009 World energy source share of electricity generation (IEA, 2011a).

2.1 Renewable energy

Renewable energy markets, investments, industries, and policies have experienced rapid changes in recent years. So, its status can't be assured without doubts. If the global economic recession, felt most acutely in Europe, that also strike the renewable energy sector due to incentive cuts announcement by several governments, is taken into account, the trend of these energy carrier systems installed capacity was supposed to decrease. However, if the three-month long oil spill in the Gulf of Mexico and the incident in nuclear power station in Japan are considered, the caused extensive damage and welfare of people in these regions led to rethink and promote the use of renewable energy carrier systems. In the following subsections, each of the renewable energy carrier systems is discussed.

2.1.1 Wind

Wind turbines are used for the conversion of wind's kinetic energy into mechanical energy and then into electricity. This form of energy produces no emissions or contamination during the system operation.

WWEA (2011) states that wind energy has reached 196.630 GW of worldwide installed capacity and 430 TWh of produced energy (2.5 % of the electricity global consumption). This sector shows a fast growth rate among the renewable energy carrier systems, but in 2010 has showed the lowest growth rate value (23.6 %) since 2004 due to the international economic situation. Although, the wind sector had a turnover of 40 billion Euro and employed 670000 persons worldwide.

The distribution of the total installed capacity has changed since 2009, as China became number one in total installed capacity (44.7 GW), dethroning the United States of America (USA) with 40.2 GW. Also, China is now the centre of the international wind industry, adding 18.9 GW within one year, accounting for more than 50 % of the world market for new wind turbines. The growth rate in European countries shows stagnation in Western ones but a strong growth in a number of Eastern European countries (Romania, Bulgaria, Turkey). Nevertheless, Germany keeps its number one position in Europe for installed capacity with 27.2 GW, followed by Spain with 20.6 GW. The highest shares of wind power in electricity supply can be found in three European countries: Denmark (21 %), Portugal (18 %) and Spain (16 %).

WWEA predicts further substantial growth of wind sector in China, India, Europe and North America. Based on the current growth rates, this agency expects by 2015 a global wind capacity of 600 GW and more than 1 500 GW by the year 2020.

2.1.2 Solar photovoltaic

The solar photovoltaic (PV) technology is a method of converting solar radiation into electricity through the photovoltaic effect. It is an environmentally friendly energy carrier system by its ability to operate noiseless and emitting no greenhouse gases.

The PV sector showed a strong growth and investment, more than doubling in 2010 (REN21, 2011) as its capacity was added in more than 100 countries during that year, ensuring that PV remained the world's fastest growing power-generation technology.

The nominal worldwide installed capacity of PV systems in 2010 was about 40 GW - more than seven times the capacity in place five years earlier. Just only in 2010, 17 GW of capacity was added worldwide. This value represents an increase of 9.7 GW comparing with the installed PV power in 2009. The number of utility-scale PV plants continues to rise, accounting for almost 25 % of total global PV capacity.

Technology cost reductions in solar PV led to high growth rates in manufacturing and cell manufacturing continued its shift to Asia.

The PV market is dominated by the European Union (EU) countries (accounting for 80% of the world total) and particularly by Germany, which owns almost of half of global market (44 %) and installed more PV in 2010 (7.4 GW) than the entire world did the previous year. The rank is followed by Spain, Japan and Italy.

2.1.3 Geothermal

The geothermal energy source comes from the sub-soil heat, several hundred meters below the surface. For every 100 meters deep, the temperature increases about 3 °C. It is possible to reach the water boiling point temperature (100 °C) at 3 km depth (Farret & Simões, 2006).

This energy can be used for direct heating or for electricity generation by producing steam to drive a turbine (Erdogmus et al., 2006). Only the latter process will be assessed in this work. In the past 25 years, the electricity production by geothermal resource has significantly grown, reaching in 2007 about 10 GW of worldwide installed capacity (Gallup, 2009). By the end of 2010, total global installations came to just over 11 GW, and geothermal plants generated about 67.2 TWh of electricity during the year (REN21, 2011). However, the availability of this energy carrier system becomes scarce due to the difficulty and costs associated with sub-soil drilling at depths that allow reaching temperatures values suited to operate a turbine.

2.1.4 Hydro

Hydropower is a clean, renewable and reliable energy carrier system, allowing for energy storage and subsequent use when needed, making it a highly available resource. The use of this renewable energy source is done by converting into electricity the kinetic energy contained in rivers and potential energy of water falls down a shaft. The energy conversion process requires directing the stored water to a hydraulic turbine in order to drive an electric generator (Varun et al., 2009). Hydro energy is a resource globally wide spread with an installed capacity of about 720 GW around the world in 2008 (Kaldellis, 2008). Since then, global hydropower production increased reaching an estimated 1010 GW. The top countries for hydro capacity are China, Brazil, the USA, Canada, and Russia, which account for 52% of total installed capacity. Brazil and Canada generate roughly 80% and 61%, respectively, of their electricity with hydropower, while many countries in Africa, likewise Norway produce close to 100% of their grid based electricity with hydro (REN21, 2011). World spread countries continue to develop hydropower on large to small scales as well as pumped storage systems.

2.1.5 Hydrogen

Hydrogen is an abundant substance on the planet due to its presence in the molecule of water that covers about 70% of the earth's surface. It is a clean energy that enables the production of electricity through hydrogen fuel cells. Fuel cells are available in units of 5 to 250 kW, being more suitable for decentralized electricity production. Two types of fuel cells will be considered in the analysis of hydrogen as a energy carrier system: the phosphoric acid and solid oxide (ceramic, zirconium oxide and yttrium oxide) fuel cells. The former represents the first generation of commercial fuel cells. However, despite its good performance, such cells showed low viability. The solid oxide fuel cells have become much more attractive. With units with capacities from 5 to 250 kW, these cells are accessible to small consumers becoming suitable for decentralized production (Afgan & Carvalho, 2004).

2.1.6 Ocean (wave and tidal)

This renewable energy carrier converts the kinetic and potential energy of ocean into electricity. Both technologies, wave and tidal, will be assessed together as one. It is a renewable energy resource with high energy potential, reaching around 320 GW along the European coast, which corresponds to 16% of the world total resource (Cruz & Sarmento, 2004, WavEC, 2004). However, both technologies used in the conversion of this energy are still in a development stage despite of the numerous devices and conversion techniques that are patented. Because they are in an emerging phase, there are only few technologies with commercial application and the information about its sustainability is still based on forecasts. Despite this fact, at 2010's end, an estimated total of 6 MW of wave (2 MW) and tidal stream (4 MW) capacity had been installed mostly off the coasts of Portugal and the United Kingdom. The world estimated power capacity of tidal barrage is around 500 MW and presently there are more than 100 ocean energy projects (exceeding 1 GW in cumulative capacity) in various phases of development (REN21, 2011).

2.1.7 Nuclear

Nuclear energy can be defined as the energy converted by the release of the binding energy of components, for example, protons and neutrons, from an atom nucleus. The source of nuclear energy is based on the well known Einstein's equation ($E = mc^2$). Thus, a small amount of mass can be transformed into a lot of energy.

The nuclear energy resource has been used along many decades and has a great potential for electricity production. Although, it usage has been always controversial due to social acceptance questions. Nuclear energy supplies about 13% (≈ 624 GW) of the world electric energy demand. The electricity generation through nuclear energy has reached a value of 2558 TWh in 2009 (WNA, 2010). The USA is the country with more electricity through nuclear power plants with 19% of the total consumption (EIA; 2010). In 2006, France has bet this value with 80% of consumed electricity produced by nuclear power plants (Beardsley, 2006). Among other countries with a remarkably steady increase in nuclear generation are China, the Czech Republic, Romania and Russia.

In 2011, there were 437 nuclear reactors operating in the world and the International Atomic Energy Agency (IAEA) currently lists 64 reactors as "under construction" in 14 countries (Schneider et al., 2011). The world installed nuclear capacity has increased slowly, and despite of seven fewer units operating in 2011 compared to 2002, the capacity is still about 8 GW higher due to a combined effect of larger units replacing smaller ones and, mainly, technical alterations at existing plants, a process known as "uprating".

2.2 Fossil fuels

Oil and coal remain the most important primary energy sources since the 70's. Coal increased its share significantly since 2000. Growth slowed in 2005 and the total share of fossil fuels dropped from 86% in 1971 to 81% in 2004 (IPCC, 2007). In 2004, around 40 % of global primary energy was used as fuel to generate 17408 TWh of electricity. Electricity generation has an average growth rate of 2.8 %/year and is expected to continue growing at a rate of 2.5 to 3.1 %/year until 2030 (IPCC, 2007). The electricity generation forecasts indicate that fossil fuels will to continue to support this energy carrier. Fossil energy resources remain abundant as proven and probable reserves of oil and gas are enough to last for decades and in the case of coal, centuries. Possible undiscovered resources (mainly in Artic) extend these projections even further.

2.2.1 Coal

Coal is the world's most abundant fossil fuel (IPCC, 2007). The fossil fuel coal here considered includes all coal, both primary (including hard coal and lignite) and derived fuels (including patent fuel, coke oven coke, gas coke, BKB, gas works gas, coke oven gas, blast furnace gas and other recovered gases). Peat is also included in this category.

Coal-fired electricity-generating plants technologies are of conventional subcritical pulverized fuel design, with typical efficiencies of about 35% for the more modern units.

Best plants with supercritical pulverized fuel design achieve efficiencies of almost 50% (IPCC; 2007). Improved efficiencies have reduced the amount of waste heat and CO_2 that would otherwise have been emitted to atmosphere.

In 2005, coal accounted for around 25% of total world energy consumption primarily in the electricity and industrial sectors. Although coal deposits are widely distributed, world's recoverable reserves are located in the USA (27 %), Russia (17 %) and China (13 %). Two thirds of the proven reserves are hard coal (anthracite and bituminous) and the remainder are sub-bituminous and lignite (IPCC, 2007). Global proven recoverable, probable and estimated additional possible reserves of all coal types are about 133 000 EJ (IPCC, 2007).

In 2009, 8119 TWh were generated in coal/peat-fired plants (IEA, 2011b). China leads world ranking for the electricity production through coal/peat fuel with 2 913 TWh. It is followed by the USA with a power of 1 893 TWh, and by India with 617 TWh (IEA; 2011). According to IPCC (2007), the demand for coal is expected to more than double by 2030 (4500 GW).

2.2.2 Natural gas

Natural gas production has been increasing in the Middle East and Asia-Oceania regions since the 1980s. During 2005, natural gas was obtained in the Middle East (11 %), Europe and Eurasia (38 %), and North America (27 %) (IPCC, 2007). Proven global reserves of natural gas are estimated to be 6500 EJ, of which almost three quarters are located in the Middle East (IPCC; 2007).

Natural gas-fired power generation has grown rapidly due to its relative superiority to other fossil-fuel technologies in terms of investment costs, fuel efficiency, operating flexibility, rapid deployment and environmental benefits. In 2009, 4301 TWh were generated in gas-fired plants (IEA, 2011). The ranking is composed by the USA (950 TWh), Russian (469 TWh) and Japan (285 TWh). Natural gas is forecast to continue to be the fastest-growing primary fossil fuel energy source worldwide. The share of natural gas used to generate electricity worldwide is projected to increase from 25 % in 2004 to 31 % in 2030 (IPCC, 2007).

2.2.3 Oil

Conventional oil products extracted from crude oil-well bores and processed by primary, secondary or tertiary methods represent about 37% of total world energy consumption with major resources concentrated in relatively few countries as two thirds of proven crude oil reserves are located in the Middle East and North Africa (IPCC, 2007). Oil comprises crude oil, natural gas liquids, refinery feedstocks and additives as well as other hydrocarbons such as oil products (refinery gas, ethane, LPG, aviation gasoline, motor gasoline, jet fuels, kerosene, gas/diesel oil, fuel oil, naphtha, white spirit, lubricants, bitumen, paraffin waxes, petroleum coke and other oil products). However, not all of these are suited as input of a oil-fired plant for electricity generation.

Similar to the operation of other conventional steam technologies, oil-fired conventional steam plants are used to generate electricity. Burning oil to generate electricity produces significant air pollution in the forms of nitrogen oxides, and, depending on the sulphur content of the oil, sulphur dioxide and particulates. Carbon dioxide and methane (as well as other greenhouse gases), heavy metals and volatile organic compounds all can come out of the smoke stack of an oil-burning power plant.

In 2009, 1027 TWh were generated in oil-fired plants (IEA, 2011). Saudi Arabia is the country with the highest production of electricity from oils that reachs 120 TWh. With electricity generated values rather low, the ranking is completed by Japan with 92 TWh and Iran with 52 TWh. Assessments of the ultimate extractable resource (proven + probable + possible reserves) have ranged between 11 500 to 17 000 EJ (IPCC; 2007). Considering that consumption rates will continue to rise (IEA; 2011b), a reasonable prediction for supply is limited between 30 to 40 years.

3. Sustainability indicators

To assess the sustainability of an energy conversion process, it is necessary to use status indicators that define and quantify the process subsystems. For properties of a system that are not directly measurable, assessment tools are used to obtain the indicators. From a sustainability point of view, indicators should quantify the technological (efficiency, lifetime and energy payback time), economical (capital cost and electricity generation cost), environmental (greenhouse gas emissions and land requirements) and social (job creation and social acceptance) performances.

3.1 Technological indicators

The technological indicators selected to quantify the sustainability of energy carriers systems into electricity were efficiency, lifetime and energy payback time. The maximum and minimum values for each one of these indicators are determined for the different energy carrier systems. The values assigned to each indicator, for each energy carrier were collected from relevant and up-to-date studies.

3.1.1 Efficiency

The efficiency of electricity generation of the selected energy conversion technologies is an indicator that quantifies the percentage of effective primary energy converted into electricity. A range of values for this indicator was acquired for each technology. The values for the respective efficiencies were collected from Hanjalic et al. (2008), Evans et al. (2009),. Roth et al. (2009), Brito & Huckerby (2010), Vob (2006), Evans et al. (2010), Graus et al. (2007) and Dones & Heck (2006). For ocean energy only one value has been considered (minimum value = maximum value) due to the lack of information on this technology. The range of values (minimum value - maximum value) of the power conversion efficiency for each technology is shown in Fig. 2.

3.1.2 Lifetime

The lifetime is an indicator that quantifies the technology durability, indicating the period of time that this energy carrier system is in full operation. The lifetime is given in years due to longevity associated to each technology. Fig. 3 shows the maximum e minimum values of lifetime for each technology based on Varun et al. (2009), Afgan & Carvalho (2004), Roth et al. (2009), Banerjee et al. (2006), Parker et al. (2007), PREGA (2005) and Wartmann et al. (2009).

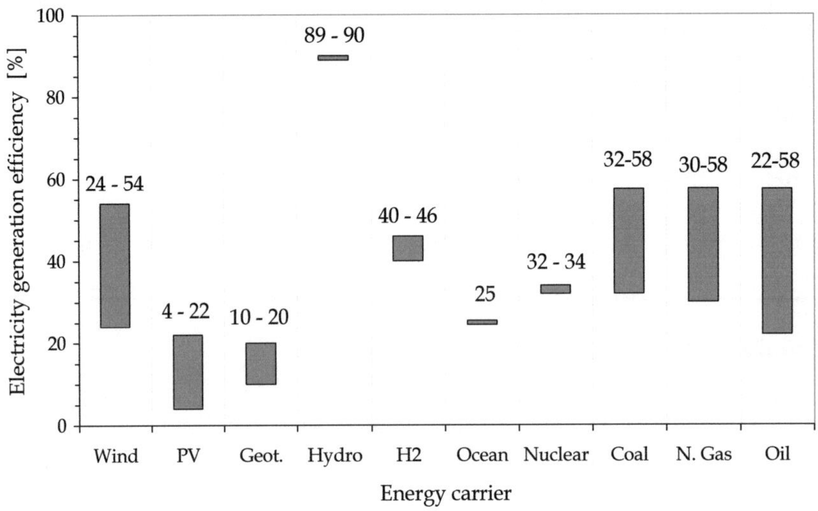

Fig. 2. Values range of electricity generation efficiency (%) for each energy carrier.

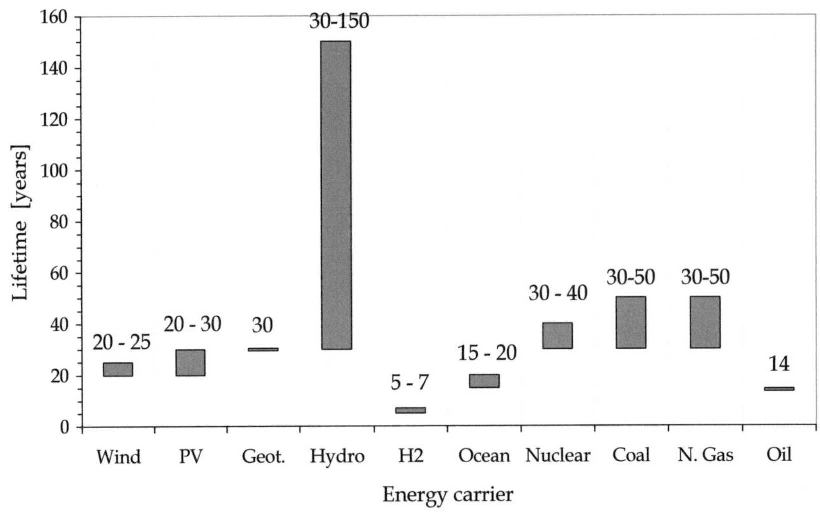

Fig. 3. Values range of lifetime (years) for each energy carrier.

3.1.3 Energy payback time

The Energy Pay Back Time (EPBT) is the time required for a technology to generate the same amount of energy needed for its manufacture and installation. This indicator is a measure of the return time. Once this period is completed, all the energy generated is profit (in energy terms). The values range of this indicator is shown in Fig. 4 is given in months. The values for E.P.B.T. were obtained by Varun et al. (2009), Banerjee et al. (2006), Parker et al. (2007), Soerensen et al. (2007), Vob (2006), WNA (2011), Randolph & Masters (2008), Mansure & Blankenship (2010), Voss (2001) and Biswas (2009).

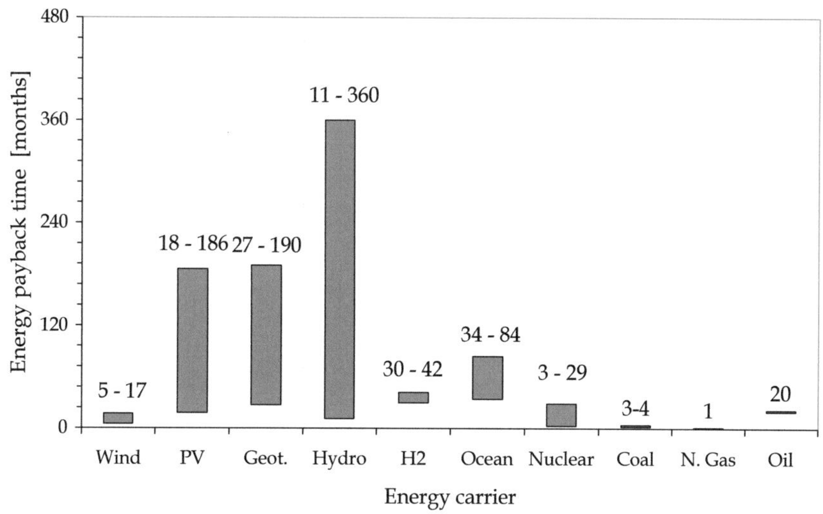

Fig. 4. Values range of energy payback time (months) for each energy carrier.

3.2 Economical indicators

The economical indicators selected to quantify the sustainability of energy carriers systems into electricity were capital cost and electricity generation cost. The maximum and minimum values for each one of these indicators are determined for the different energy carrier systems. The values assigned to each indicator, for each energy carrier were collected from relevant and up-to-date studies.

3.2.1 Capital cost

The capital cost is an economic indicator that measures the cost of installing technology for energy conversion considering its electricity generation capacity, i.e. its rated power. This indicator is given in €/MW. Fig. 5 shows the values range of this indicator for each technology. These values were obtained in Afgan & Carvalho (2004), Denny (2009), WNA (2011), Lako (2010), EIA (2010), Afgan & Carvalho (2002) and ESMAP (2007).

3.2.2 Electricity generation cost

This indicator quantifies the unit cost associated with the electricity production. It is given in the $€_{cent}$/kWh. The values were collected in Varun et al. (2009), Afgan & Carvalho (2004), Evans et al. (2009), Roth et al. (2009), Banerjee et al. (2006), Parker et al. (2007), Dalton et al. (2010), Dunnett & Wallace (2009), Allan et al. (2011), Lee et al. (2007), Evans et al. (2010) and PREGA (2005). Fig. 6 shows the range of values corresponding to this indicator.

3.3 Environmental indicators

The environmental indicators selected to quantify the sustainability of energy carriers systems into electricity were green house gas emissions and land requirements. The

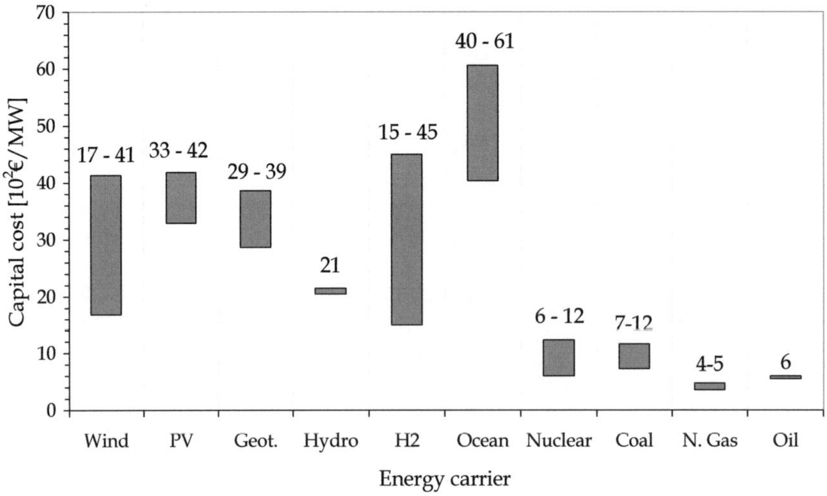

Fig. 5. Values range of capital cost (10^2€/MW) for each energy carrier.

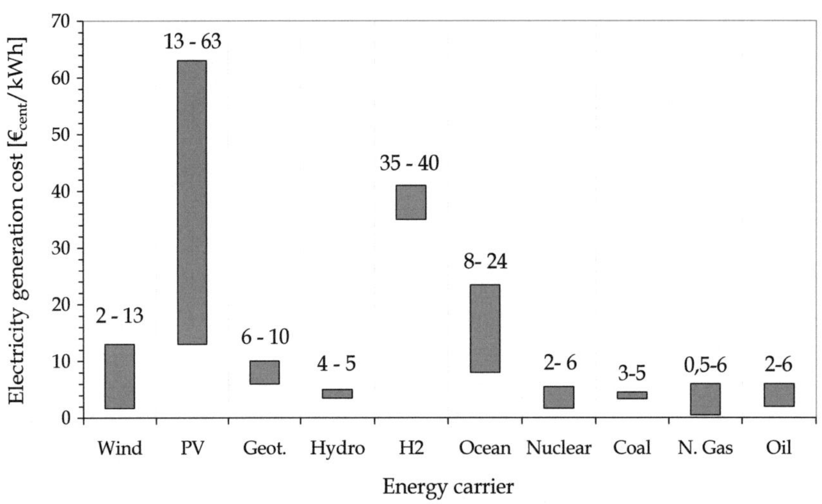

Fig. 6. Values range of electricity generation cost (€$_{cent}$/kWh) for each energy carrier.

maximum and minimum values for each one of these indicators are determined for the different energy carrier systems. The values assigned to each indicator, for each energy carrier were collected from relevant and up-to-date studies.

3.3.1 Greenhouse gas emissions (CO_2)

Greenhouse gases emissions have been the main reason why it became necessary to resort to the use of renewable energy. It is an indicator that quantifies the environmental impact of greenhouse gas emissions during the use of energy conversion devices. Sustainable energy carrier systems require minimal values of greenhouse gases emissions in order to not affect adversely the environment as it has been done by using fossil fuels combustion for electricity generation. There are several substances that are harmful greenhouse gases. However, carbon dioxide (CO_2) is the substance with more weight in the greenhouse gases composition, being the only one considered in this assessment. The CO_2 emissions are considered during the manufacturing and installation of the conversion technologies. This indicator is given in gCO_2/kWh of electricity generated and the values used to quantify it were obtained in Varun et al. (2009), Afgan & Carvalho (2004), Evans et al. (2009), Roth et al. (2009), Gagnon et al. (2002), Kannan et al. (2006), Sherwani et al. (2010), Varun et al. (2009), Raugei & Frankl (2009), Evans et al. (2010), ABB (2011), NEI (2011a) and Voss (2001). The values range of this indicator is shown in Fig. 7.

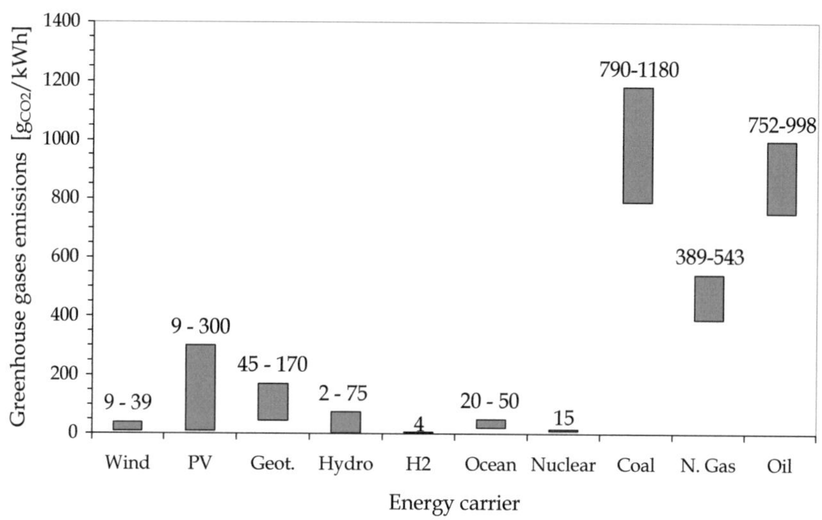

Fig. 7. Values range of greenhouse gases emissions (gCO_2/kWh) for each energy carrier.

3.3.2 Land requirements

This environmental indicator quantifies the area occupied by the installed technology. If the footprint is high, there may be harmful consequences on the environment due to the destruction of ecosystems. For this reason, the area occupied by the infrastructure of the energy conversion technology should be as small as possible. This indicator describes the area required to produce a given amount of energy per year.

The occupied land is referred to the field area used by the technology structure expressed in $km^2/TWh/year$. The values range for this indicator shown in Fig. 8 were collected from Afgan & Carvalho (2004), Evans et al. (2009), Gagnon et al. (2002), Evrendilek & Ertekin (2003), Rourke et al. (2010), NRC (2011), ESMAP (2007) and Wackernagel & Monfreda (2004).

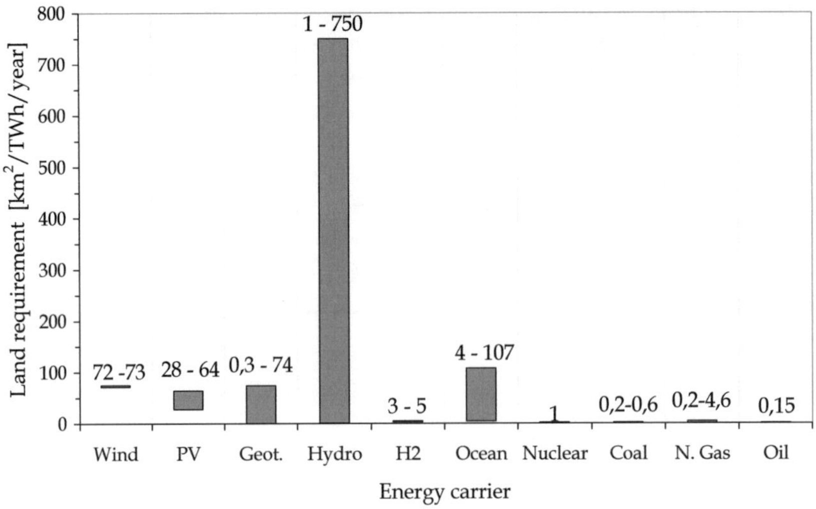

Fig. 8. Values range of land requirements (km²/TWh/year) for each energy carrier.

3.4 Social indicators

The social indicators selected to quantify the sustainability of energy carriers systems into electricity were job creation and social acceptance. The maximum and minimum values for each one of these indicators are determined for the different energy carrier systems. The values assigned to each indicator, for each energy carrier were collected from relevant and up-to-date studies.

3.4.1 Job creation

Job creation is a social indicator that quantifies the number of jobs created by the construction of a technological system of energy conversion. The energy carrier systems impact in society is of significant importance for their sustainability, since its operation can be blocked by the population if the conversion technology is not socially advantageous. Thus, job creation becomes a good indicator to assess the social impact as indicates the capacity of employment of each installed technology. Its unit is Jobs n.^{er}/MW. The used values for this indicator are from Dalton & Lewis (2011), Rinebold et al. (2009), Peters (2010), Plowman (2004), NEI (2011b) and EFL (2011). The values range for this indicator is shown in Fig. 9.

3.4.2 Social acceptance

Social acceptance of any energy source involves both the general attitude towards the technology to capture energy as well as systems deployment decisions at local, regional or even national levels. In this context appears the latest formulation of "energy social acceptance" concept, named "triangle model". This model identifies three key dimensions of social acceptance: the socio-political acceptance, community acceptance and market acceptance (EWEA, 2009). The socio-political acceptance refers to the energy conversion technology acceptance and policies to a more general level. This component is not limited to levels of acceptance by the public in general, but includes the acceptance by stakeholders

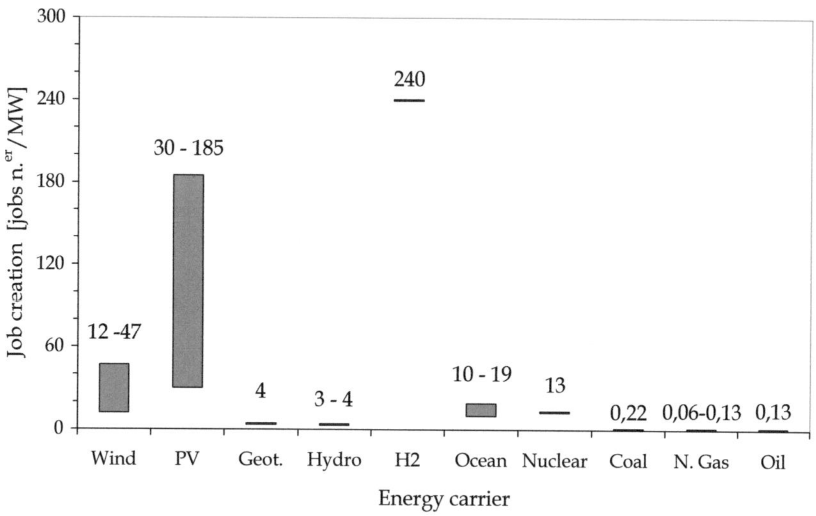

Fig. 9. Values range of job creation (Jobs n.er/MW) for each energy carrier.

and policy makers. The various political and private agents involved in the discussion are crucial in planning and promoting local initiatives. The community acceptance refers to the acceptance of specific projects at the site of potentially affected populations, of the main local stakeholders and local authorities. It is this part where the social debate around energy capture system suitable for a particular region emerges and develops. Market acceptance refers to the process by which commercial parties adopt and support (or not) an implementation of a particular technological system of energy capture.

The values range is divided into the percentage of individuals in favour (minimum value) and the sum of individuals in favour and undecided (maximum value) concerning the installation of the technological system to capture a particular energy source. It considers the public support levels for different types of energy sources obtained through polls and attitude surveys conducted by the Survey Standard Eurobarometer (EB) on the population of the European Union (EU) (EC, 2006, 2007). In the case of geothermal energy, Ungemach (2007) indicates that this energy source has a low social acceptance. So, it is considered a value of 40%, which does not affect or benefit the index. By other side, van Bree & Bunzeck (2010) indicate that the hydrogen energy conversion systems have a high social acceptance, because it is a clean energy source without any kind of controversy, although it is reported a low level of knowledge about the hydrogen technology. In this sense, is considered a value of 60% of global acceptance, that it will neither detriment nor benefit this energy capture system.

Globally, the social acceptance indicates the population approval to install and explore a certain technology power plant. This indicator encompasses the social-politic, community and market approval. This indicator quantification relies on statistical studies which were provided by EWEA (2009), Ungemach (2007), van Bree & Bunzeck (2010), Evans et al. (2010).

The values range for this indicator is shown in Fig. 10.

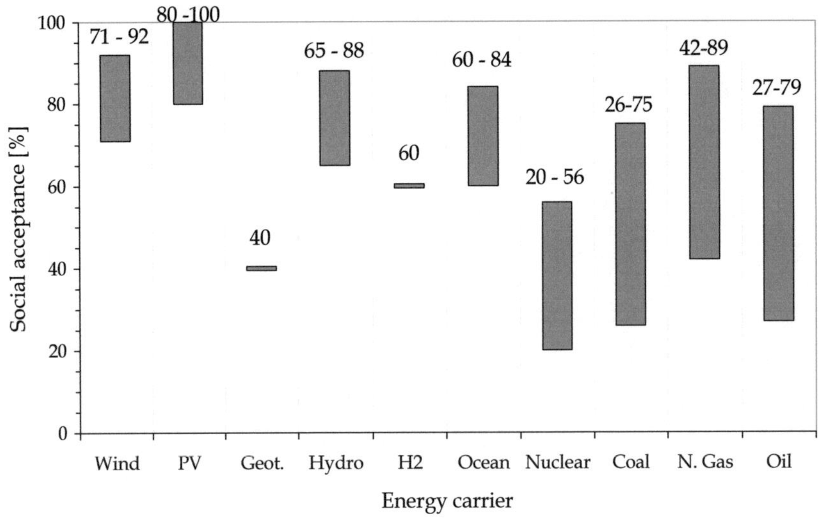

Fig. 10. Values range of social acceptance (%) for each energy carrier.

4. Sustainability assessment

The sustainability assessment is performed through a mathematical formulation that non-dimensional indicators, within the maximum and minimum values of all energy carriers. This procedure is performed for all the indicators, technological, economical, environmental and social ones. This mathematical formulation allows quantifying the sustainability indicators within the maximum variation range among all energy carriers. Two global sustainability indexes will be determined, a maximum and a minimum one. The former uses the maximum values among all energy carrier systems to obtain the non-dimensional relation. This procedure is made for each indicator. Afterwards, weighting functions are applied to perform a multi-criteria decision analysis. This procedure is performed likewise using the minimum values among all energy carrier systems within each indicator. This sensitivity analysis allows verifying which are the most relevant indicators to assess energy carriers sustainability.

4.1 Mathematical formulation

After choosing the indicators, an index is formulated to quantify the sustainability of each energy carrier system. It is necessary to relate them so they can be expressed by a quantifiable single value. This relationship is obtained through mathematical expressions that use a membership function for each indicator. The procedure is performed for the minimum and maximum values of each indicator in order to obtain of values range for the global sustainability index.

For each indicator:

- Select the maximum, $max(x_i)$, and minimum, $min(x_i)$, for each indicator separately for minimum and maximum values ranges.
- Evaluate whether the function $q(x_i)$ increases or decreases with the increase of x_i. Depending on the variation of function $q(x_i)$, select the proper expression.

If membership function $q(x_i)$ increase with x_i indicator, their relationship is expressed by:

$$q(x_i)=\begin{cases} 0 & \text{if } x_i \le \min(x_i) \\ \dfrac{x_i-\min(x_i)}{\max(x_i)-\min(x_i)} & \text{if } \min(x_i)<x_i<\max(x_i) \quad (1) \\ 1 & \text{if } x_i \ge \max(x_i) \end{cases} \tag{1}$$

If membership function $q(x_i)$ decreases with x_i indicator, their relationship is expressed by:

$$q(x_i)=\begin{cases} 1 & \text{if } x_i \le \min(x_i) \\ 1-\dfrac{x_i-\min(x_i)}{\max(x_i)-\min(x_i)} & \text{if } \min(x_i)<x_i<\max(x_i) \; (2) \\ 0 & \text{if } x_i \ge \max(x_i) \end{cases} \tag{2}$$

The global sustainability index (Q) is the sum of the various indicators taking into account the weight (w_i) that each one has in the mathematical expression of the index. Considering m indicators for the process characterization, the final mathematical expression is given by:

$$Q(q;w)=\sum_{i=1}^{m} w_i q_i \tag{3}$$

4.2 Multi-criteria decision analysis based on weighting function variation

The multi-criteria decision analysis is performed with the weighting function variation. Several case scenarios are developed considering different weights for each indicator. This condition allows the analysis of the most relevant indicators on the sustainability assessment.

According to values range obtained through extensive bibliographic review, two index values are calculated, one referred to minimum values and other to maximum ones. For the indicators that have just a single value, this one is used in both situations. Several case studies are performed. In first case scenario (CS#1), all indicators are considered of equal importance, i.e. having the same weight (w_i = 11 %).

Efficiency, energy payback time, capital cost, electricity generation cost, greenhouse gases emissions and social acceptance are decisive indicators in the quantification of energy carrier systems sustainability (Afgan & Carvalho, 2004). According to Afgan & Carvalho (2004) and Hanjalic et al. (2008), where similar studies are presented, the most important indicator's weight coefficient is within the range of 60-70%. The mean value will be considered, i.e., a 65% weight coefficient for the most important indicator in each case study. Thus, in CS#2 to CS#7, per case, one indicator is considered more important than the others. This indicator will have a higher weight (w_i = 65 %) and the remaining an equal lower weight coefficient (w_i = 4,375 %).

For last, CS#8 considers that decisive indicators most relevant than the remainders. So, these indicators will have a higher weight (w_i = 14 %) and the remaining an equal lower weight coefficient (w_i = 7,5 %).

The respective weights for case studies are shown in Table 1. The membership function given by Equation (1) or (2) is used depending on the energy carrier system sustainability increase or decrease with the indicator, respectively (see Table 2).

In expressions (1) and (2) are used as maximum and minimum, respectively the largest and smallest value found for this indicator, given the range of values corresponding to all energy carriers. Thus, each indicator will be represented by a membership function that varies between 0 and 1.

Indicator, q_i		Weight, w_i (%)							
		CS#1	CS#2	CS#3	CS#4	CS#5	CS#6	CS#7	CS#8
q_1	Efficiency	11,1	65	4,375	4,375	4,375	4,375	4,375	14
q_2	Electricity generation cost	11,1	4,375	65	4,375	4,375	4,375	4,375	14
q_3	Capital cost	11,1	4,375	4,375	65	4,375	4,375	4,375	7,5
q_4	Lifetime	11,1	4,375	4,375	4,375	4,375	4,375	4,375	7,5
q_5	Greenhouse gases emissions	11,1	4,375	4,375	4,375	65	4,375	4,375	14
q_6	Land requirement	11,1	4,375	4,375	4,375	4,375	4,375	4,375	7,5
q_7	Job creation	11,1	4,375	4,375	4,375	4,375	4,375	4,375	7,5
q_8	Energy payback time	11,1	4,375	4,375	4,375	4,375	65	4,375	14
q_9	Social acceptance	11,1	4,375	4,375	4,375	4,375	4,375	65	14

Table 1. Weight of each indicator for the case studies.

Indicator		Sustainability	Equation used
q_1	Efficiency	Increases	Equation 1
q_2	Electricity generation cost	Decreases	Equation 2
q_3	Capital cost		
q_4	Lifetime	Increases	Equation 1
q_5	Greenhouse gases emissions	Decreases	Equation 2
q_6	Land requirement		
q_7	Job creation	Increases	Equation 1
q_8	Energy payback time	Decreases	Equation 2
q_9	Social acceptance	Increases	Equation 1

Table 2. Sustainability variation with indicator's value.

5. Analysis and discussion of results

Based on the mathematical formulation results, where the maximum and minimum values for the global sustainability index are obtained for each test scenario, the most relevant indicators are defined as well as the mix of energy carrier systems that should be considered in the electricity supply portfolio. In the following sections the results obtained with the case studies are discussed.

5.1 Case scenario n.^{er} 1 (CS#1): equal weighting factors for indicators

The minimum and maximum values of the sustainability index for each energy carrier system when it is considered an equal weighting factor (w_i = 11,1 %) for all indicators is shown in Fig. 11. The following considerations can be highlighted:

1. The high efficiency and lifetime of hydro systems contribute significantly to its global sustainability index;
2. The reduced E.P.B.T. and low cost of electricity generation associated with wind and nuclear systems contribute to their overall levels of sustainability;
3. The capital cost needed to generate energy by photovoltaic systems, geothermal and ocean (wave and tidal) penalizes their sustainability indexes;
4. The reduced CO_2 emissions associated with wind systems, hydro, hydrogen and nuclear contribute to their sustainability indexes;
5. The reduced land requirements of geothermal, hydrogen, nuclear and ocean energy and all fossil fuels conversion systems also influence their global sustainability index;
6. The number of jobs generated by conversion systems of hydrogen into electricity and in a lesser extent by photovoltaic systems potentates their index;
7. Social acceptance is less controversial in the wind, photovoltaic and hydro systems;
8. The global sustainability indexes for fossil fuel energy carrier systems (coal, natural gas and oil) rely on the reduced electricity generation cost, capital cost and land requirements when compared with other technologies;
9. The lifetime of fossil fuel energy carrier infrastructures benefit their global sustainability indexes;
10. The land requirement for a fossil fuel-fired plant to generate electricity (without taking into account natural resources extraction) is lower comparing to renewable energy plant, which promotes its global sustainability index;
11. Among the fossil fuel energy carrier systems, those of natural gas emit lower quantity of greenhouse gases, which increases their indicator value;
12. The number of jobs created in a fossil fuel-fired plant is higher than in a renewable one, enlarging this indicator, and consequently, promoting the increase of global sustainability indexes of fossil fuel energy carrier systems.

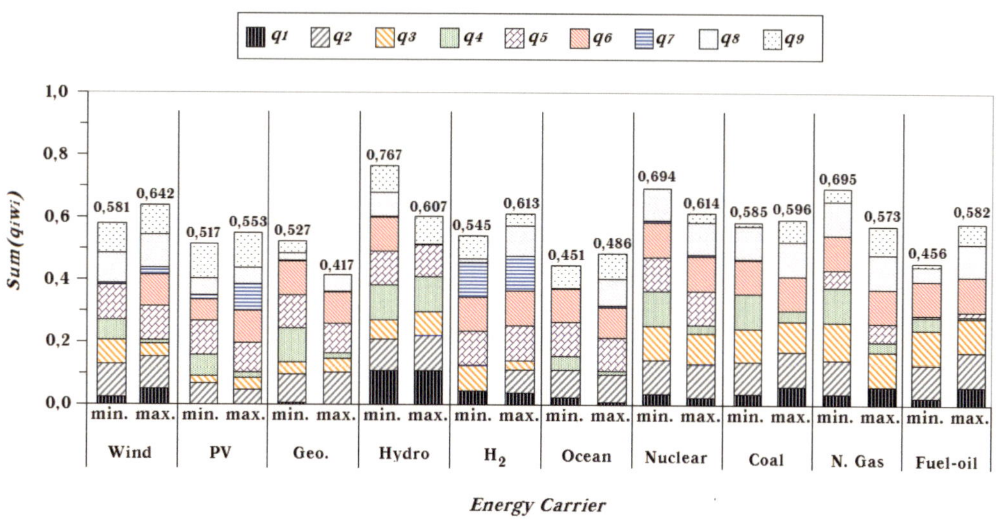

Fig. 11. Maximum and minimum values range of sustainability index for each energy carrier system - Case Scenario 1 (Equal weighting factors for indicators).

The analysis of this figure allows us to suggest a hierarchy of sustainability ranking of conversion technologies and propose a mix of technologies to convert energy into electricity. Thus, hydro, nuclear, natural gas and wind energy carrier systems are the ones which stand in front of a sustainable future for the electricity supply. Among these energy carrier systems, only one is based on fossil fuel, natural gas-fired plants. However, if nuclear energy is excluded from this comparison due to the controversy generated by this energy carrier system, coal and hydrogen conversion system are included in the front line of most sustainable energy conversion systems. Although fossil fuel-fired plants are a huge damage to ecology and its availability is very limited, natural gas and coal energy carriers are among those more globally sustainable due to the same weighting factor for all indicators.

Nevertheless, this analysis is refined considering the following case studies where the global sustainability is analysed from a single indicator view point. This kind of analysis is important to figure which are the indicators more significant on sustainability.

5.2 Case scenario n.^{er} 2 (CS#2): highest weighting factor for efficiency indicator

Fig. 2 shows the sustainability index for each technology considering energy conversion efficiency as the core indicator (Case Scenario n.^{er} 2: CS#2). The weighting factor for this indicator was considered as w_1 = 65 %. For this case scenario, hydro, natural gas and coal energy carrier systems are the most sustainable. The following renewable energy carrier systems most sustainable use hydrogen and wind as resource. By other hand, photovoltaic and geothermal energy carrier systems are the less sustainable.

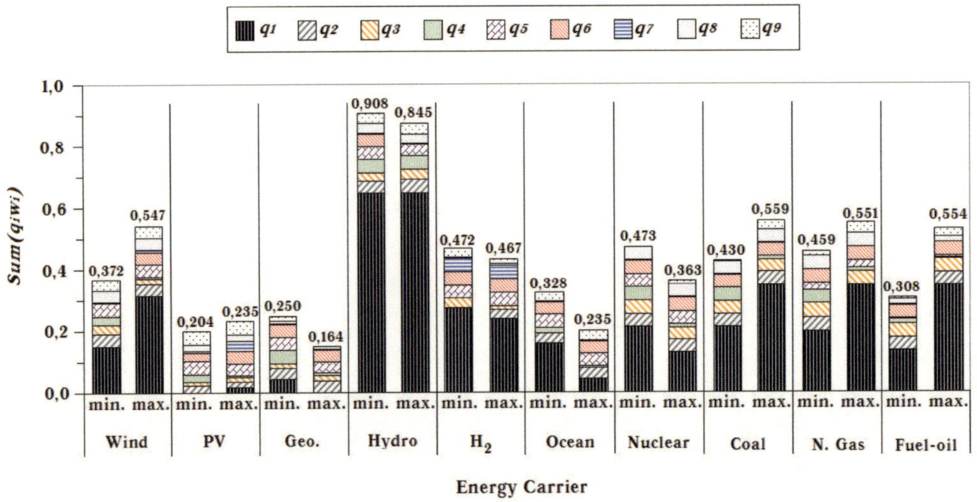

Fig. 12. Maximum and minimum values range of sustainability index for each energy carrier system - Case Scenario 2 (Highest weighting factor for efficiency indicator: q_1).

5.3 Case scenario n.^{er} 3 (CS#3): highest weighting factor for electricity generation cost indicator

In this case scenario, the sustainability ranking is modified. Nuclear, hydro, coal and wind energy carrier systems are the most sustainable when electricity generation cost is considered the most relevant indicator to assess sustainability. Likewise in previous case scenarios, at least one fossil fuel energy carrier system is included among the most sustainable. Photovoltaic and hydrogen energy carrier systems are the less sustainable.

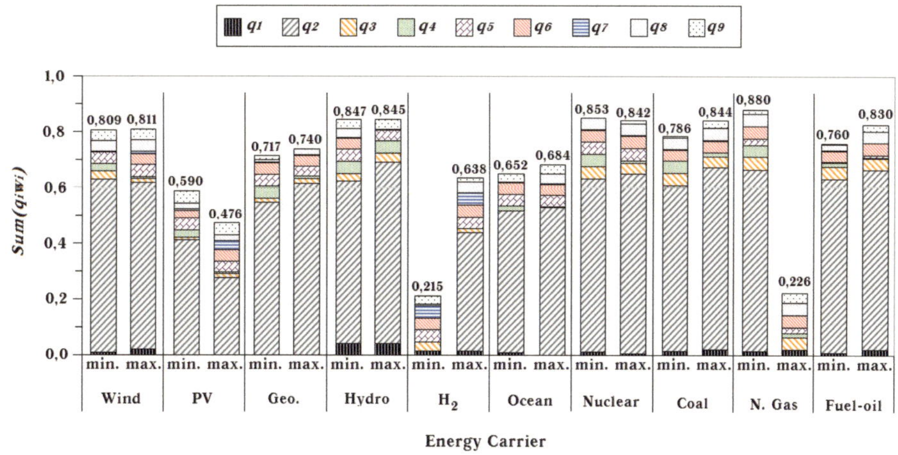

Fig. 13. Maximum and minimum values range of sustainability index for each energy carrier system - Case Scenario 3 (Highest weighting factor for electricity gen. cost indicator: q_2).

5.4 Case scenario n.er 4 (CS#4): highest weighting for capital cost indicator

Considering case scenario n.er 4 where capital cost is the most relevant indicator, the most relevant energy carrier systems are mainly based on fossil fuel resources (natural gas, nuclear, oil and coal). Following these systems appear the energy carrier systems based on renewable resources: hydro and wind. In this case study, ocean and photovoltaic energy conversion system are the less sustainable. These results mean that the capital cost for constructing and operating a fossil fuel-fired plant is less than a renewable one.

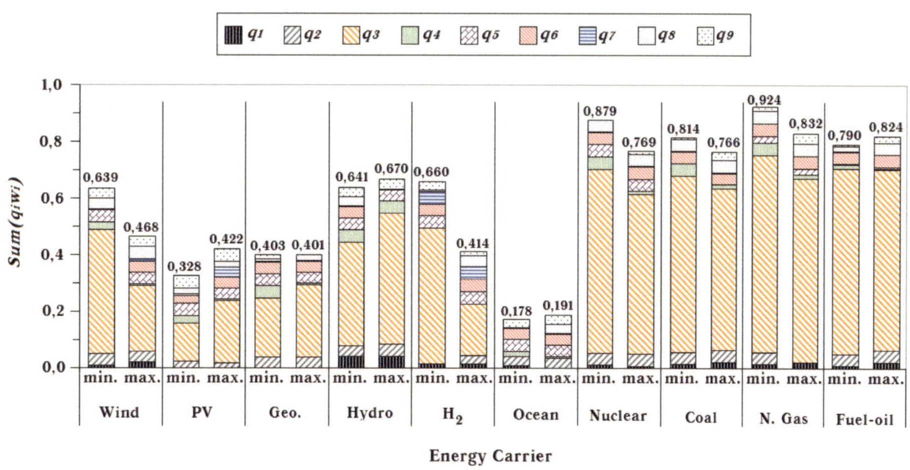

Fig. 14. Maximum and minimum values range of sustainability index for each energy carrier system - Case Scenario 4 (Highest weighting factor for capital cost indicator: q_3).

5.5 Case scenario n.[er] 5 (CS#5): highest weighting for greenhouse gases emissions indicator

For the case scenario where greenhouse gases (GHG) emissions is considered as the most relevant indicator (case scenario n.er 5), the ranking is composed by renewable energy carrier systems: nuclear, hydro and wind. All fossil fuel-fired power plants, i.e. natural gas, oil and coal energy carrier systems are the less sustainable. Due to the combustion emissions, these results were expected. Geothermal and photovoltaic energy conversion systems are the less sustainable among the renewable ones.

5.6 Case scenario n.[er] 6 (CS#6): highest weighting for energy payback time indicator

Case study n.er 6 considers E.P.B.T. as the core indicator. In this case, natural gas, nuclear and coal energy carrier systems are ahead in the ranking of most sustainable. Taking into account the results of previous case scenarios, fossil fuel-fired plants include the leader positions when economic and efficiency aspects are combined to define a mix for electricity supply. Geothermal and ocean energy conversion systems are the less sustainable.

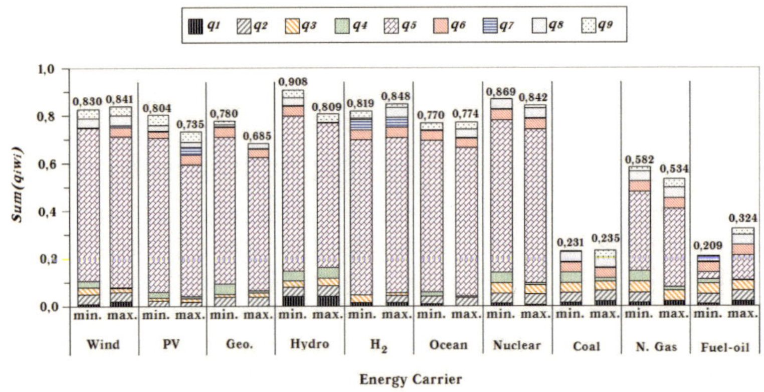

Fig. 15. Maximum and minimum values range of sustainability index for each energy carrier system - Case Scenario 5 (Highest weighting factor for GHG emissions indicator: q_5).

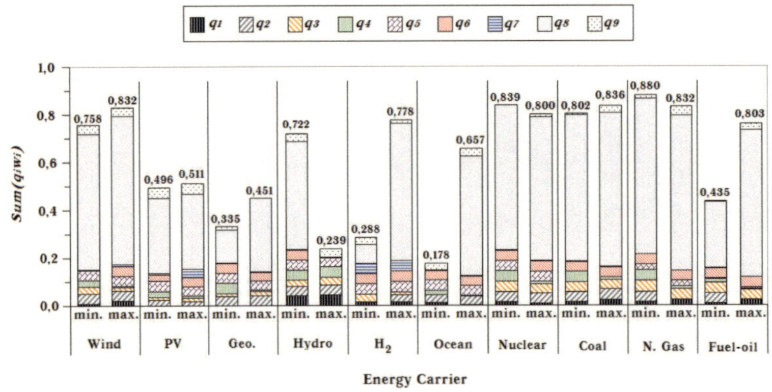

Fig. 16. Maximum and minimum values range of sustainability index for each energy carrier system - Case Scenario 6 (Highest weighting factor for energy payback time indicator: q_8).

5.7 Case scenario n.er 7 (CS#7): highest weighting for social acceptance indicator

Being social acceptance considered a meaningful indicator, there is a shift on leadership of the sustainability global index. Photovoltaic systems are considered now the most sustainable. Hydro and wind systems also include the leading group. Oil, nuclear and geothermal energy conversion systems are the less sustainable. The latter have reduced social acceptance due to the lack of reliable data. By other hand, nuclear and oil conversion systems possess low social acceptance worldwide due to the recent incidents.

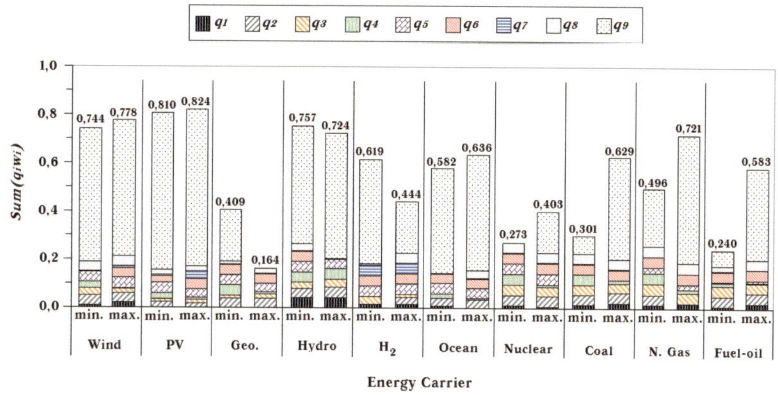

Fig. 17. Maximum and minimum values range of sustainability index for each energy carrier system - Case Scenario 7 (Highest weighting factor for social acceptance indicator: q_9).

5.8 Case scenario n.er 8 (CS#8): Higher weighting for selected indicators

A higher weighting factor is considered for selected indicators, i.e. efficiency, electricity generation cost, greenhouse gases emissions, energy payback time and social acceptance are considered as the relevant indicators to assess sustainability. Hydro, wind and nuclear energy carrier systems take the leadership for the mix on the electricity supply. Ocean, oil and geothermal energy carrier systems are the less sustainable assuming these conditions.

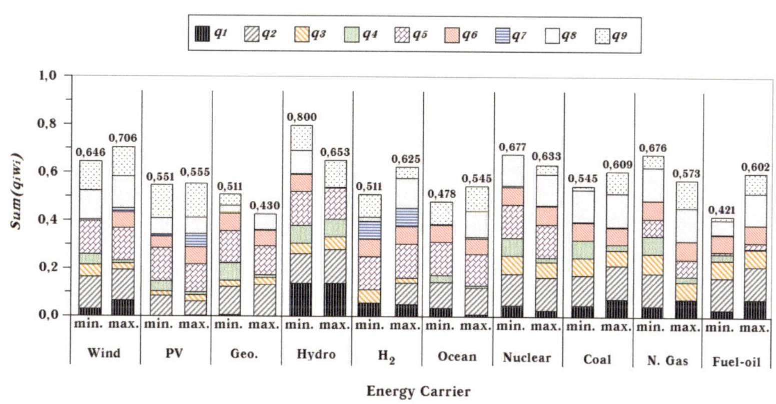

Fig. 18. Maximum and minimum values range of sustainability index for each energy carrier system - Case Scenario 8 (Highest weighting factor for selected indicators: q_1, q_2, q_5, q_8, q_9).

Table 3 includes a summary of the case scenarios. It is considered a ranking varying from 1 to 10 (equal to the number of the energy carrier systems assessed), corresponding respectively to an order from the less sustainable to the most sustainable. The sum of the global sustainability index for each energy conversion system, taking into account the case studies where one indicator is considered as the most relevant, provide an insight of the electricity supply portfolio mix more sustainable (see *total 1* in Table 3). Independently of the case study, analysing the different faces of a global sustainability, hydro, nuclear, wind and natural gas-fired power plants are those that should be considered as the most relevant on the electricity supply portfolio taking into account the individual significance of each indicator. It must be pointed out that among these four energy carrier systems, only one (natural gas) is of fossil fuel kind. Geothermal and ocean energy conversion systems are the opposite counterpart, i.e., are the less sustainable.

Case Scenario		Wind	PV	Geo.	Hydro	H2	Ocean	Nuclear	Coal	N.Gas	Oil
#1	Equal weight	**7**	4	2	**10**	5	1	**9**	6	**8**	3
#2	Efficiency	6	2	1	**10**	**7**	3	4	**8**	**9**	5
#3	Electricity gen. cost	**7**	2	5	**9**	1	4	**10**	**8**	3	6
#4	Capital cost	5	2	3	6	4	1	**9**	**7**	**10**	**8**
#5	GHG emissions	**8**	5	4	**9**	**7**	6	**10**	1	3	2
#6	E.P.B.T.	**7**	4	1	3	5	2	**9**	**8**	**10**	6
#7	Social acceptance	**8**	**10**	1	**9**	5	**7**	2	4	6	3
#8	Selected indicators	**9**	4	1	**10**	5	3	**8**	6	**7**	2
Total 1: $\sum_{i=2}^{7} Q(q;w)_{CS\#i}$		**41**	25	15	**46**	29	23	**44**	36	**41**	30
Total 2: $\sum_{i=1}^{8} Q(q;w)_{CS\#i}$		**57**	33	18	**66**	39	27	**61**	48	**56**	35

Table 3. Summary of global sustainability index variation.

Despite of the accordance between the ranking results on the energy carrier systems sustainability for electricity supply for the different case scenarios, it must be taken into account that all case scenarios were developed for a worldwide scale. For a particular national context, the results of the comparative assessment of energy supply options will be significantly different as some constrains would be considered. Examples of these constrains on a national level are described as follows:

- PV systems create the number of jobs depicted in Fig. 9 only in countries that do manufacture them, but not to the same extent in countries that import them;

- Social acceptance criteria is strongly locally dependent. For instance, Fig. 10 shows quite high acceptance for nuclear, but likewise other countries, it is known that in Germany the nuclear option is already abandoned. In the particular context of this country, the nuclear option must be neglected, i.e., the weighting factor must be zero;
- The possibility of using geothermal energy in large extend to supply electricity demand is also strongly locally dependent. In 1999, were identified 39 countries as having the potential to meet 100% of their electricity needs through domestic geothermal resources, although significant power production had been developed in only nine: Costa Rica, El Salvador, Guatemala, Iceland, Indonesia, Kenya, Nicaragua, Papua New Guinea, and the Philippines (Holm et al., 2010). Nowadays, geothermal plants around the world (in 24 countries) generated 0.33% (67.2 TWh) of the electricity demand (20 055 TWh). Based on this example, it must be pointed out that not all possible energy sources are uniformly distributed across all countries.

Many other constrains can be set for the indicators used to quantify sustainability. Although this can be seen as a limitation of the model, it is important to clarify that this comparative assessment of energy supply options can be developed in a national context, requiring precise values range of the indicators, null weighting values when a specific indicator can not be considered as well as the availability of the energy sources. Using these values, the energy carrier systems sustainability can be assessed in a national context, providing additional case scenarios to those that were developed along this chapter.

6. Summary

In a global sustainability context, now and near future electricity supply must be supported by an energy carrier systems mix that provide affordable, abundant, and reliable electricity while minimizing impacts on the environment. This chapter provides a road-map to develop sustainability analysis and the specific results for the energy carrier systems' context.

A wide range of indicators is used to characterize the technological, economical, environmental and social dimensions of current energy carrier conversion systems into electricity. Minimum and maximum values of selected indicators were collected from specialized and specific literature for each energy conversion system. Firstly, the same weight is given to all indicators in order to evaluate a global sustainability index from an equality point of view. Then, indicators are used separately to assess sustainability. For this evaluation, the weighting factor of a selected indicator is higher than the others. Finally, the weighting factors of indicators assumed as more relevant, are higher compared to the weighting factors of the remainder indicators.

A hierarchy ranking is outlined from the results of this multi-criteria analysis. Hydro, nuclear, wind and natural gas-fired power plants mix stand out for a sustainable future for the electricity supply. Notice that social acceptance of nuclear technology was based on data collected prior to the disaster in Fukushima power station. Nowadays, the social acceptance of this technology is probably lesser, affecting its overall level of sustainability.

In the opposite side, geothermal and ocean energy conversion systems are and will continue to be the less sustainable. This condition arises from the specific needs for the location of geothermal power plants as well as from the low values for each indicator when comparing with the others energy conversion systems. Ocean energy conversion system also includes this group mainly due to its technological infant stage. This energy conversion system still requires a lot of research and development to be competitive.

An update on variation range of different sustainability indicators is provided. The implementation of a particular system type over another change continuously, due to usual technology improvements. These improvements increase the energy conversion efficiency and reduce the greenhouse gases emissions, as well as the installation and operation costs. Additionally, these improvements can lead to changes in society mentality. However, is must be taken into account that all test scenarios were developed on a worldwide basis. In a particular national context, some constrains must be evaluated for each indicator. Additionally, the values range of each indicator must be determined locally.

This work aims to contribute on the debate on current and future electricity supply from energy carrier systems, taking into account that we will need to continue to use fossil fuel to supply the worldwide increasing demand on electricity.

7. References

ABB (2011). *Trends in global energy efficiency 2011.* Enerdata and Economist Intelligence Unit, The ABB Group.

Afgan, N. H. & Carvalho, M. G. (2002). Multi-criteria assessment of new and renewable energy power plants. *Energy,* Vol.27, No.8, pp. 739-755.

Afgan, N.H. & Carvalho, M.G. (2004). Sustainability assessment of hydrogen energy systems, *International Journal of Hydrogen Energy,* Vol.29, No.13, pp. 1327-1342.

Allan, G. et al. (2011). Levelised costs of Wave and Tidal energy in the UK: Cost competitiveness and the importance of "banded" Renewables Obligation Certificates, *Energy Policy,* Vol.39, pp. 23-39.

Banerjee, S., Duckers, L. J., Blanchard, R. & Choudhury, B. K. (2006). *Life Cycle Analysis of Selected Solar and Wave Energy Systems,* Advances in Energy Research, 2006.

Beardsley, E (2006). *France Presses Ahead with Nuclear Power,* NPR.

Biswas, P. (2009). Water Use and Re-Use in Energy Technologies in a Carbon Constrained World. *Proceedings of First National Expert and Stakeholder Workshop on Water Infrastructure Sustainability and Adaptation to Climate Change,* Arlington, Virginia, USA, January 6-7.

Brito, A. & Huckerby, J. (2010). *Implementing Agreement on Ocean Energy Systems - Annual Report,* International Energy Agency (IEA), Paris, France.

Cruz, J. & Sarmento A. (2004). *Energia das ondas: Introdução aos aspectos tecnológicos, económicos e ambientais,* Instituto do Ambiente. (in portuguese)

Dalton, G. J. & Lewis, T. (2011). Metrics for measuring job creation by renewable energy technologies, using Ireland as a case study, *Renewable and Sustainable Energy Reviews,* Vol.15, pp. 2123-2133.

Dalton, G. J. et al. (2010). Case study feasibility analysis of the Pelamis wave energy convertor in Ireland, Portugal and North America, *Renewable Energy,* Vol.35, pp. 443-455.

Denny, E. (2009). The economics of tidal energy, *Energy Policy,* Vol.37, No.5, pp. 1914-1924.

Dones, R. & Heck, T. (2006). LCA-Based evaluation of ecological impacts and external costs of current and new electricity and heating systems. *Proceedings of Materials Research Society Symposium*, Vol.895, pp. 37-50.

Dunnett, D. & Wallace, J. S. (2009). Electricity generation from wave power in Canada, *Renewable Energy*, Vol.34, pp. 179-195.

EFL (2011). *How Transitioning to Renewable Energy Can Create Jobs in the Dominican Republic.* Energy for Life (EFL), Milan, Italy.

EIA (2010). *Summary status for the US*, U.S. Energy Information Administration (EIA).

EIA (2010). *Updated Capital Cost Estimates for Electricity Generation Plants*, U.S. Energy Information Administration (EIA), U.S. Department of Energy, Washington, USA.

Erdogmus, B. et al. (2006). Economic assessment of geothermal district heating systems: A case study of Balcova-Narlidere, Turkey. *Energy and Buildings*, Vol.38, No.9, pp. 1053-1059.

ESMAP (2007). *Technical and economic assessment of off-grid, mini-grid and grid electrification technologies.* Energy Sector Management Assistance Program (ESMAP), Technical Paper 121/07, The International Bank for Reconstruction and Development (The World Bank), Washington, DC, USA.

Evans, A. et al. (2009). Assessment of sustainability indicators for renewable energy technologies. *Renewable and Sustainable Energy Reviews*, Vol.13, No.5, pp. 1082-1088.

Evans, A., Strezov, V. & Evans, T. (2010). Comparing the sustainability parameters of renewable, nuclear and fossil fuel electricity generation technologies. *Proceedings of 21st World Energy Congress - Montreal 2010*, Montreal, Canada, September 11 - 16.

Evrendilek, F. & Ertekin, C. (2003). Assessing the potential of renewable energy sources in Turkey. *Renewable Energy*, Vol.28, No.15, pp. 2303-2315.

EWEA (2009). *Wind Energy – The Facts: Chapter 6: Social acceptance of wind energy and wind farms.* European Wind Energy Association (EWEA), Earthscan Ltd., March.

Farret, F.A. & Simões, M. G. (2006). *Integration of alternative sources of energy*, Wiley-IEEE Press, 2006.

Frantzis, L. (2010). Job Creation Opportunities in Hydropower. National Hydropower Association Annual Conference.

Gagnon, L. et al. (2002). Life-cycle assessment of electricity generation options: The status of research in year 2001. *Energy Policy*, Vol.30, pp. 1267-1278.

Gallup, D. L. (2009). Production engineering in geothermal technology: A review, *Geothermics*, Vol.38, No.3, pp. 326-334.

Graus, W.H.J., Voogt, M. & Worrell, E. (2007). International comparison of energy efficiency of fossil power generation. *Energy Policy*, Vol.35, pp. 3936–3951.

Hanjalic, K., van de Krol, R. & Lekic, A. (2008). *Sustainable Energy Technologies, Options and Prospects*, Springer, 2008.

Holm, A., Blodgett, L. Jennejohn, D. & Gawell, K. (2010). *Geothermal Energy: International Market Update*, Geothermal Energy Association, May 2010.

IEA (2011a). *2011 Key World Energy Statistics*, Organisation for Economic Co-operation and Development (OECD)/International Energy Agency (IEA), Paris, France.

IEA (2011b). *World Energy Outlook 2011*, Organisation for Economic Co-operation and Development (OECD)/International Energy Agency (IEA), ISBN , Paris, France.

IPCC (2007). *IPCC Fourth Assessment Report: Climate Change 2007.* Working Group III - Intergovernmental Panel on Climate Change (IPCC), B. Metz, O.R. Davidson, P.R. Bosch, R. Dave, L.A. Meyer (eds), Cambridge University Press, Cambridge, United Kingdom and New York, NY, USA.

Kaldellis, J. K. (2008). Critical evaluation of the hydropower applications in Greece, *Renewable and Sustainable Energy Reviews*, Vol.12, No.1, pp. 218-234.

Kannan, R. et al. (2006). Life cycle assessment study of solar PV systems: An example of a 2.7 kWp distributed solar PV system in Singapore. *Solar Energy*, Vol.80, pp. 555-563.

Lako, P. (2010). *Marine Energy*, International Energy Agency (IEA), Technology Brief E13 November 2010.

Lee, T. J. et al. (2007). Strategic environments for nuclear energy innovation in the next half century, *Progress in Nuclear Energy*, Vol.49, pp. 397-408.

Mansure, A. J. & Blankenship, D. A. (2010). Energy Return in Energy Investment, An Important Figure-Of-Merit For Assessing Energy Alternatives. *Proceedings of 25th Workshop on Geothermal Reservoir Engineering*, Standford University, California, USA, February 1-3, 2010.

Mendes, R. P., Gaspar, P. D. & Gonçalves, L. C. (2011b). Sustainability quantification of renewable energy conversion systems for electricity supply. *Proceedings of the International Conference on Engineering ICEUBI 2011*, University of Beira Interior, Covilhã, Portugal, November 28-30.

Mendes, R. P., Gonçalves, L. C. & Gaspar, P. D. (2011a). Comparação da sustentabilidade das tecnologias de conversão de energia a partir de fontes renováveis por indicadores tecnológicos, económicos, ambientais e sociais. *Proceedings of X Congresso Ibero - Americano em Engenharia Mecânica*, Porto, Portugal, September 4-7.

NEI (2011a). *Protecting the Environment.* Nuclear Energy Institute.

NEI (2011b). *New Nuclear Plants: An Engine for Job Creation, Economic Growth.* Nuclear Energy Institute (NEI).

NRC (2011). *Generic Environmental Impact Statement for License Renewal of Nuclear Plants.* U.S. Nuclear Regulatory Commission (NRC), Staff report NUREG-1437.

Parker, R. et al. (2007). Energy and carbon audit of an offshore wave energy converter, *Proceeding of the Institution of Mechanical Engineers, Part A: Journal of Power and Energy*, Vol. 221, pp. 1119-1130, 2007.

Peters, N. (2010). *Promoting Solar Jobs: A policy framework for creating solar jobs in New Jersey.* New Jersey.

Plowman, C. (2004). *U.S. Job Creation Due to Nuclear Power Resurgence in United States.* U.S. Department of Energy.

PREGA (2006). *Fuel switching from oil to gas for power generation.* Promotion of Renewable Energy, Energy Efficiency and Greenhouse Gas Abatement (PREGA).

Randolph, J. & Masters, G. M. (2008). *Energy for Sustainability: Technology, Planning, Policy,* Island Press, 2008.

Raugei, M. & Frankl, P. (2009). Life cycle impacts and costs of photovoltaic systems: Current state of the art and future outlooks. *Energy*, Vol.34, No.3, pp. 392-399.

REN21 (2011). *Renewable 2011 - Global Status Report*, Renewable Energy Policy Network for the 21st Century, Paris, France.

Rinebold, J. M., Drajer, T. & Brzozowski, A. (2009). *A National "Green Energy" Economic Stimulus Plan Based in Investment in the Hydrogen and Fuel Cell Industry*. The Connecticut Center for Advance Technology, Inc.

Roth, S. et al. (2009). Sustainability of electricity supply technology portfolio. *Annals of Nuclear Energy*, Vol.36, No.3, pp. 409-416.

Rourke, F. O. et al. (2010). Tidal energy update 2009. *Applied Energy*, Vol.87, No.2, pp. 398-409.

Schneider, M., Froggatt, A. & Thomas, S. (2011). *The world nuclear industry status report 2010–2011: Nuclear Power in a Post-Fukushima World, 25 Years After the Chernobyl Accident*, Worldwatch Institute, Washington, D.C., U.S.A.

Sherwani, A. F. et al. (2010). Life cycle assessment of solar PV based electricity generation systems: A review. *Renewable and Sustainable Energy Reviews*, Vol.14, pp. 540-544.

Soerensen, H. C., Naef, S., Anderberg, S. & Hauschild, M. Z. (2007). Life Cycle Assessment of the Wave Energy Converter: Wave Dragon, *Proceedings of International Conference on Ocean Energy*, Bremerhaven, 2007.

Ungemach, P. (2007). Conclusions of the Workshop: Increasing policy makers awareness and public acceptance. *Proceedings of ENGINE Workshop 6 "Increasing policy makers 'awareness and the public acceptance", ENhanced Geothermal Innovative Network for Europe (ENGINE)*, Athens, Greece, September.

van Bree, B. & Bunzeck, I. (2010). WP6*: D6.1: Social acceptance of hydrogen demonstration projects*. NextHyLights Industry and Institute partners, Supporting Action of the Fuel Cell and Hydrogen Joint Undertaking (FCH JU), New Energy World JTI framework, European Commission, August.

Varun et al. (2009). Energy, economics and environmental impacts of renewable energy systems, *Renewable and Sustainable Energy Reviews*, Vol.13, No.9, pp. 2716-2721.

Varun, Bhat, I.K. & Prakash, R. (2009). LCA of renewable energy for electricity generation systems--A review. *Renewable and Sustainable Energy Reviews*, Vol.13, No.5, pp. 1067-1073.

Vob, A. (2006). A Comparative Assessment of Electricity Generation Options from an ecological and sustainability viewpoint," *Proceedings of the International Conference on Nuclear Power Plants for Poland* (NPPP), Warsaw, Poland, June 1-2, 2006.

Voss, A. (2001). LCA and External Costs in Comparative Assessment of Electricity Chains. Decision Support for Sustainable Electricity Provision? *Proceedings of IEA Conference - Energy Policy and Externalties: The Life Cycle Analysis Approach*, November 15-16.

Wackernagel, M. & Monfreda, C. (2004). Ecological Footprints and Energy. *Encyclopedia of Energy*, Vol.2., Elsevier Inc.

Wartmann, S., Jaworski, P., Klaus, S. & Beyer, S. (2009). *Scenarios on the introduction of co2 emission performance standards for the eu power sector*. Report commissioned by the European Climate Foundation. EcoFys Germany GmbH, Niederlassung Nürnberg, Germany.

WavEC (2004). *Potential and Strategy for the Development of Wave Energy in Portugal*, Wave Energy Centre (WavEC).

WNA (2010). *Another drop in nuclear generation*, World Nuclear Association (WNA) News.

WNA (2011). *Energy Analysis of Power Systems*, World Nuclear Association (WNA).

WNA (2011). *The Economics of Nuclear Power,* World Nuclear Association, url: http://www.world-nuclear.org/info/inf02.html acessed: [28-04-2011].

WWEA (2011). *World Wind Energy Report 2010,* World Wind Energy Association (WWEA), Bonn, Germany.

12

The Need for Efficient Power Generation

Richard Vesel and Robert Martinez
ABB Inc.
USA

1. Introduction

This chapter makes the business case for energy efficient plant auxiliary systems and discusses some trends in electricity markets and power generation technologies. The information in these colored sections is specific to power generation industries and/or process plants with large on-site power and/or steam heat generation.

2. Trends in power demand and supply

Currently growing 2.6 percent per year, world electricity demand is projected to double by 2030. The share of coal-fired generation in total generation will likely increase from 40 percent in 2006 to 44 percent in 2030. The share of coal in the global energy consumption mix is shown in the figure below. This share is now increasing because of relatively high natural gas prices and strong electricity demand in Asia, where coal is abundant. Coal has been the least expensive fossil fuel on an energy-per-Btu basis since 1976.

China expanded coal use by 11 percent in 2005 and surpassed the U.S. as the number one coal user in 2009. Coal is the most abundant fossil fuel, with proven global reserves at the end of 2005 of 909 billion metric tons, equivalent to 164 years of production at current rates (International Energy Agency, 2006).

In the U.S., coal-fired plants currently provide 45%, down from 51% just a few years ago, of total generating capacity (Woodruff, 2005), or about 400 GW, from about 600 power plants. Total electrical generation capacity additions are estimated to be 750 GW by 2030 (International Energy Agency, 2006). Of that new capacity, 156 GW is projected to be provided by coal plants (Ferrer, Green Strategies for Aging Coal Plants: Alternatives, Risks & Benefits, 2008). Other estimates put capacity addition to 2030 at 280 coal-fired 500MW plants (Takahashi, 2007).

In North America, declining natural gas prices are again creating a trend toward more energy efficient and lower emission plant designs, a trend now expected to continue at least thru 2020. The generating costs of combined-cycle gas turbine (CCGT) plants, which use natural gas, are expected to be between 5–7 cents per kWh, while coal-fired plants are in the range 4–6 cents/kWh (International EnergyAgency, 2006). Integrated gasification combined cycle (IGCC) plants are not yet competitive as of 2008 (which is why government is subsidizing many such projects). Their low relative costs make coal-fired plants competitive in the U.S. with other large central generating plants.

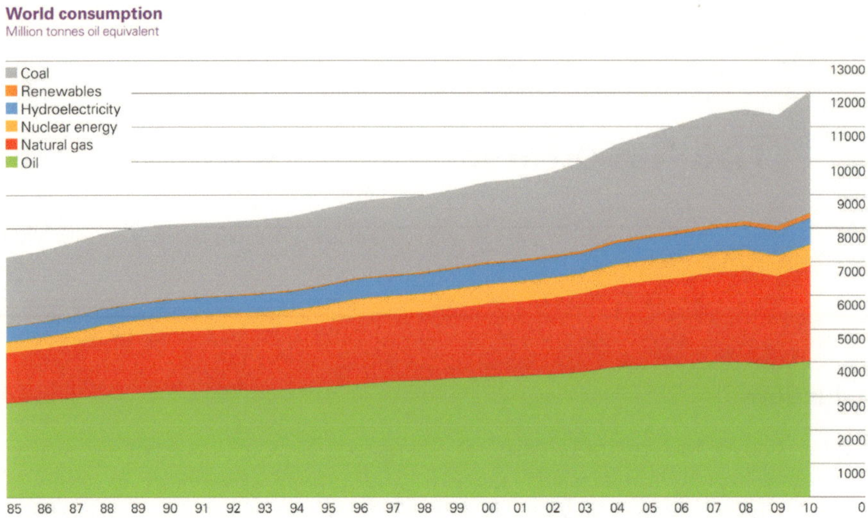

World primary energy consumption grew by 5.6% in 2010, the strongest growth since 1973. Growth was above average for oil, natural gas, coal, nuclear, hydroelectricity, as well as for renewables in power generation. Oil remains the dominant fuel (33.6% of the global total) but has lost share for 11 consecutive years. The share of coal in total energy consumption continues to rise, and the share of natural gas was the highest on record.

Fig. 1. Trends in energy consumption, (2011 BP Statistical Review of World Energy).

Many new coal plants were being planned or constructed as of 2008, but with some uncertainty regarding the future trend due to carbon footprint and other environmental concerns over current coal-fired plant technology. Regulations imposing carbon dioxide emissions charges will eventually change the economics in favor of CCGT and other more efficient fossil plant types. Even without emissions taxes, the licensing of new plants is threatened by growing grass-roots opposition at local and state levels. According to the US Department. of Energy (DoE), 59 of 151 planned new coal plants were either refused licenses or abandoned in 2007, and 50 plants are being challenged in court. Environmental groups have successfully challenged these new plants by arguing that the additional capacity could be gained through energy efficiency and renewable sources of power. With the industry facing a possible moratorium on new plants, it is more important than ever to make existing plants as energy efficient as possible.

Whether limited by emissions or supplies, the fossil-fuel power generation industry must sooner or later reduce the carbon per unit energy produced. The prominence of coal means that it will play an important role in the transition to a low-carbon future. Dr. Amory Lovins, a leading US energy analyst, anticipated the need for such a transition many years ago when he said; "It is above all the sophisticated use of coal, chiefly at modest scale, that needs development. Technical measures to permit the highly efficient use of this widely available fuel would be the most valuable transitional technologies." (A. Lovins, Energy Strategy: The Road Not Taken 1976)

3. Trends in steam plant designs and efficiency

Large fossil-fuel-fired steam plants use a closed steam cycle in which water is converted to steam in a boiler. This steam is then superheated and then expanded through the blades of a turbine whose shaft rotates an electrical generator. The steam exits the turbine and condenses to water, which is pumped back up to boiler pressure.

3.1 Sub-critical plant types

The most common type of plant using this design is alternatively referred to as 'drum boiler' or 'subcritical,' because water is circulated within the boiler between a vessel (the drum) and the furnace water-wall tubing where it absorbs combustion heat, but does not exceed critical pressure. Existing subcritical pulverized coal (PC) boiler steam power plants can theoretically achieve up to 36–40 percent efficiency at full load. Due to major process design changes such as supercritical boilers and other technology improvements, the average efficiencies of the newest coal-fired plants are up to 46 percent compared to 42 percent for new plants in the 1990s (IEA CoalOnline, 2008).

Energy efficiency improvements of several percentage points in new plants have resulted from improved designs of the main components and auxiliaries in steam power plants: including auxiliary drivepower:

- Improvements in turbine blade design
- Improvements in fans and flue gas treatment methods
- Reduction of furnace exit gas temperature
- Increase of feed water temperature
- Reduction of condensing pressure
- Use of double reheat on main steam flow
- Optimization and reduction of the consumption of auxiliary drivepower

3.2 Super-critical coal-fired steam plants

Supercritical plants, also called 'once-through' plants because boiler water does not circulate multiple times as it does in drum-boiler designs, have efficiencies in the mid-40 percent range. New 'ultra critical' designs using pressures of 4,400 psi (30 MPa) and dual stage reheat are capable of reaching about 48 percent efficiency (IEA Coal Online - 2, 2007). Plant availability problems with the first generation of large supercritical boilers led to the conclusion that pulverized coal-fired electricity generation was a mature technology, with an efficiency limited by practical and economic considerations to around 40 percent. However, improvements in construction materials and in computerized control systems led to new designs for supercritical boilers that have overcome the problems of the earlier plants (IEA Coal Online - 2, 2007). Although most new coal-fired plants are expected to use drum steam boilers, the share of supercritical technology is rising gradually (International Energy Agency, 2006).

3.3 Combined-Cycle Gas Turbine (CCGT)

A combined-cycle gas turbine (CCGT) power plant uses a gas turbine in conjunction with a heat recovery steam generator (HRSG). It is referred to as a combined-cycle power plant because it combines the Brayton cycle of the gas turbine with the Rankine cycle of the HRSG. The thermal efficiency of these plants has reached a record heat rate of 5690 Btu/kWh, or just under 60 percent.

3.4 Some steam plants are lagging

At the beginning of the 21st century, it was believed that a single-cycle coal-fired power station with an efficiency of more than 50 percent would be possible by 2015 (Kjær and

Boisen, 1996 in IEA Coal Online - 2, 2007). The efficiency of some new design plants may be high, but almost 75 percent of the existing coal-based fleet of plants in the U.S. is over 35 years old, with an average net plant efficiency of only slightly above 30 percent (Ferrer, 'Green Strategies for Aging Coal Plants,' 2008).

In addition to the less efficient design of core equipment, these older plants suffer an additional efficiency handicap due to plant aging; they become less reliable and generally less efficient due to leakage, fouling, and other mechanical factors. Another trend which lowers efficiency is the change in fuel supply systems toward off-design coals for which the boiler has not been optimized (IEA Coal Online - 2, 2007). Fuel supplies may be subject to further tweaking as generating companies seek to reduce their carbon footprint by substituting a portion of the coal they use with biomass.

Another important reason that older plants are lagging in efficiency is that many of them are operating at 30–50 percent below their rated capacities, where efficiencies of all sub-systems are lower. The realities of a more deregulated and competitive marketplace, with renewable and distributed energy sources and new system operating reserve requirements, have led to previously baseloaded plants being operated as dispatchable plants; an unforeseen operating regime (ABB Power Systems, 2008). One view of this latter issue is the global distribution of load factor of nominally baseloaded steam turbine plants less than 500MW for the period 2001–2005. The following figure shows that the median load factor is only 64 percent.

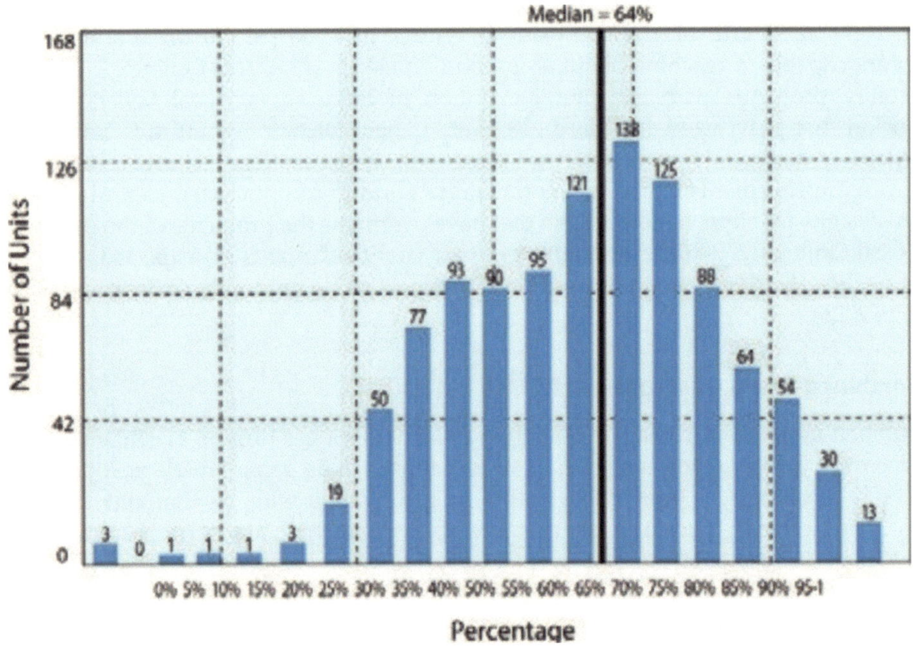

Fig. 2. Distribution of load factor of base-loaded plants, (World Energy Council, 2007).

3.5 Plant auxiliary power usage is on the rise

The share of total plant auxiliary electrical power in the fleet of fossil-fuel steam plants has been increasing due to these main factors:

- Addition of anti-pollution devices such as precipitators and sulfur dioxide scrubbers which restrict stack flow and require in-plant electric drive power. About 40 percent of the cost of building a new coal plant is spent on pollution controls, and they use up about 5 percent of gross power generated (Masters, 2004).
- Additional cooling water pumping demands to satisfy environmental thermal discharge rules.
- A trend away from mechanical (e.g. condensing steam turbine) drives toward electrical motors as the prime mover for in-plant auxiliary pump and fan drives.

For PC power plants, the auxiliary power requirements are now in the range of 7–15 percent of a generating unit's gross power output for PC plants. Older PC plants with mechanical drives and fewer anti-pollution devices had auxiliary power requirements of only 5 to 10 percent (GE Electric Utility Engineering, 1983). These figures are for traditional drum boiler type plants, but the auxiliary power requirements of supercritical boilers are not any lower. The feedwater pump power required to reach the much higher boiler pressure is approximately 50 percent greater than in drum boiler designs. Increased demand for auxiliary power increases a plant's net heat rate and reduces the amount of salable power.

4. Plant auxiliary energy efficiency improvements

In-plant electrical power, when taken from the generator bus, may be priced artificially low in some utility companies' auxiliary lifecycle calculations. A process industry customer, however, must always pay high commercial rates (and sometimes penalties), thus providing a strong incentive to improve their auxiliary energy efficiency. Price dis-incentives, regulations permitting cost-pass thru, and other non- technical barriers are discussed in the handbook section on Barriers to Increased Energy Efficiency.

These barriers may result in sub-optimal energy designs for power plant auxiliaries, most commonly in oversized motors, fans and pumps. These design decisions have particularly negative consequences when the base-loaded plant then moves to a new operating mode at 50–70 percent capacity (see previous section for a discussion of this trend). Auxiliaries such as pumps and fans that use constant speed motors and some form of flow restriction for control will waste much more power when operating under such partial-load conditions. Other plant systems will also run less effectively below their design points. Boilers at partial loads, for example, run with relatively higher excess air to achieve complete combustion, which lowers efficiency; these topics are discussed in greater detail in the handbook sections on Drivepower and Automation.

5. The potential for energy efficiency

5.1 Technical efficiency improvement potential

A recent study by the International Energy Agency (IEA) suggests a technical efficiency improvement potential of 18–26 percent for the manufacturing industry worldwide if the best available (proven) technologies were applied. Most of the underlying energy-saving measures would be cost-effective in the long term. Another study, by the U.S. Dept. of Energy, focused on the energy efficiency opportunity provided by automation and electric power systems in process industries. An improvement potential of 10–25 percent was suggested by industry experts, who were asked to consider improvements within the context of operational or retrofit situations. The results of that study are shown in the figure below.

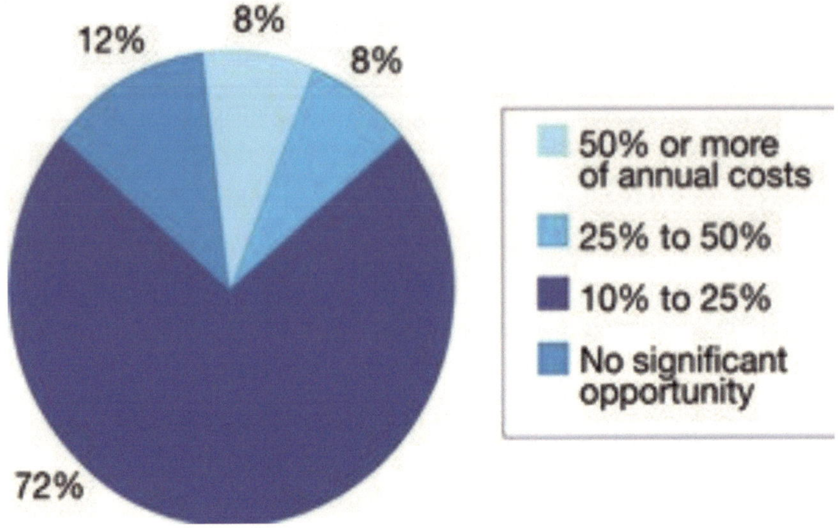

Fig. 3. Process industry survey results on potential of energy efficiency, (US DoE, 2004).

5.2 Potential revealed through performance benchmarking

Access to power generation plant performance data is important for identifying areas for improvement and for showing the results of best practice. Market fragmentation and the increased competitiveness of de-regulated markets in the past have made access to data difficult. There has also been a lack of standards or practices for measuring performance.

The World Energy Council (WEC), through its Performance of Generating Plant (PGP) Committee, is now gathering and normalizing such data so that valid comparisons can be made across countries and markets.

Similar performance benchmarking efforts are done in the U.S., but through industry-funded organizations like EPRI. Standardization efforts are best represented by IEEE Std 762-2006 IEEE Standard for Definitions for Use in Reporting Electric Generating Unit Reliability, Availability, and Productivity.

Interestingly, the WEC found that 'new drivers geared toward profitability, cost control, environmental stewardship, and market economics are shifting the focus away from traditional measures of technical excellence such as availability, reliability, forced outage rate, and heat rate' (World Energy Council, 2007). Their PGP database has added individual unit design and performance indices that can be used to compare efficiency and reliability across designs. The published performance data will help industry improve practices, and will put a spotlight on under-performing plants and companies.

5.3 Efficiency potential revealed by country comparisons

The potential for energy efficiency, at least from a U.S. perspective, is also indicated in a recent (2007) comparison of fossil-fuel-based power generation efficiencies between nations that together generate 65 percent of worldwide fossil-fuel-based power. The Nordic countries, Japan, the United Kingdom, and Ireland were found to perform best in terms of fossil-fuel-based generating efficiency and were, respectively, 8 percent, 8 percent and 7 percent above average in 2003. The United States is 2 percent below average. Australia, China, and India perform 7 percent, 9 percent and 13 percent, respectively, below average. The energy savings potential and carbon dioxide emissions reduction potential if all countries produce electricity at the highest efficiencies observed (42 percent for coal, 52 percent for natural gas and 45 percent for oil-fired power generation), corresponds to potential reductions of 10 exajoules of consumed thermal energy and 860 million metric tons of carbon dioxide, respectively (Graus, 2007).

The IEA analysis mentions that more than half of the estimated energy and carbon dioxide savings potential is in whole-system approaches that often extend beyond the process level (Gielen, 2008). 'Integrative Design' is this handbook's approach to the most challenging energy efficiency issues in plant auxiliary design.

6. Energy efficiency is attracting interest and investment

The previous sections showed an engineer's view of the importance of energy efficiency. What are the views and plans of corporate energy decision makers and investors?

6.1 From corporate energy managers

According to a recent survey on energy efficiency of corporate and plant-level energy managers at more than 1,100 North American companies (Johnson Controls, 2008):

- 57 percent expect to make energy-efficiency improvements during the same time period, devoting an average of 8 percent of capital expenditure budgets on energy-efficiency projects.
- 64 percent anticipate using funds from operating budgets, allocating 6 percent to energy-efficiency improvements.
- 40 percent have replaced inefficient equipment before the end of its useful life in the past year.
- 70 percent have invested in educating staff and other facility users as a way to increase support for increasing internal energy efficiency.

6.2 From industry investors

When 18 U.S. investment organizations were surveyed about energy efficiency, the results indicated that the technologists should have no trouble funding their projects. According to that study (Martin, 2004), the energy technology attracting the greatest investment interest is energy intelligence (smart instruments, advanced control, and automation). The handbook sections on Instruments, Controls & Automation discuss these technologies and how they can be used to improve plant energy efficiency.

6.3 Carbon dioxide emissions must be reduced

According to a 2005 report from the World Wide Fund for Nature (WWF), coal-based power stations are at the top of the list of least 'carbon efficient' power stations in terms of the level of carbon dioxide produced per unit of electricity generated. Based on current developments in Europe and in the U.S., regulations which limit or tax carbon dioxide emissions seem inevitable for all Western economies. A carbon charge of $25 per metric tonne (carbon dioxide) is a conservative estimate used in IEA scenarios. The impact of carbon pricing on fossil-fuel plant generating costs, shown in the figure below, is dramatic compared to most other generation methods. At prices above $20 per metric tonne coal-based plants become the most expensive type to operate at current non-optimized cost levels.

China and India account for four-fifths of the incremental demand for coal, mainly for power generation. For the first time, China's carbon dioxide power emissions in 2008 exceeded the United States' emissions; the lower quality coal used in India and other rapidly expanding economies, decreases plant efficiency and leads to increased carbon dioxide emissions per unit electricity (International Energy Agency, 2006).

6.4 Energy efficiency is key to CO_2 mitigation

The IEA Energy Technology Perspectives model is a bottom-up, least-cost optimization program. The model was developed to describe the global potential for energy efficiency and carbon dioxide emissions reduction in the period to 2050, particularly in the industrial sector. In the 'accelerated technology scenario' (ACT), the potentials for carbon dioxide reduction on all power consumption are shown in the figure below. This figure illustrates the scenario in which carbon dioxide emissions are stabilized globally in 2050 to 2005 levels, and the world narrowly avoids a costly climate crisis.

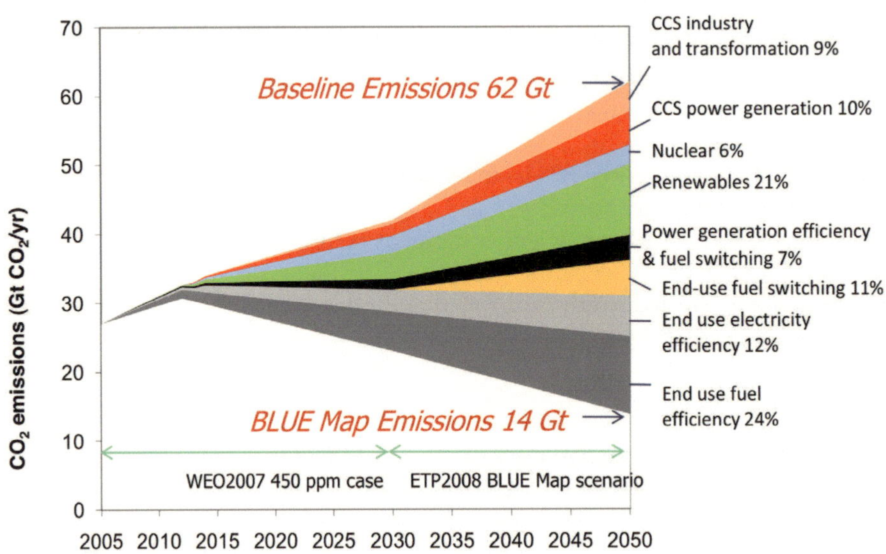

Fig. 4. Relative share of CO2 mitigation efforts, all consumption, (International Energy Agency, 2006).

7. The role of power generation in reducing emissions

The IEA's ACT scenario suggests that power generation efficiency can contribute significantly to the overall global effort to stabilize carbon dioxide emissions by 2050 at or near 2005 levels. Surprisingly, the model shows that power generation efficiency alone, which includes improved auxiliaries and other measures, has a larger climate impact than even nuclear power.

When the model is applied to process industries alone, the impact of energy efficiency is proportionately larger. The figure below shows the 'blue' scenario, which uses the same ACT scenario describe above, but with a higher carbon dioxide charge of $50 per (metric) tonne, instead of $25/tonne (Taylor, 2008).

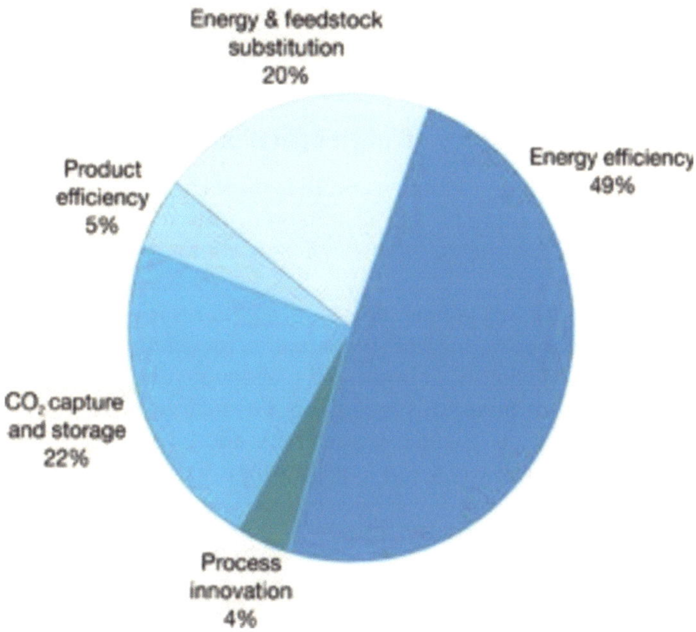

Fig. 5. Relative share of CO2 mitigation efforts in process industries, (Taylor, 2008).

Applying this model to the power generation sector in particular suggests that its carbon dioxide emissions are cut by 36 percent using all of the approaches shown. Half of those savings (18% of total) can be attributed to relatively low- technology energy efficiency measures alone.

Energy efficiency measures are the most important of all the carbon dioxide mitigation approaches for process industries, contributing to almost half of the impact on emissions (Martin, 2004). Although these predictions apply to process industries, the relative potentials are likely to be valid for the steam power generation sub-sector as well.

8. Multiple benefits of energy efficiency

The primary benefits of a increased plant energy efficiency are reduced emissions and energy or fuel costs.

Power plants which operate partially or wholly at full load will have more salable power. At less than capacity, the fuel savings are significant. In coal-fired steam power plants, fuel costs are 60-70% of operating costs.

The following is a more complete list of benefits accompanying energy efficiency design improvements for plant auxiliaries:

8.1 Operational benefits

- Improved reliability/availability. As has been found with stricter safety design regulations, any extra attention to the process is rewarded with improved uptime.
- Improved controllability: energy is wasted in a swinging, unstable process, partly through inertia in the swings, but mainly because operators in such situations do not dare operate closer to the plant's optimum constraints.
- Reduced noise and vibration, reduced maintenance costs.

8.1.1 Results of improved efficiency on plant operations and profitability

- Better allocation: under deregulation, as utilities dispatch plants within a fleet, heat rate improvement can earn plants a better position on the dispatch list (Larsen, 2007).
- Avoiding a plant de-rating due to efficiency losses after anti-pollution retrofits or other plant design changes.
- Improved fuel flexibility—by efficiently using a wider variety of fuels (coal varieties) and, in some cases, increasing the firing of biomass, for example.
- Improved operational flexibility 1) Improved plant-wide integration between units will reduce startup-shutdown times; this benefit applies mainly to de- regulated markets. 2) The heat rate versus capacity curve is made flatter and lower, which allows the plant to operate more efficiently across a wider loading range.

8.2 Plant investment benefits

- Avoiding forced retirement due to pollution non-compliance: An ambitious retrofit programme may save some older plants from early retirement due to non- compliance with regulations.
- Tax credits take advantage of newer policies such as EPACT 2005, which may provide tax credits for efficiency efforts. Similar policies are in effect in the EU and China.
- Mainstream industry authority Engineering News-Record's influential Top Lists rankings now include "Top Green Design Firms" and "Top Green Contractors":
- 'The market for sustainable design has passed the tipping point and is rapidly becoming mainstream' (http://enr.construction.com/).
- Increasingly, shareholders and capital markets are rewarding companies who treat their environmental mitigation costs as investments (Russel, 2005).

Retrofitting may save some older plants from early retirement due to non- compliance with regulations such as the EU's Large Combustion Plant Directive on pollution (nitrous oxides, sulfur dioxide, mercury, and particulates) (International Energy Agency, 2006). In the US, increased compliance may smooth permitting of new units or plants.

All of the 'dirty dozen' in Carbon Monitoring For Action's (CARMA) list of top carbon dioxide emitting sources in the U.S. are coal-fired power plants, emitting an average of about 20 million tonnes of carbon dioxide per year per plant. 'Blacklists' like these, which include rankings by company as well, are increasingly being consulted by large institutional investors and sovereign wealth funds. With tightened credit markets, there is therefore an even greater incentive for top management to watch carbon dioxide emissions. See the section on Benchmarking for other global efforts toward increased transparency.

Non-Technical Barriers to Energy Efficient Design Despite all of the benefits and incentives, and the low-capital-cost improvement potential described in previous sections, the implementation of integrative, energy efficient design and operation is still hindered by several obstacles. Methods for improved design are known and the required technologies are widely available 'off the shelf.' Individual components are generally available in high-efficiency variants. So why are power and industrial process plants energy inefficient in their design as a whole? One clue, is the fragmentation found in engineering disciplines, vendor equipment packages, and even in the way projects are executed.

The current situation with energy efficiency is analogous to the status of safety in process industries a decade or two ago. Operational safety was acknowledged as important and was codified, but there were no standards on how safety could be managed during the design process on how it could be 'designed-in' from the start. The recent Functional Safety standards IEC-61508 and 61511 point the way forward for energy design and management standards evolution.

Many of the barriers listed below are managerial or procedural rather than technical in nature. These important non-technical aspects are discussed elsewhere. The discussion here is generic for most large power and process facilities, but a specific industry will have additional competitive and regulatory pressures.

Local, State, National and International Regulatory Authorities Authorities provide the regulatory framework for the activities of all the other stakeholders. The efforts of authorities are closely linked with those of the standards organizations. These factors, however, may contribute to inefficient plant designs:

- Regulations often permit pass-thru of all fuel-related costs directly to the rate base. This financially discourages any economization efforts related to fuel consumption, i.e. efficiency.
- Lack of clarity, unity and commitment to emissions charging makes investors wary of long-term investments in energy efficiency and/or carbon dioxide emissions reduction.
- Deregulation and the ensuing volatility in fuel and energy prices may also discourage the long-term thinking necessary to make some efficiency and carbon dioxide emissions reduction schemes justifiable.

8.2.1 Shareholders & investors

The observations in this paragraph regarding shareholders and investors apply mainly to new construction or large-scale redevelopment projects. See the following paragraphs for barriers more applicable to facility owner/operators of older plants and retrofit project contexts. Shareholders and investors often influence project schedules, contract clauses, functional specs for new construction and major retrofits of plants. These factors, however, may contribute to plants that are ultimately energy inefficient:

- Project schedules are compressed; front-end design and concept studies are underfunded or curtailed.
- Scope of redevelopment projects is narrow because investors 'generally want to avoid changes to the long remaining lifespan of the standing capital stock' (International Energy Agency, 2006).
- Designs are 'frozen' early by a pre-established milestone date, even if important data may be missing.
- Cost analysis methods are too crude, or not coupled tightly enough to the conceptual process design, or have wrong initial assumptions regarding risk, return, and lifetime; calculations may ignore significant indirect costs and savings such as substitution costs, maintenance savings, and peak energy prices.
- Operational energy costs may be treated as a fixed cost and therefore receive much less attention than a variable cost.
- Low-bid, fixed-price contracting without strong, well-defined and enforceable energy performance guarantees, at the plant, unit and equipment levels.
- Purchasing managers seek multiple suppliers to reduce cost; this strategy leads to increased design and data fragmentation. Purchasing managers may still prefer individual vendors versus full-service/system integrators.
- Energy-expert consulting companies are usually the last to be hired, and therefore have much less influence over the conceptual design.
- Drawings are issued 'for construction' before even the first vendor drawing is seen, much less approved (Mansfield, 1993), leading to hasty, often energy- inefficient re-design at the interfaces.
- Capital scarcity might favor smaller plants with lower efficiency (Gielen, 2008).

8.2.2 Facility operators

Facility Operators craft the original specifications, validate the design during commissioning and acceptance trials, determine operational loading and maintenance of facility, and usually initiate and manage retrofit projects. These factors, however, may contribute to plants that are ultimately energy inefficient:

- Retrofit projects to improve energy efficiency are funded from operating budgets, not from larger capital expenditure budgets; payback expectations and discount rates are all generally much higher than in green-field projects.
- Managers focus on optimizing process productivity, in which energy is only one of several other cost functions and may not receive the consideration it deserves; many modern plant-wide optimization systems optimize for productivity, which only indirectly improves energy efficiency.
- There is a war for money between process improvement and energy efficiency camps in a typical plant; process improvement teams and their measures seem to 'get more respect.'
- An increasing number of plants are centralizing purchasing, which means less engineer involvement in purchasing decisions. 'Since purchasing centralization… we've seen companies shift away from using a lifecycle cost model, which seems very short-sighted to us. Some of the decisions customers have been making are committing them to a stream of ongoing expenses that could have been reduced.' (Control Engineering article 8/15/2005).

- Facility operators receiving a new/retrofitted plant/unit are under-pressure to begin operations as soon as possible so they are therefore less critical with respect to energy targets during acceptance tests.
- Facility operators do not or cannot operate plant at design capacities due to changes in market or other factors.
- Plant engineering and maintenance teams are losing experienced older staff; facility operators do not provide adequate training for staff on energy efficiency.
- Power plant/power house energy managers (superintendents or maintenance directors) lack the necessary communication and salesmanship skills to push through good energy efficiency proposals. (a post on J. Cahill's blog, 2007).
- Reluctance to admit non-optimal, energy inefficient, operation to upper management – the perception is that this reflects badly on plant management and their plant operations team.
- Fear of production disruptions from new equipment or new procedures to improve efficiency (International Energy Agency, 2006); doubts about safety, controllability or maintainability
- Expansion projects will simply duplicate an existing unit on the same site, repeating many of the same design mistakes, to reduce the up-front engineering hours; low-labor copy-and-paste projects may also overlook opportunities for rationalization & integration with the existing unit(s).

8.3 Design and engineering companies

Design and engineering companies determine design specs of facility, select components and execute the design. These factors, however, may contribute to energy inefficient plants:

- A tendency to oversize pumps, fans, and motors by one rating, and oversize them again after handoff to another discipline, and then again by project leaders:
- Bottom limits in standards already have a safety margin, but these limits are interpreted as a bare minimum (from fear of litigation) and an additional safety margin is added.
- Overload maximums received from process engineer are interpreted by mechanical and electrical teams as continuous minimums; fat margins are added in lieu of detailed loading study.
- Additional margins are then added for future, but unplanned, capacity increases.
- Engineers on auxiliary systems are inordinately fearful of undersizing and risk being singled out as the bottleneck that prevents operation at full design capacity of other, more expensive, hardware
- Large, commodity motors and fans are commercially available only in discrete sizes. After all the margins, an engineer will choose the next size up if the design point falls between two sizes.
- A tendency to aggressively reduce engineering hours to increase margins on fixed-priced contracts and to avoid selecting premium components for such contracts.
- Trade-offs between floor space and pipe/ductwork efficiency are not life-cycle cost-estimated; civil and architectural concerns are the default winners due to their early head start in most projects.
- Trade-offs between reliability and energy efficiency are not life-cycle cost- estimated. Higher energy costs are seen as insurance against large, but virtual opportunity costs.
- Lack of energy design criteria and efficiency assessment steps in the standard engineering workflow. There is typically a design optimization step for cost, safety, reliability, and other concerns, but not for energy efficiency.

- Shortage of engineers in key industries; junior and outsourced engineers are making higher-impact decisions.
- A reluctant to deviate from their 'standard design' templates, especially on expansion projects where the design has been delivered on previous units. This leads to short cuts and uncritical copying-and-pasting of older, non-optimal designs.
- Engineers work mainly within the confines of their discipline and do not see opportunities for inter-disciplinary optimization of the total design. For example:
- Mechanical engineers miss out on optimizations from chemical engineering to use waste heat and to optimize plant thermodynamics or create useful by-products.
- Process engineers do not leverage the full potential of automation, selecting instead familiar equipment like valves to perform control tasks better suited to a variable frequency drive.
- Electrical engineers do not fully understand the process needs for power, such as duty cycles, and therefore do not fully optimize their designs.
- None of the engineers mentioned above are typically very quick to leverage advances in materials science, which enable higher operating parameters.

Equipment vendors and design tool providers

Equipment Vendors and Design Tool Providers determine component energy efficiencies. The vendor's tools directly affect the engineer's workflow, models, and documentation. These factors, however, may contribute to energy inefficient plants:

- Vendors provide black-box components with closed/proprietary/rigid interfaces, which are not easily optimized for the whole system; this is the result of a trend toward 'commoditization.'
- Proliferation of design tools and data formats which are non-integrated and their design model is non-navigable between vendor tools; this hinders integrative design.
- Lack of full-scope energy-optimization functionality in the leading design and modeling tools
- As components become commodities, salesmen are replacing sales engineers, and misapplications are increasing (Plant Services.com, 2008)

8.4 Professional and standards organizations

Professional and standards organizations provide basic education standards and best practice certifications. These factors, however, may contribute to energy inefficient designs:

- No widely accepted standards specifically for energy efficient designs of entire plants; some operational energy management standards are in development, however.
- No widely accepted certification for energy design for whole plant systems; an energy manager certification is available in the USA, however.
- Lack of mandatory international labeling system for industrial motors, transformers, and other equipment to enable comparison.
- Protectionism and turf wars limit the global adoption of a single set of standards; the divisions between English and SI units are a cause for some confusion, design errors, and incompatibilities.

8.4.1 Educators and academia

Educators and academia provide basic skills and certification (by diploma) of the next generation of designers and engineers. These factors, however, may contribute to inefficient designs:

- Educators tend to focus on the abstract and theoretical, as opposed to best practice design using state-of-the art commercially available engineering components.
- Systems engineering courses are not mandatory or sometimes not even offered in the average curriculum.
- Electrical engineering curricula increasingly favor more modern topics of electronics and discrete logic at the expense of courses on old-fashioned power engineering; courses on power station design have been dropped.
- Programs or degrees toward industrial or engineering management are too general: the specifics of each discipline cannot be made more abstract.
- Some engineering schools offer no 'capstone' design course that encourages synthesis of all the disciplines toward a single design task.

Standards, Best Practice, Incentives, and Regulations Standards are the designer's and engineer's best design guidelines. Standards also offer customers and authorities an objective measure for applying regulation and incentives. 'Best practices' encompass more than standards, and include case studies, more application details, and some costing information.

8.4.2 Role of standards in energy efficiency

Energy efficiency is an invisible quality and is subject to various interpretations; it is important, therefore, for engineers and managers to be able to have some common definitions and methodologies when assessing efficiency performance. International standards for energy design and management are emerging and some countries, including the U.S., have some standards in these areas. It is likely that these standards will be closely linked to future carbon dioxide compensation schemes, whether at national or international levels (ISO, 2007).

Common benchmarking for performance is good, but at a deeper level, standards can provide the equipment and system inter-operability that can enable a higher performance design. Highly efficient components which are mismatched or poorly integrated make for an inefficient overall system. A joint ISO/IEA technical committee that was recently formed to identify gaps in industrial standards coverage recommended more emphasis on the systemic approach and encouraged a focus on energy efficiency of overall systems and processes as well as retrofitting and refurbishing. This expert committee also recommended that standards should address efficiency improvements through industrial automation.

8.4.3 Standards and best practice

The standardization efforts relevant to plant auxiliaries' energy performance cover a wide variety of disciplines. The list far below refers to existing standards relevant to the systems in this handbook that specify design, application, labeling and minimum energy performance standards. The list focus is on U.S. standards, but some important international standards are also mentioned, in italicized text. The premier, official sources of unbiased standards are the national standards bodies such as American National Standard Institute (ANSI) for the U.S., CEN/CENELEC (for the EU) and the international bodies such as the IEC and the ISO. A convenient way to search for U.S. and global standards is by using the ANSI NSSN search engine at www.nssn.org. Search by title 'power station design' or 'power plant.'

Other sources of objective standards are the professional societies and industry associations, although the latter may show more bias toward their industry in certain situations:

- Institute of Electrical and Electronic Engineers (IEEE)
- Instrumentation, Systems, and Automation Society (ISA)
- The Hydraulic Institute (HI)
- National Fluid Power Association (NFPA)
- Air Movement and Control Association (AMCA)
- American Society of Mechanical Engineers (ASME)
- National Fire Protection Association (NFPA)

Many standards for steam-water cycle design of cycle equipment can be found in the various ASME and NFPA codes, but these are not within the scope of this handbook. The ASME test codes for determining efficiency, however, are of interest. Energy is a political as well as a technical subject; some 'associations' (not those mentioned above) promoting best practice are actually lobby groups with strong, but not obvious, links to commercial or political entities with various agendas. These sources can be useful if their advice is taken together with the objective sources listed above. Some of these unofficial sources of design guidance are listed in the Reference section of this handbook. The following list is not a comprehensive list of all relevant standards; appearing here are only those that have some relevance to plant auxiliaries' energy performance and design.

Power Plant Facilities

- IEEE Std 666-2007 IEEE Design Guide for Electric Power Service Systems for Generating Stations (Revision of IEEE Std 666-1991)
- IEEE Std 762-2006: Standard for Definitions for Use in Reporting Electric Generating Unit Reliability, Availability, and Productivity
- ANSI/ISA S77.43.01-1994 (R2002) : Fossil Fuel Power Plant Unit/Plant Demand Development (formerly ANSI/ISA S77.43-1994)
- ANSI/ASME PTC 46-1996: Overall Plant Performance codes
- ASME PTC 47-2006: Integrated Gasification Combined-Cycle Plants
- IEEE 803.1-1992 : Recommended Practice for Unique Identification in Power Plants and Related Facilities - Principles and Definitions
- ISO 13600 series (1997–2002): Technical energy systems. Methods for analysis of technical energy systems, - enabling the full costing and life cycle analysis

Best Practices

- DoE EERE Best Practice guides for Steam, Pumping Systems, Fans www1.eere.energy.gov/industry/bestpractices
- EPRI studies and reports – there is a large population of useful reports
- ABB Electrical Transmission and Distribution Reference Book (the 'T&D' manual)

Pump and Fan Systems

- ANSI/HI 1.3-2007 : Rotodynamic (Centrifugal) Pumps for Design and Application

- ANSI /HI Pump Standards : Available through the Hydraulic Institute, a standards partner (www.pumps.org/)

Best Practices for Pump and Fan Systems

- ANSI/HI Optimizing Pumping Systems Guidebook
- US DoE Sourcebook (2006). Improving Pump System Performance, from EERE Industrial Technologies Program:
- US DoE Sourcebook (2006). Improving Fan System Performance. from EERE Industrial Technologies Program
- Air Movement and Control Association (AMCA) International

Motors and Drives

- NEMA MG 1 : Motors and Generators
- NEMA ANSI C50.41:2000 : Polyphase induction motors for power generating stations
- IEEE Std 958-2003 : Guide for Application of AC Adjustable-Speed Drives on 2400 to 13,800 Volt Auxiliary Systems in Electric Power Generating Stations
- IEC 60034-3 Ed. 6.0 b:2007 Revises IEC 60034-3 Ed. 5.0 b:2005 Rotating electrical machines - Part 3: Specific requirements for synchronous generators driven by steam turbines or combustion gas turbines
- IEC 60034-2-1: Motor efficiency testing (September 2007); published as EN 60034-2-1 at CENELEC level.

Best Practices for Motors and Drives

- DoE Motor System Best Practices

The Energy Independence and Security Act of 2007 (EISA) calls for increased efficiency of motors manufactured after December 19, 2010.

Electric Power Systems

- IEEE 493 Design of Reliable Industrial and Commercial Power Systems, also has useful equipment reliability data.
- C57.116-1989 IEEE Guide for Transformers Directly Connected to Generators
- IEEE Std C37.010™, IEEE Standard Application Guide for AC High-Voltage Circuit Breakers Rated on a Symmetrical Current Basis.64, 65
- IEEE 519-2006 Harmonic voltage and current distortion limits
- IEEE Std 946-2004 IEEE Recommended Practice for the Design of DC Auxiliary Power Systems for Generating Stations
- IEEE Std C37.21-2005 IEEE Standard for Control Switchboards
- 525-1992 IEEE Guide for the Design and Installation of Cable Systems in Substations
- C62.92-1993 IEEE Guide for the Application of Neutral Grounding in Electrical Utility Systems, Part III-Generator Auxiliary Systems
- IEC 60076-1 Power Transformers (VDE 0532 Part 101)
- IEC 62271-1 Ed. 1.0 b:2007 High-voltage switchgear and controlgear
- IEC 61000-2-4 (Worldwide) Harmonic voltage and current distortion limits

Best Practices for Electric Power Systems

- ABB Switchgear Manual, 11th edition, 2006 (available online)

- ABB Transformer Manual, 2007

Energy & Environmental Management

- ISO 14064 and ISO 14065 provide a methodology to help organizations assess carbon footprints and implement emissions trading schemes
- ISO 14001 is an internationally recognized framework for environmental legislation, regulation, management, measurement, evaluation, and auditingassessing.
- ISO 13600 series provides guidelines on technical energy systems

Instrumentation & Control Automation Systems

- ANSI/ISA-77.44.01-2007 - Fossil Fuel Power Plant - Steam Temperature Controls
- ANSI/ISA-RP77.60.05-2001 (R2007) - Fossil Fuel Power Plant Human-Machine

Interface: Task Analysis

- ANSI/ISA-77.42.01-1999 (R2006) - Fossil Fuel Power Plant Feedwater Control System – Drum-Type
- ANSI/ISA-77.20-1993 (R2005) - Fossil Fuel Power Plant Simulators - Functional Requirements
- ANSI/ISA-77.41.01-2005 - Fossil Fuel Power Plant Boiler Combustion Controls
- ANSI/ISA-RP77.60.02-2000 (R2005) - Fossil Fuel Power Plant Human-Machine

Interface: Alarms

- ANSI/ISA-77.70-1994 (R2005) - Fossil Fuel Power Plant Instrument Piping Installation
- ANSI/ISA-77.43.01-1994 (R2002) - Fossil Fuel Power Plant Unit/ Demand

Development-Drum Type

- ANSI/ISA-77.13.01-1999 - Fossil Fuel Power Plant Steam Turbine Bypass System
- 502-1985 IEEE Guide for Protection, Interlocking, and Control of Fossil-Fueled Unit-Connected Steam Stations
- ASME PTC PM-1993 Performance Monitoring Guidelines for Steam Power Plants
- ISO 13380:2002 Condition monitoring and diagnostics of machines - General guidelines on using performance parameters
- ISO/TS 18876-1:2003 Industrial automation systems and integration - Integration of industrial data for exchange, access and sharing

Best I&C Practices

- ISA Instruments and Automation Society, http://isa.org/, both a standards and industry organization with sources on best practice

9. Power generation regulations and incentives

The regulatory environment for coal-fired plants appears likely to change significantly before 2010. Some US states (2008) are considering a moratorium on new coal plant construction, and may slow or stop permitting of plants under construction. A US Supreme Court ruling in 2007 determined that CO2 is an air pollutant; this raises the possibility that CO2 will soon be regulated as such under the Clean Air Act. Some of the most relevant existing regulations are listed below:

- EPACT: Energy Policy Act (1992)
- EPACT: Energy Policy Act (2005)
- CAAA: Clean Air Act (1970, 1990) and National Ambient Air Quality Standards (NAAQS)

EPACT 2005: tax credits for the construction of coal-fired generation projects requisite on meeting efficiency and emissions targets. (International Energy Agency, 2006): According to the IEA, this leads to an increased share of IGCC and 'clean coal' projects, but may also have impact on traditional coal-fired plant designs and operation.

The recent legislation in the Energy Independence and Security Act of 2007 (EISA) will also have an impact on the design and operation of fossil-fuel fired power plants. EISA calls for increased efficiency of motors manufactured after December 19, 2010, for example.

Engineering Basic Standards

- IEEE 280: Standard Letter Symbols for Quantities Used in Electrical Science and Electrical Engineering
- ISO 15926: A meta-structure for information concerning engineering, construction and operation of production facilities.
- ISO 31: Quantities and units, International Organization for Standardization, 1992, now being superseded by the harmonized ISO/IEC 80000 standard.

Efficiency and Lifecycle Cost Calculations

Efficiency Calculations

Efficiency is a measure of how effective a system or component can convert input to output. Efficiency is normally given in units of percentage, or as a value from 0 (0 percent) to 1.0 (100 percent). Energy efficiency can be calculated using either energy (kW/h) or power (kW)

Efficiency percent = (Useful Power Out (kW) / Power In (kW)) x 100

10. Energy and power calculations

Energy must always be defined relative to a given time period or to a given volume, etc. The energy consumed by a system or components during a given time period is determined by multiplying its input power over a time period. The common term 'losses' means wasted energy. Losses can be treated as energy (kWh) in all the calculations in this section. The common term 'loads' means output power. Most energy calculations are based on a year's time, and a year is conventionally assumed to be only 8,000 hours to account for system downtime, when energy consumption is 0. (There are otherwise 8,760 hours in a full year.)

Annual energy consumption (kWhr) = 8,000 (hrs/year) x Power (kW) Load Profile

In practice, power levels (or 'loads') are not constant, as assumed in the formula above. Loads vary over a given period due to changes in the process or ambient conditions. This variation is described by the component's 'load profile,' which describes the percentage of time (in hours per year) at each loading level (as a percentage of full load) as shown in the sample load profile below:

%	Hours	% of Full Load
5%	400	100
10%	800	90
15%	1200	80
20%	1600	70
20%	1600	60
15%	1200	50
10%	800	40
5%	400	30
0%	0	20
100%	8,000 hrs	Weighted Avg 65

Table 1. Load profile.

A more accurate view of annual energy consumption for the above component's profile is the sum of the energies at each load level:

Annual Energy (kWhr) = (#hrs) at load level(i) x (%) full load at load level(i) x full load (kW)

Duty cycle is similar to load profile, but is used to refer to shorter time periods (days or hours) and for cycling (on-off) loads, rather than more continuously variable loads.

11. Energy and power units

Energy has many forms and can be described using many units. These are the three most commonly used units in the global power generation industry.

1 horsepower (hp) = 0.7457 kW = 2546 Btu/hr

11.1 Savings calculations

Savings calculations are used to determine the difference in energy and cost between two components or systems.

By combining the formulas above, one can compare the annual savings of energy for two components or systems of varying efficiency E1(%) and E2(%). The result is an energy saving (Se) in kW per year (assume 8,000 hrs in absence of data):

Annual Energy Savings (kWhr) = 0.746(kW/hp) x P(hp) x 8,000 x 100(%) x (1/E2 - 1/E1)

One can then multiply by the cost of energy (in $/Kwh) to determine the financial (or capitalized) cost of the annual energy savings calculated above, in $:

Annual Dollar Savings ($) = Se (kWh) x Q ($/kWh), where Q is the price per kWh of electricity

In these calculations the price (Q) of energy is assumed to be constant. In fact, energy prices may change as often as every 15 minutes in a de-regulated market, with much higher prices during peak periods. The average annual price of electricity shows a rising trend. See the section on present value for methods to account for this change.

11.2 Lifecycle costing methods

Life-cycle costing (LCC) is a method of calculating the cost of a system over its entire lifespan. LCC is calculated in the same way as 'total cost of ownership' (TCO). A technical accounting of systems costs includes initial costs, installation and commissioning costs, energy, operation, maintenance and repair costs as well as down time, environmental, decommissioning and disposal costs. These technical costs, for an example transformer, are listed below.

$$LCC = C_A + C_E + C_I + \sum_{0}^{n} (CP_M + CC_M + CO_P + CO_0 + CR) + C_n$$

where C_A = cost of apparatus

C_E = cost of erection

C_I = cost of infrastructure

CP_M = cost of planned maintenance

CC_M = cost of corrective maintenance

CO_P = cost of operation (load and no-load losses)

CR = cost of refurbishment or replacement

C_D = cost of disposal

n = years of operational life span

Fig. 6. Life-cycle costing method.

Additional, non-technical costs that should be accounted for in budgetary estimates include insurance premiums, taxes, and depreciation.

All costs in an LCC calculation should be discounted to present value (PV) dollars using the present value formulas in the following section. A very simplified LCC calculation with fewer terms considers only the cost of apparatus and the cost of operation, and does not consider inflation or variation in price of energy per kWh. The operational cost term in an LCC formula is typically the annual energy costs calculated using the formulas above, discounted to PV dollars. See the section Motor System Calculations for a numerical example.

For systems that directly emit carbon dioxide or other pollutants, the cost of operation should include remediation costs, and the taxes which authorities charge (or may charge) per unit of emissions. For electrical loads powered from a fossil-fuel- based source, the carbon dioxide amounts (in tons) are still relevant, but the carbon dioxide tax (in $) should not be added to that component's operational costs if the tax has already been factored into the price of the consumed electricity.

11.3 Carbon dioxide cost calculations

For coal-fired power plants, 1.3 tons of carbon dioxide is emitted per MW hour (C.P. Robie, P.A. Ireland, for EPRI, 1991). A conservative estimate for a future carbon dioxide tax is $25 per metric ton, globally and in the U.S. The tax may take many forms, ether as a direct tax or a traded quota, etc. A metric ton (1,000kg = 2240 lbs) is also written 'tonne.'

The energy and dollar savings calculations can now be applied, using the above data, to give a carbon dioxide (tons) saving and a carbon dioxide dollar savings ($) for reducing power from a fossil-fuel-based source:

Annual carbon dioxide savings (tonnes) = 0.746(kW/hp) x P(hp) x 8,000 x 100(%) x (1/E2 – 1/E1) x 1,300

Annual carbon dioxide tax savings ($) = $25/tonne x annual carbon dioxide savings (tonnes)

A rule of thumb for coal-fired plants: a 2 percent steam cycle efficiency improvement can reduce carbon dioxide emissions by up to 5 percent (Ferrer, 'Small-Buck Change Yields Big-Bang Gain,' 2007).

11.4 Limitations of LCC methods

LCC analyses often count only single benefits, such as the electricity directly saved by a new motor's higher nameplate efficiency. In fact, there are numerous other benefits to reduce electricity consumption on the size and wear of upstream, power system components. Other benefits that are hard to quantify in LCC analysis include reduced maintenance via the elimination of the control valve, for example. In a detailed LCC calculation it is important to consider substitution cost.

11.5 Present value formulas

Most of the costs shown in the LCC calculation accrue in the future. These payments must be translated into present values using the time-value of money formulas given here.

Present value (PV) of a future amount (FV) at period 'n' in the future at 'i' interest rate is:

$PV = FVn \times 1/(1 + i)n$

Present value of a uniform series of payments, each of size US (for Uniform Series):

$= US \times ((1 + i)n - 1)/i(1 + i)n$

Where 'i' is the interest rate from 0-1 (for a 6% rate, i = 0.06)

The formula for PV of a uniform series can be used to determine the value of annual energy savings, where the annual cost is calculated as shown at the start of this section.

If the average annual price of electricity rises at p% per year, then the flat rate Q must be multiplied by the following rising price factor 'f':

$$f = (qn - 1) / (q - 1)$$

Where: $q = 1 + p/100$

And p is the price increase in %

Using the formula for a 1 kW loss after 20 years shows an accumulated cost which is 41 times the cost of the first year if the average annual increase in the energy price is 7 percent (ABB Ltd,Transformers, 2007).

11.6 Payback calculations

If the PV of the energy savings over 'n' periods (years) exceeds that of the investment cost (X), then the investment should be made. The number of periods required for PV to equal X is the 'payback' period. For a given value of X, therefore, the payback period 'n' can be calculated.

The monetary value of energy losses, called the capitalized loss value, is defined as the maximum amount of money the user is willing to invest to invest to reduce losses by 1 kW.

11.7 Levelized cost calculations

For non-uniform payments, use the levelized cost (LC) method to determine the levelized amount. This method simply uses the PV formula on each amount to determine the total PV of the stream, then applies the inverse of the PVus formula to determine a levelized amount for each period. To evaluate projects, one can use either the total PV or the LC method. Both will reach the same conclusion, except that the LC shows a comparison by period. In evaluating energy efficiency project alternatives, it may be useful to calculate the 'capital equivalent cost' (CEC).

The CEC is found by adding the capital cost to the PV of all the operating costs over the unit's lifetime. This calculation provides a sound basis for comparing bids.

11.8. Limitations of PV MethodsPresent value methods make assumptions regarding lifetime (number of periods 'n') and discount (interest) rate 'i' which have a large impact on the calculated value. In evaluating energy efficiency projects or components, the conventional assumptions tend to undervalue the savings. High-quality, high efficiency motors, for example, may have a longer lifespan ('n') than standard motors. Also, the lower risk of energy efficiency projects should be reflected in a lower discount rate, especially in common comparisons with new capacity. This comparison is between 'negawatts' (energy efficiency) and Megawatts (new capacity).

11.8 Plant heat rate calculations

In power plants, efficiency is often expressed as 'heat rate,' which is the amount of energy generated (kWh) per unit of fuel heating value (Btu = British Thermal Units).

Energy Value: 1kWh= 3414.4 Btu = 3.6 MJ

For a plant with Net Plant Heat Rate of 10,000 Btu/kWh (10.54 MJ/kWh), then the thermal efficiency = 34.14 percent.

Note that heat rate is the inverse of efficiency; a reduction in heat rate is an improvement in efficiency. Sub-critical steam plants use the fuel's higher heating value (HHV) as basis for heat rate and efficiency calculations, whether the fuel is coal, oil, or gas. Combined-cycle gas turbine plants are usually evaluated on the basis of the lower heating value (LHV) of their fuel. This can lead to the differences in apparent efficiency being somewhat greater than they actually are (Eng-tips.com, Fowler, 2006).

Coals vary considerably in their composition, which determines their heating value and carbon dioxide emissions during combustion. A typical coal has a heating value of about 8,000 Btu/pound, a carbon content of about 48 percent by weight and a moisture content of about 20 percent by weight and is combusted with up to 10 percent excess combustion air.

The ASME performance test codes 6 and 6A for steam turbines describe the method to determine steam turbine efficiency in existing plants. Whole plant international test codes are ASME PTC 46 and ISO 2314.

12. Energy accounting for reliability

12.1 Reliability concepts

The methods and terminology in this section are common to the field of quantitative reliability analysis. Reliability (R) is the probability that a unit is still operational after one year, based on the unit's mean time between failure (MTBF) specification. Reliability is expressed as failure rate on per year basis.

$$R = e(-8760hr/MTBF)$$

Availability (A) of a unit can be calculated as: A = MTBF / (MTBF + MTTR)

Where:

MTTR = mean time to repair

Energy Cost of Plant Trips

The total cost of a plant trip is composed of many parts, including opportunity cost of lost power sales, cost of substitute purchased power, ISO fines, trip-induced repairs, and energy. The energy wasted per year due to trip events is therefore R multiplied by energy wasted during startup/shutdown procedures (R x Ess). The wasted energy due to a complete shutdown and cold restart (Ess) is composed of two parts:

E(shutdown energy) + E(startup energy) =Ess

Where:

E(startup energy) = hours duration of startup x energy input/hr
E(shutdown energy) = rotational energy in all machinery + chemical energy in process lines

E for shutdown is more difficult to measure and calculate. As a rough estimation, therefore, E shutdown is assumed to be ¾ of the E startup. So ultimately, the annual energy costs of plant trips :

R x Ess = R x 1.75 x hours duration of startup x energy input (MMBtu)/hr x Energy price ($/MMBtu)

13. References

International Energy Agency (IAE) , www.iae.org, Global data on coal applications
IEA Clean Coal Centre, http://www.iea-coal.org.uk,Unbiased info on coal tech; IEA Coal Online, www.coalonline.org
ISO, http://www.iso.org/iso/hot_topics_energy, Int'l standards organization, with emerging energy mgmt. standards
World Energy Council , http://www.worldenergy.org/, Pubs on power plant efficiency, energy policies (World Energy Council, 2007)
Consortium for Energy Efficiency (CEE) , http://www.cee1.org/
US Dept. Of Energy: Energy Efficiency & Renewable Energy, http://www1.eere.energy.gov/industry/, Industrial Technologies Program
Electric Power Research Institute , http://my.epri.com, Non-profit R&D and consulting organization, utility & industry funded
Leonardo Energy, http://www.leonardo-energy.org, Global community of sustainable energy professionals

13

Energy Efficiency Initiatives for Saudi Arabia on Supply and Demand Sides

Y. Alyousef[1*] and M. Abu-ebid[2]
[1]*Energy Research Institute, King Abdulaziz City for Science and Technology, Riyadh,*
[2]*AEA Technology plc, Didcot,*
[1]*Saudi Arabia*
[2]*United Kingdom*

1. Introduction

The Kingdom of Saudi Arabia (KSA) is blessed with an abundance of energy resources. It has the world's largest proven oil reserves, the world's fourth largest proven gas reserves, has abundant wind and solar renewable energy resources, and is the world's 20th largest producer and consumer of electricity. Saudi Arabia makes negligible use of its renewable energy resources and almost all its electricity is produced from the combustion of fossil fuels. Despite attempts to diversify the economy, the oil and gas industry still accounts for approximately 75% of budget revenues, 45% of GDP, and 90% of export earnings. Exploitation of the natural resources has allowed the Saudi government to keep energy prices low through a system of direct and indirect subsidies. The nation has benefited greatly from these policies, but together with increased prosperity and sophistication, a culture of wasteful energy usage has become established.

KSA is experienced rapid economic growth over recent years. Since 2000, the energy consumption per capita has increased by more than 30%. This increase in primary energy consumption has occurred during a period of declining oil exports. In 2008, the total primary energy consumption has approximately reached 800 million barrels of oil equivalent (BOE), of which more than 60% was oil. The consumption of primary energy within the Kingdom is expected to double in 2030 leading to diminishing oil exports based on current trends (Ministry of Water and Electricity, 2009).

There is widespread recognition within KSA that with growing internal demand for primary energy there will be a declining proportion of oil for export. Consequently, the national government has identified energy efficiency as a key national priority, reflecting the rapid increase in domestic consumption of petroleum products, related GHG emissions and the associated opportunity cost of lost export revenues. There is also a strategic national push to develop an energy efficiency and renewable technology R&D and manufacturing base in an attempt to diversify the economy away from fossil fuels.

---

*Corresponding Author

2. Fossil fuel production and consumption

2.1 Oil production and consumption

Saudi Arabia is the largest producer and net exporter of oil in the world with more than 10 million barrels/day produced in 2007. The state-owned oil company, Saudi Aramco, is the world's largest oil company. The country has around 100 major oil and gas fields and more than 1500 wells. Recently, the Saudi Arabia's Ministry of Petroleum and Mineral Resources (MPMR) announced their plan to increase the production capacity to 12.5 million barrels/day by 2009 but these plans have been delayed due to the collapse of oil price at the end of 2008 (Alowaidh, et.al, 2010). In 2008, KSA exported an estimated 8.4 million barrels/day of petroleum liquids, the majority of which was crude oil. Increasing oil exports is a national priority which can positively influence economic development and prosperity in the country. However, while Saudi Arabia has the necessary infrastructure to double its export capacity, oil exported over the past few years has been gradually decreasing due to increasing internal consumption. Figure 1 shows that consumption of oil within the country has been gradually increasing due to population growth, strong economic, industrial growth and subsidised prices for electricity and transport fuel.

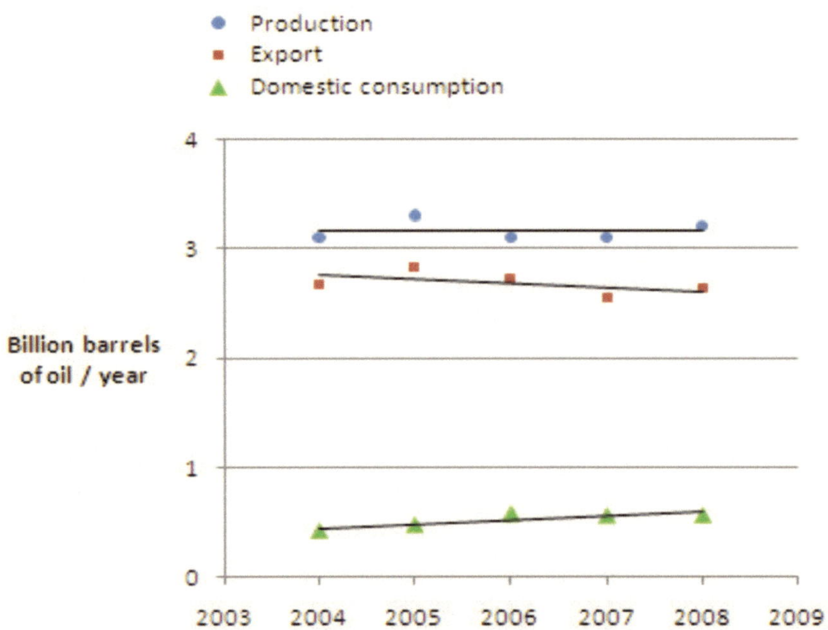

Fig. 1. Oil production, export and consumption in Saudi Arabia.

Approximately 3200 million barrels of oil were produced in 2008, of which 560 million barrels (952 TWh) was consumed in the country. About 9.6% of oil consumed within KSA was used as feedstock and the rest as primary energy. As primary energy, oil was mainly used for transport (43%), power generation (39%), in co-generation desalination plants (8%), and for other uses (10%) such as for example the oil and petrochemical industry(Alowaidh, et.al, 2010).

2.2 Natural gas production and consumption

In 2008, Saudi Arabia produced 86 billion m^3 (550 million barrels of oil equivalent or about 900 TWh) of natural gas of which 13% was lost in flaring. All natural gas produced was consumed within KSA as feedstock (38%), in power generation (34%), in desalination plants (11%) and in industry (17%). The production of natural gas has been increasing to fuel the growth in the petrochemical, power generation, and water desalination sectors(Alowaidh, et.al, 2010). While the majority of oil produced is exported, all natural gas produced in Saudi Arabia is used within the country. According to Saudi Aramco, the production of natural gas is expected to double to 150 billion m^3in 2030.

2.3 Potential for oil savings

According to independent analysis quoted in industry reports, demand for oil is expected to rise by 8 to 10% in 2010 mostly in the area of power generation. A plan which allows KSA to cut down on its oil consumption in the power generation sector and to re-direct that additional oil for export can bring about many economic and environmental benefits.

Based on current growth rates, oil consumption is expected to reach 800 million barrels/year by 2030. A 10% annual reduction in oil consumption within KSA in 2030 will result in the release of 80 million barrels of oil/year for export. At today's price of oil, this corresponds to additional revenue from oil export of $6 billion/year. This 10% reduction in oil consumption is a realistic target for Saudi Arabia and can be achieved through

i. Energy efficiency improvement on the supply side (i.e. in power stations and industry) as well as on the demand side (i.e. reducing electricity consumption),
ii. Energy and resource conservation (e.g. reducing water demand reduces energy demand in desalination plants), and
iii. The utilisation of renewable energy sources such as solar and wind energy.

In general, increased oil savings can be achieved by gradual fuel switching (combined with energy efficiency improvement) in the power generation and co-generation desalination sectors. The share of natural gas in these two sectors is currently 47%. If the share of natural gas in these two sectors alone increases to 60% in 2030, the potential annual oil savings can amount to 120 million barrels/year leading to additional revenue of $9 Billion/year at current oil prices. This switch from oil to gas is also associated with environmental benefits since natural gas has lower CO_2 emissions than oil.

3.Overview of primary energy flow

Saudi Arabia's annual primary energy consumption increased from about 10 MWh/capita in 1971 to 47 MWh/capita in 2008(Electricity and Cogeneration Regulatory Authority,2008). Currently KSA is one of the top 15 countries in the world in terms of primary energy use on a per-capita basis. This significant increase in energy demand is due to rapid economic and industrial growth in Saudi Arabia over the past few decades. In 2008, the total oil and gas consumption exceeded 1 billion barrels of oil equivalent (BOE), of which approximately 23% was used as feedstock and the rest as an energy source in power generation plants, co-generation desalination plants, transport and industry. This consumption of primary energy (oil and natural gas) is expected to double over the next two decades as shown in Figure 2.

In 2030, oil and natural gas demand is expected to increase to more than 1500 million barrels of oil equivalent/year.

As evident from Figure 2, the current share of both oil and natural gas is 63% and 37% respectively. The future share of oil and natural gas in the fuel mix in KSA will depend on policies regarding additional gas production and usage. The projection in Figure 2 assumes the same current primary energy mix in 2030. It is expected, however, that natural gas demand in KSA will double by 2030 thus displacing some oil usage and adding to oil exports. The total fossil fuel consumption in KSA in 2008 was approximately 1,766 TWh, 23% of which was used as feedstock in industry, and in other sectors (e.g. LPG in homes). The fuel used as feedstock is not part of the primary energy use and so it is beyond the scope of this paper. The remaining fossil fuel (77%) is used as primary energy in the power generation, water desalination, industrial and transport sectors. This paper does not include energy conservation and oil saving in the transport sector where 28% of the primary energy in KSA is consumed (mainly as oil). About 55% of the primary energy in KSA is used in power stations and desalination plants for generating electricity and desalinated water. Of this, 44% is used in power stations owned by the Saudi Electricity Company (SEC) and the remaining 11% is used in co-generation desalination plants. The remaining primary energy is consumed in the industrial sector (11%) and other sectors (6%) such as agriculture and construction(Saudi Electricity Company,2009).

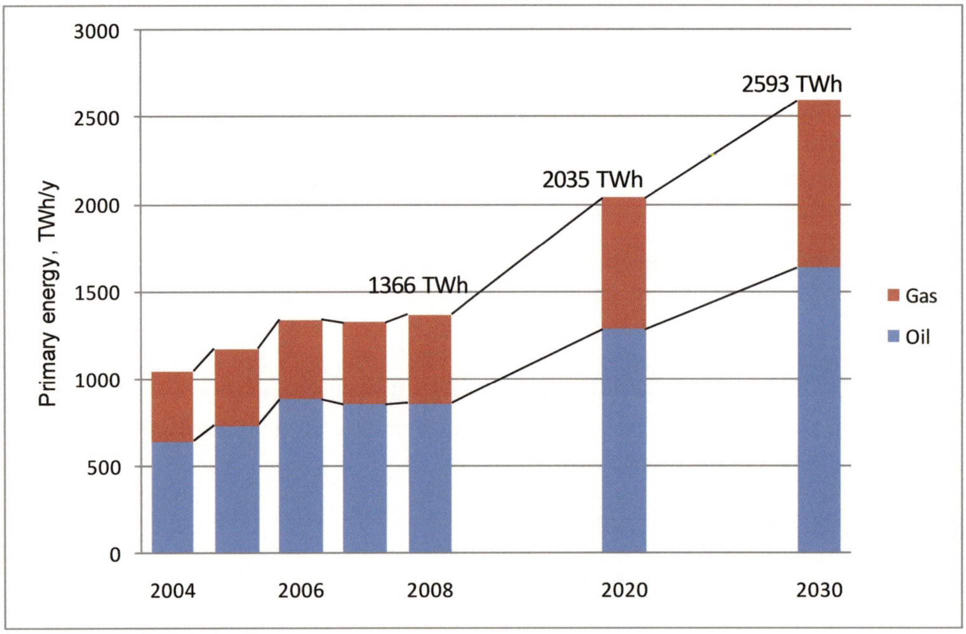

Fig. 2. Growth of primary energy demand in KSA (without feedstock).

3.1 Primary energy consumption in the power generation sector

The Saudi Electricity Company (SEC) controls the electricity sector and owns a total of 70 power generation stations. The current power generation capacity is around 39 GW of which 89% is owned by SEC, 6% from desalination co-generation plants and 5% from on-site generation (mainly at ARAMCO's sites). The breakdown of power stations under the control of SEC in terms of capacities is shown in Figure 3(Saudi Electricity Company, 2010).

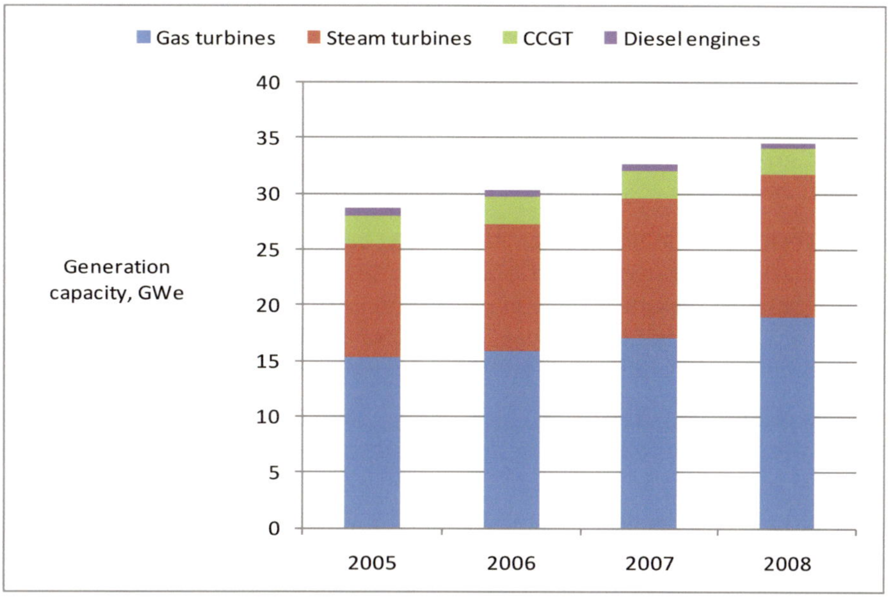

Fig. 3. Breakdown of actual generation capacity (GW) in SEC power stations (2004–2008).

In 2008, the more-efficient combined cycle gas turbines (CCGT) accounted for only 7% of the total capacity with steam and gas turbines making up the majority of the generating capacity. The electricity generated from the different types of power generating stations is shown in Figure 4.

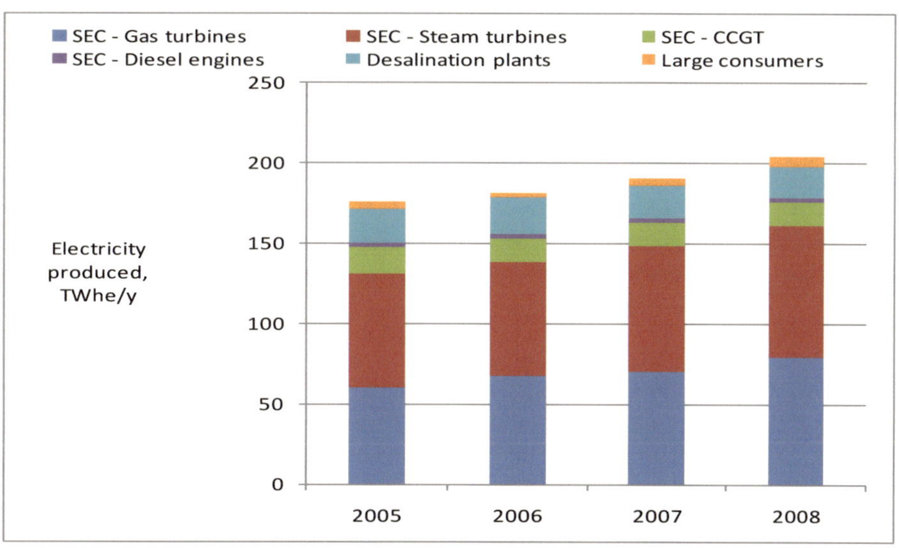

Fig. 4. Breakdown of electricity produced from different generation technologies (2004-2008).

The fuel consumed in power stations amounted to 604 TWh (355 million barrels of oil equivalent) with 55% being from oil and 45% from natural gas. Based on 2008 fuel consumption and electricity production figures, a nominal power generation efficiency of 29.5% is obtained for all power generation in Saudi Arabia. This efficiency is low in comparison to world averages. For example, the average power generation efficiency for UK power generation is 38.6%(Electricity and Cogeneration Regulatory Authority,2009). If retiring power stations in KSA are replaced by CCGT, higher efficiencies from the power sector can be expected, as modern CCGT power stations are capable of delivering seasonal efficiencies in the order of 45%-50%, based on KSA climate conditions.

While fuel consumption in SEC power stations has been gradually increasing, the share of natural gas has decreased from over 50% in 2004 to 45% in 2008. Thus, there could be a great potential in the power sector for gas to replace oil which will then lead to additional oil exports and so contributing to economic and environmental benefits. The switch from oil dominated electricity generation sector to more natural gas, will lead to more oil becoming available for export, improved generation efficiency and reduction in CO_2 emissions.

In order to satisfy future growth, it is predicted that an additional 35 GWe of electricity generation capacity, with an additional capital investment of $120 billion, is required in Saudi Arabia by 2030(Elhadj,E.,2004). If the current power generation mix is maintained, fuel consumption in the power generation sector will be 48% of the total primary energy consumption (2,593 TWh in 2030). If a scenario, where all newly-built power stations are CCGT, is considered, the nominal power generation efficiency could increase to 37% leading to about 20% reduction in predicted fuel consumption and also reducing the need for investment in new generation capacity. This shows the importance of increasing the share of CCGT in the power generation mix.

3.2 Primary energy consumption in the desalination sector

In Saudi Arabia there exists a strong link between water and energy consumption because a large portion of water consumed is desalinated water which is transported for long distances. A summary of water production and demand in Saudi Arabia is given in Figure 5(Abdel-Jawad, M., 2001).

Desalination is an energy intensive process. In Saudi Arabia, in 2008, a total of 153 TWh (57% gas and 43% oil) were used to produce 1135.6 m^3 of water and 19 TWh_e of electricity. The two main desalination methods used in KSA (and in the Middle East in general) are multi-stage flash (MSF) distillation and reverse osmosis(World Bank,2007).

Power generation produces significant amounts of heat which, if not utilised, will be dumped to the atmosphere. Heat from power generation can be utilised in desalination. Combining power generation with desalination has higher energy efficiency than generating electricity or desalinating water separately, with energy efficiency improvement reaching 10-20% better. So an effective policy is one which promotes the construction of co-generation desalination plants where heat can be recovered and used while at the same time generating electricity.

In order to satisfy future growth in water demand, an additional 20-30 desalination plants with total capital investment of $50 billion will be needed by 2030. If the volume of desalinated water doubles in 2030 as expected, and if desalination plants maintain their current efficiencies, fuel consumption by these plants will also double to more than 300 TWh.An effective policy for reducing primary energy consumption in the co-generation desalination sector is to introduce standards for minimum co-generation efficiencies from such plants.

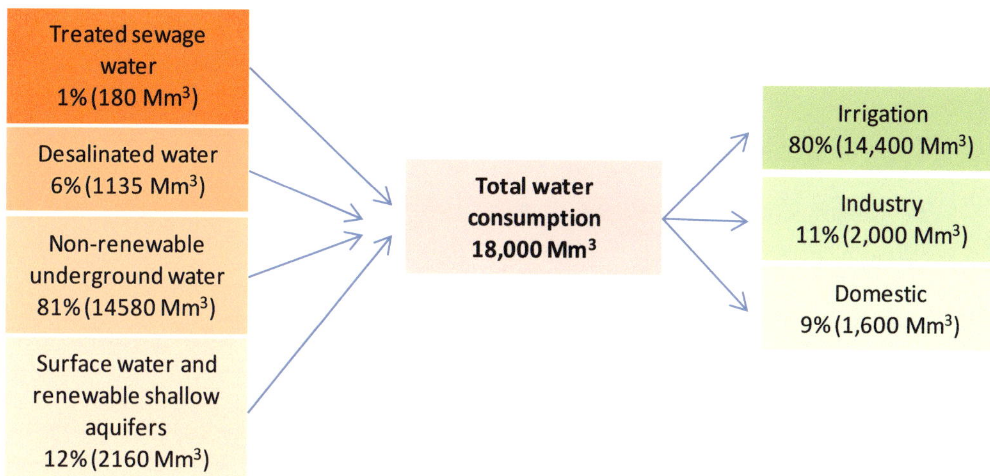

Fig. 5. Water production and demand in Saudi Arabia.

3.3 Primary energy consumption in the industrial sector

The industrial sector in KSA consists of oil refining, petrochemicals, iron and steel, cement in addition to other sectors. The average growth rate in the industrial sector is 4.3% with a total GDP of more than $240 Billion. The petrochemical sub-sector is one of the fastest growing sectors and has a 1.2% share in GDP. In 2008, primary energy consumption in the industrial sector reached 150 TWh representing 11% of total primary energy use in KSA. Assuming a growth rate of 4.3%, this is expected to increase to 379 TWh in 2030.

4. Electricity demand in Saudi Arabia

In 2008, electricity produced in Saudi Arabia was 204 TWh, 11% of which was lost in transmission and distribution. The total electricity consumption in 2008 was about 181 TWh mainly in the residential sector (53%) with a consumption of 97 TWh at the user end which corresponds to approximately 285 TWh (168 million barrels of oil equivalent) at the power station inlet. Between 2004 and 2008, electricity consumption in the Kingdom has grown at an annual average of 5.1%. As a result of population and economic growth in Saudi Arabia, electricity consumption is expected to double to about 360 TWh in 2030.The breakdown of electricity delivered is shown in Figure 6(Saudi Electricity Company,2010).

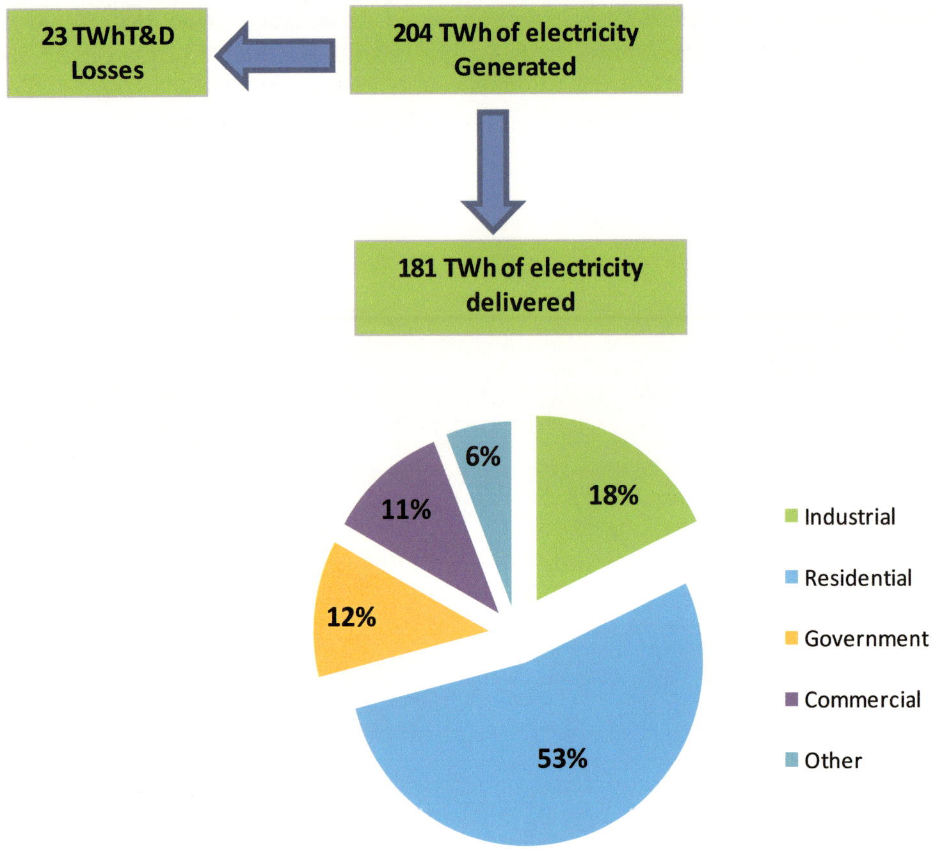

Fig. 6. Electricity flow in the Kingdom of Saudi Arabia in 2008.

5. Growth of primary energy and electricity demand

Figure 2 above shows the expected primary energy consumption in 2020 and 2030. Based on the discussion in Sections 2, 3 and 4 above, future primary energy demand in different sectors is estimated as shown in Table 1. Table 1 gives a summary of the assumptions undertaken in obtaining the future primary energy demand in KSA in 2030 for the power generation, desalination and industry sectors following a business-as-usual (BAU) scenario. As stated above, primary energy consumption in the transport and the "other" (including agriculture, construction industry, etc.) sectors are excluded from this study.

According to the BAU scenario, the share of the total primary energy demand for these three sectors increases from the current 66% to 74% in 2030. Based on the BAU scenario, the annual growth rate for primary energy demand in Saudi Arabia as a whole is 3% and in comparison with 1.6%/annum worldwide growth demand projected by the IEA(International Energy Agency,2009). For the three sectors in Table 1, a growth in demand of 3.4%/annum is estimated indicating the importance of implementing primary energy saving measures in these sectors in particular.

Sector	Projection	Current, TWh/y	CO_2 emissions, T CO_2/y(1)	2030, TWh/y	CO_2 emissions, T CO_2/y(3)
Power generation	Fuel consumption in the power generation sector is approximately 600 TWh. The power generation capacity and power output are expected to double. Assuming the same generation mix (i.e. CCGT, steam turbines, gas turbines) is maintained, fuel consumption will also double.	604	328	1208	655
Desalination	Municipal water demand will increase from the current 1.6 billion m^3/year to 3.4 billion m^3/year in 2030. Assuming the same share of desalinated water in the municipal sector is maintained (i.e. 70%), an additional 1.3 billion m^3 of water from desalination processes will be required. The trend in KSA is to build cogeneration desalination plants, so a doubling of primary energy consumption is almost required. Using the ratio of desalination capacity in 2030 and currently, the energy use in 2030 is 325 TWh.	153	77	325	179
Industry	Using a growth rate similar to the current rate (i.e. 4.3%), the demand by industry in 2030 is 379 TWh	150	54(2)	379	135(2)

(1) Based on emission factors of 350 g/kWh for gas and 700 g/kWh for oil.
(2)The smaller emissions from industry are due to the much higher natural gas consumption in comparison to oil.
(3) Assuming same energy mix and same emissions factors

Table 1. Summary of current and future (business-as-usual) primary energy demand by sector.

6. Energy efficiency challenges and barriers

The challenge electricity-generating utilities face is the provision of secure and stable supplies of electricity to their customers. A major tool in meeting this challenge is energy efficiency; however, both challenges and barriers face multi-faceted cultural, economic, technical, and institutional problems, which may require mandatory and educational initiatives to overcome them, as well as institutional reorganisation

6.1 Cultural barriers

The historical low fuel and electricity prices, along with decades of increasing prosperity, have led to an endemic culture of profligate energy usage. The high standards of living taken for granted by new generations of Saudi youths entering the workforce depend on secure and stable energy, yet these figures strongly indicate that unless energy-efficiency measures are swiftly incorporated at all levels of the supply-demand equation, economic development may not be able to meet expectations. The problem is compounded by the twin growth profiles of population and per-capita energy consumption. From 2002 to 2006, the population has grown by an average of 2.6% p.a. and per-capita energy consumption by an average of 6.1% p.a. Over the same period as shown in Figure 7, SEC's generating capacity has grown by an annual average of 5.4% yet total peak loads have grown by an annual average of 7.6%. Total consumption increased from 128,629 GWh in 2002 to 163,147 GWh in 2006, an average increase of 6.7% p.a. Although in absolute terms the increase in residential consumption during the period was greater than the combined total consumption of the agricultural and commercial sectors, expressed as percentages, the greatest average annual increase was in the commercial sector at 13.4%, followed by the residential sector at 7.9%. The smallest increase in consumption was in the industrial sector, which only increased by industrial consumption by 2.8% p.a. (Ministry of Water and Electricity, 2006).

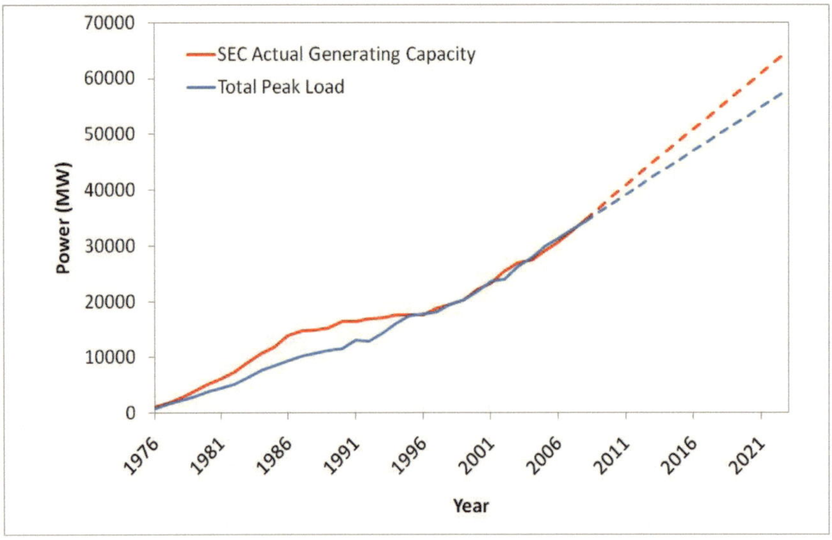

Fig. 7. Growth of Saudi electricity utility generating capacity and total peak loads in Saudi Arabia. Solid lines (1976-2007) are MOWE statistics; dashed lines (2008-2023) are projections.

Despite the large annual increases in electricity consumption there is currently limited interest in the subject of energy efficiency among consumers from any sector, and little awareness of the opportunities available to them or of the benefits of specific energy efficiency technologies and practices. The recent government decision to abolish electricity price subsidies may provide an impetus to change this into a culture of energy efficiency; however, for this to happen, consumers need to be energy aware and tariffs need to actually reflect the true costs of generation, transmission, and distribution.

6.2 Economic barriers

The Saudi Electric Company needs to raise at least SR 380 billion in investments for the period 2009-2017 (Saudi Electricity Company,2009) if it is to ensure security of supply and meet its capacity increase targets. Although there are proposals to open the electricity generating industry to private sector involvement, the high capital costs and low tariffs mean that it would not be an economically viable investment. This is especially the case in sparsely-populated rural areas where the low customer density profile means that the cost of generation and distribution to utility companies may be significantly higher than in urban areas where customer densities are much higher.

Saudi financial institutions have yet to make major investments in energy efficiency projects. Despite many projects offering 2-3 year payback periods, committing capital to investments in new energy technologies is still considered an adverse risk. A similar attitude is found in all electricity consumer sectors. Not only are they also reluctant to make investments in energy efficient technologies, but the direct and indirect medium- and long-term economic gains of energy efficiency are often overlooked in favour of the short-term capital-cost savings available with cheaper equipment and work practices.

Although the Government recently decided to abolish electricity price subsidies, costs to consumers have risen, and environmental costs are still not internalised in energy tariffs. To make sustainable reductions in energy use, consumers must see clear financial benefits: Subsidized energy prices mean consumers have little incentive to save energy. Even assuming that consumers had accurate information about the benefits of energy efficiency measures, had a range of energy-efficient appliances from which to choose, and had the financial means to purchase them, there would still be little economic reason for them to do so because electricity tariffs are lower than the expected avoided costs of electricity. In addition, despite import duties on electrical goods, the market is flooded with cheap imported products.

6.3 Technical barriers

Perhaps the biggest problem facing the energy supply sector is the large seasonal variation in electricity consumption. In the hot summer season, there is increasing energy demand for air conditioning, especially by the residential and commercial sectors. Figure 8 shows a chart of the daily variation in peak load and the daily temperature profile measured in Riyadh on 09 September 2006. Peak load follows temperature throughout the day, with minimum demand and minimum temperatures in the early morning and maximum demand and maximum temperatures in the early afternoon. Peak load power is provided by gas turbines; they have the advantages of fast start-up times and can quickly respond to changes in demand, but have the disadvantage of being relatively inefficient. An additional disadvantage is that during peak hours, when demand is at a maximum, ambient temperatures are also at a maximum, which can degrade turbine efficiencies by up to 20%.

Another problem is the low reserve available generation margin. In Figure 7, the generating capacity of the electric utilities from 1976 to 2007 is shown along with the total peak load. Each increase in generation capacity is met by an increase in consumption. In 2002, generating capacity increased by 9.6%, but the following year, 2003, saw consumption increase by 9.8%. In 1996, 2001, and every year since 2004, the electric utilities have had insufficient capacity to meet the peak load.

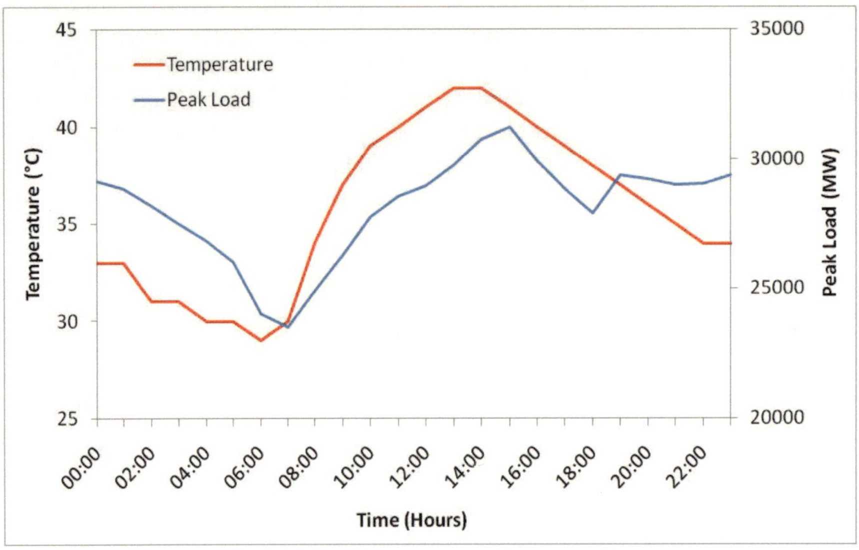

Fig. 8. Daily variation in peak load and temperature (temperatures for Riyadh, 09 Sept. 2006).

6.4 Institutional barriers

Perhaps because of its unique position as a major oil and gas producer, and its resultant low domestic energy prices, Saudi Arabia lags behind many other developing nations in implementing energy efficiency strategies. As a consequence, there is no central institution responsible for energy efficiency at a national level, responsibility for oil, natural gas, and electricity is divided amongst several ministries and institutions, and there are no explicit end-user policies.

The Ministry of Petroleum and Mineral Resources (MPMR) is responsible for the activities of oil companies, the largest of which is Saudi Aramco. Saudi Aramco produces almost all the country's crude oil and all of its natural gas liquids. It is involved in exploration, production, and refining of crude oil, as well overseas marketing and domestic distribution of petroleum products. The Ministry is also responsible for the gas industry. Saudi Aramco produces, transports, and markets natural gas to large gas consumers such as power plants, desalination plants, and process industries. It also provides gas to the Saudi Gas Company, which delivers and sells natural gas to small consumers.

The electricity industry answers to the Ministry of Water and Electricity (MOWE), although the Electricity Cogeneration Regulatory Authority (ECRA) regulates the industry and ensures high quality electric power and services are provided at reliable levels and reasonable prices; ECRA is also responsible for energy conservation. Power generation, transmission, and distribution are carried out the Saudi Electric Company (SEC). As an entity, SEC has been operational since 2000 and was formed by ministerial decree through the merger of ten regional power companies. SEC is a joint stock company; however, the majority stockholder is the Saudi government.

Although many national organisations deal with limited aspects of energy efficiency and have sophisticated skill sets, their work is often duplicated, their strategies and policies are limited in scope, and their activities are not coordinated. Ultimately, none of the initiatives taken by these bodies, or those taken by major commercial and industrial concerns, are enforceable outside their own organisations or have much effect on a national scale due to lack of transparency on the part of governmental organisations and commercial confidentiality concerns by commerce and industry. There are a few sustainable energy policies and standards at a national level, but they are not enforced by law and are not legally binding. The Saudi Arabian Standards Organisation (SASO), for example, has published standards for consumer electrical appliances, which if implemented, should decrease the availability of cheap imports and increase the uptake of energy efficient appliances; however, the organisation is not able to enforce the standards, and despite import tariffs, the market is flooded with cheap energy inefficient products (National Energy Efficiency Program, 2010).

7. Energy efficiency initiatives

A number of energy efficiency and conservation initiatives have been taken by various agencies in response to the increasing demand for electric power and to help ensure security of supply. Steps to reduce energy demand have been taken by the Saudi Arabian Standards Organization (SASO) with standards for the use of advanced insulating materials in the construction of new commercial buildings, and by the restructuring of electricity tariffs by the Saudi Electric Company.

At the same as expanding conventional generation capacity, MOWE is attempting to reduce peak demand by setting limits on the maximum power delivered to large electricity consumers, and by establishing a demand-side management programme. MOWE is also enabling major consumers in industry and commerce to reduce peak demand time consumption and to shift their usage patterns to off-peak times, while agricultural consumers are required to reschedule their irrigation to off-peak times. In addition, new developments designed with large air conditioning loads are required to incorporate thermal energy storage systems. Another initiative is the proposal to privatize and reorganize the electricity sector into three independent sectors: generation, transmission, and distribution, and to allow private-sector participation in new power-generation projects. An estimated 837 MW peak load reduction was achieved in 2001 through these initiatives in the governmental sector and by 34 MW in the industrial sector; the commercial sector, however, increased their peak load by 126 MW during the same period.

Although each of these initiatives is important in its own right, they have not been coordinated by a central authority and may have limited long-term effect. For this reason, perhaps the most important initiative has been the establishment of the National Energy Efficiency Programme.

During the 5 years that NEEP has been active, it has coordinated a varied programme of activities including:

- A pilot time-of-use tariff programme for major industrial and commercial customers with SEC
- Drafting energy efficiency labels for three major classes of electrical appliances with SASO
- Walk-through energy audits of governmental, commercial, and industrial facilities
- Detailed energy audits of governmental, commercial, and industrial facilities
- An energy efficient equipment leasing programme
- A programme of energy efficiency information and awareness
- The promotion of an energy service industry
- Technical and managerial training through workshops and seminars

7.1 Tariff restructuring

Historically, tariffs have been heavily subsidised. The subsidy being the difference between the average cost of production and the tariff. Prior to 1984 there was single flat-rate tariff of 0.05 SR/kWh for industrial customers and 0.07 SR/kWh for residential customers. For tariff purposes, industrial customers also include agriculture, hospitals, and charities; while government is included with residential customers. Over the next decade, the flat-rate industrial tariff was unchanged, while residential usage was assessed according to a three-tier tariff. During this time, despite rising production costs, rather than raising tariffs or adjusting rates downward, they were adjusted upward, first in 1985 which saw the usage level at which the maximum tariff started double from 2000 to 4000 kWh/month, and again in 1992 when it was increased to 6000 kWh/month (Saudi Electricity Company, 2010).

In 1995, a two-tier tariff was introduced for industrial customers, which effectively doubled charges as the higher tariff of 0.10 SR/kWh started at 2000 kWh/month. At the same time, the maximum charge for residential users was increased from 0.15 SR/kWh to 0.20 SR/kWh with no adjustment in the level at which the highest tariff started.

The major change in tariff structure occurred in 2000, with the introduction of 11 tiers in the tariffs for residential customers ranging from 0.05 SR/kWh to 0.38 SR/kWh, and the reintroduction of a flat rate tariff for industrial customers at the higher rate of 0.12 SR/kWh. The residential tariffs were in force for 7 months before being lowered to 0.05 SR/KWh to 0.26 SR/KWh; however, the number of tiers and usage was not changed.

7.2 Time-of-use tariffs

In addition, to restructuring industrial tariffs, a voluntary pilot time-of-use-tariff (TOU) programme was introduced for industrial and commercial customers with consumptions exceeding 600 MWh per annum. The pilot programme was limited to customers in Riyadh, which is part of SEC's Central Operating Area and was conducted by SEC in collaboration with NEEP.

The TOU programme divides the year into two: summer, which runs from June to September, and winter, which runs from October to May. The summer has two periods, on-peak and off-peak. On peak applies on weekdays (i.e. Saturday to Wednesday) from one o'clock in the afternoon until 5 o'clock in the evening, this being the time of maximum domestic usage as well the time of least gas-turbine efficiency due to high ambient temperatures. Off-peak applies on weekdays at all other times, and at weekends. The winter is off-peak at all times.

Generally, the more customers shift or curtail their usage to a TOU tariff during peak periods, the greater the net benefits to both the utility companies and its customers should be. Since the pilot TOU tariff program was voluntary, to ensure a high rate of customer participation, a peak to off-peak rate ratio was selected to maximise its appeal to a wide range of customers, rather than to maximise the benefit to the utility company. Several peak to off-peak rate ratios were proposed to large customers; most expressed interest in 4:1 and 2:1 ratios, with the strongest preference being for a TOU rate with a 4:1 peak to off-peak rate ratio.

The TOU tariffs were calculated using Long Run Marginal Cost differentiated by time-of-use. The peak TOU tariff was set at 0.35 SR/kWh for industrial customers and 0.76 SR/kWh for commercial customers, and the off-peak TOU tariff was set at 0.09 SR/kWh and 0.19 SR/kWh for industrial and commercial customers respectively. The outcome of the 2006 pilot TOU tariff programme is summarised in Table 3. From June to September 2006, the maximum on-peak reduction was 15.3 MW. About 10% of the total peak energy used, 2494 MWh, was shifted from peak to off-peak periods, 1717 MWh by industrial customers and 777 MWh by commercial customers. The total reduction in bills for all customers was 351510 SR, industrial customers saved 78160 SR, and commercial customers 273350 SR. Industrial customers contributed 69% to the total energy shift and their bill reduction was 22% of the total. In 2007 the number of participating industrial customers increased to 145 and commercial customers to 37; the maximum peak-load reduction was 90 MW with total savings of 64 million SR. In order to attract more participants, the on-peak tariff in 2008 was reduced to 0.09 SR for industrial customers and 0.19 SR for commercial customers.

	Industrial	Commercial
Tariffs		
Normal (SR/kWh)	0.12	0.05 - 0.26
On-Peak TOU† (SR/kWh)	0.35	0.76
Off-Peak TOU‡ (SR/kWh)	0.09	0.19
Outcome		
Customers	26	9
Max Peak-Load Reduction (MW)	11.9	3.4
Total Energy Saving (MWh)	1717	777
Total Bill Reduction (thousand SR)	78.16	273.35

†On-Peak TOU: June-Sept, Sat-Wed, 13:00-17:00
‡Off-Peak TOU: All other periods
Source: NEEP (2006)

Table 3. Pilot 2006 Time-of-Use (TOU) tariff programme.

7.3 Energy efficiency labels and standards for new equipment

Substantial energy savings are possible by replacing energy-inefficient electrical appliances with energy-efficient appliances. A cost-effective method of doing so is to determine which classes of appliances consume the most energy and to develop minimum energy efficiency standards to which they have to comply. The standards describe testing and classification procedures for compliance. The appliances then bear energy-efficiency labels to provide consumers with information about their energy performance. These can be expressed in terms of energy use per hour, energy efficiency as a percentage, a coefficient of performance, or as an estimated cost of usage. In addition to such technical data, labels can also show a comparative index to simplify purchasing decisions for the general public; typical forms are 'energy-stars' or 'good-better-best' systems.

Testing, performance-measuring procedures, and energy efficiency and labelling standards have been drafted by SASO in collaboration with NEEP for three of the major energy-consuming classes of electrical equipment: Ducted and non-ducted air-conditioners; household refrigerators, refrigerator-freezers, and freezers; and household clothes washing machines. Prior to this programme, the only appliances in Saudi Arabia to have energy efficiency regulations were air conditioners; however, they were ineffectively enforced. The existing standards for refrigerators, freezers, and washing machines, as well as for air-conditioners, were for health and safety and for physical and technical compatibility. Locally manufactured equipment tends to have low energy efficiencies and imported electrical equipment is subject to import duties. These factors, in addition to low levels of awareness amongst consumers, mean that energy use is not generally a significant factor in consumer purchases. Initiatives to change purchasing priorities include a comprehensive programme of education and awareness aimed at all sectors of the supply-demand chain, and financial incentives in the form of rebates for the purchase of energy-efficient appliances.

The energy efficiency labels display manufacturer specific details, the appliance's electrical specifications and operational capacity, and its performance under standard tests (Saudi Arabia Standards Organization, 2006). The label also shows a number of stars indicating its claimed energy efficiency at rated conditions. An example energy efficiency label for washing machines is shown in Figure 9; labels for other appliances differ only in their appliance-specific requirements in the main panel. A maximum of six stars can be displayed; the more stars, the greater the energy efficiency. The number of stars is derived differently for each class of appliance: For air-conditioners, it is the ratio of output cooling in Btu/hr to the input power in Watts at a given operating point; for domestic washing machines, it is primarily a function of the energy used per load and its rated load capacity; and for refrigerators and freezers, it is the reduction in energy consumption for their class expressed as a percentage. The derivation of energy star ratings for these appliances is in shown in Table 4.

In addition to the testing and certification procedures, the programme is also addressing market transformation issues and the ability of the local manufacturing base to meet the standards. The likely costs for new product design, new technology licenses from foreign partners, production line re-tooling costs, and associated lead times to comply with the standards has been estimated for these appliances and a market transformation strategy is being developed. Information and awareness programmes have been developed and education and advertising campaigns targeted to industry, business, and consumers have been designed. Future assessments will be made to determine whether, as in developed countries, market forces will drive the adoption of more efficient models or whether mandatory standards will be required(Saudi Arabia Standards Organization, 2007).

Fig. 9. Energy efficiency label for washing machines.

Washing Machines Function of input energy and load capacity (dimensionless)	**Refrigerators & Freezers** Amount energy consumption < class limit (%)	**Air Conditioners** Ratio of output cooling rate to input power (Btu/h/w)	**Star Rating**
≥ 6.0	30	> 10	6
5.0 to 5.9	25	9.5 to 10	5
4.0 to 4.9	20	9 to 9.5	4
3.0 to 3.9	15	8.5 to 9	3
2.0 to 2.9	10	7.5 to 8.5	2
< 2.0	5	< 7.5	1

Source: SASO (2005, 2006, 2007)

Table 4. Derivation of energy star rating for washing machines, refrigerators and freezers, and air-conditioners.

7.4 Energy audits

NEEP conducted a programme of energy audits to focus attention on the concept and benefits of energy efficiency by providing demonstrable 'proof of concept'. The energy audits increased corporate awareness of energy efficiency measures and technical and economic recommendations were made available for policy makers.

The energy audits were conducted at various levels; the simplest identified areas where consumers could reduce their energy usage. It involved walk-through surveys of governmental, commercial, and small industrial premises and provided immediate feedback of zero- and low-cost energy-saving possibilities. The audits examined electric utility equipment, HVAC systems, lighting, and water heating systems. As well examining the actual components of each system, the audits also examined the types of components used, locations, capacities, control systems, set points, replacement schedules, operational profiles, and historical records of electricity consumption. The results showed energy savings of at least 15% in educational buildings, 10% in shopping malls, and 10% in the industrial sector, with some smaller commercial premises showing potential savings of up to 35%. Savings in HVAC systems were obtained by rescheduling chiller and exhaust operation; all chillers were stopped overnight with maximum venting during the early morning to remove accumulated heat, and daytime break-period venting was minimised. Lighting systems phased in lower wattage fluorescent tubes, replaced electro-magnetic ballasts with electronic ballasts, and phased in halogen lamps with 50% lower wattage high-pressure sodium lamps; in addition, better use of natural lighting was made. Occupancy-based lighting and air-conditioning schedules were introduced, with occupancy sensors fitted where appropriate, and in larger facilities HVAC fresh-air streams were pre-cooled by exhaust streams.

More-detailed energy audits were performed at medium- and large-scale industrial and commercial facilities to identify specific areas in which electrical and thermal energy savings could be made. Industrial facilities included refineries, power plants, and chemical and processing plants; and commercial facilities included hotels, hospitals, shopping centres, private schools, large mosques, and office buildings. In addition to the systems mentioned above, the detailed audits also examined more specialised equipment and plant such as compressed air systems, electric motors and pumps, steam systems, and boilers and furnaces. The audits indicated that energy consumption in KSA could typically be reduced by up to 20% compared to current levels, with commensurate reductions in CO_2 emissions, and that in some older industrial facilities energy savings between 40% and 70% could be obtained by rescheduling plant and equipment operation and introducing energy performance monitoring systems. Where retrofit or replacement was recommended, payback periods between 3 months and 5 years were estimated.

The most-detailed energy audits were of boilers, furnaces, and steam systems. These were conducted in conjunction with intensive training courses in boiler and furnace efficiency. After implementing all zero- and low-cost measures guidance was given in preparing a loan request package to finance high capital-cost measures. Recommended energy efficiency improvements ranged from simple tune-ups to complete retrofits with potential energy savings between 5% and 10%.

Data from the programme are being used to target marketing efforts for detailed energy audits and feasibility studies, and are also being used for national energy efficiency planning and policy analysis. Although most facilities were well maintained, much of the maintenance was reactive, not proactive, and in some facilities cooling and lighting was operated continuously regardless of occupancy. Most importantly, however, despite management awareness of the concept of energy conservation, few energy accounting systems were in place. Although the audits highlighted energy saving potentials, long-term energy savings can only be realized through total energy management programmes, the foremost aspect of which is commitment at a corporate level to energy conservation.

7.5 Promotion of an energy service industry

Energy Service Companies (ESCOs) have been successfully operating in industrial countries for more than 50 years. ESCOs specialize in reducing their customers' energy consumption through a combination of engineering expertise and financial services. After conducting energy audits, ESCOs help customers upgrade their facilities and reduce energy costs by investing future cash flows from energy savings using a type of project financing called 'performance contracting', in which cost-saving measures are implemented at no initial cost to the customer and the energy savings are used to pay back the initial investment. ESCOs in Saudi Arabia face the same barriers that other energy efficiency initiatives and projects have encountered. Most of these barriers are due to unfamiliarity with the concept and mode of their operation by clients and financing intuitions; the financial barriers, however, relate to the fact that the ESCO industry is new in Saudi Arabia. In countries with well-established bank structures, ESCOs guarantee energy savings large enough to cover their clients' debt obligations and banks carry the credit risks. In countries with new ESCO markets, such as Saudi Arabia, banks are reluctant to carry credit risks and ESCOs must carry both performance and credit risks; the clients carry the business risk in both cases (National Energy Efficiency Program, 2006).

The annual energy efficiency market in Saudi Arabia has been estimated at SR 1.2 billion for the commercial, governmental, and industrial sectors alone [20]. Many organizations in Saudi Arabia have old and inefficient facilities, and although they would like to implement energy-conservation measures, they have neither the expertise nor the initial capital to do so, despite potential energy savings of up to 35%. There are presently five ESCOs in Saudi Arabia, and in an initiative designed to increase this number and to help those already established, NEEP organised two business advisory seminars addressing all aspects of the energy services industry. The first presented 25 topics over three days, and the second, 32 topics over five days. The seminars gave potential ESCOs advice about how to overcome contractual and legal start-up barriers, and how to identify and exploit energy-efficiency opportunities on a commercial basis. They also provided hands-on experience in business development, marketing, customer relations, contracts, staffing, and international joint ventures. In the short-term, however, the institutional, financial and commercial barriers are likely to prevent them from exploiting the opportunities.

7.6 Energy efficiency information and awareness

Due to the very low levels of awareness of the advantages and possibilities of energy efficiency on the part of both the energy service industry as well as that of energy end users, NEEP initiated an information and awareness programme to determine current levels of awareness of energy efficient opportunities, equipment, and financing mechanisms. The first part surveyed the energy service industry, and the second part surveyed energy end-users. The majority of planned energy efficiency measures target high energy consumers. The data obtained will help formulate information management and distribution strategies, and to assist energy service companies identify and exploit market opportunities.

The information and awareness programme compiled detailed information on the breakdown of electricity consumption. The Western Region accounts for 38% of residential electricity consumption, the Central Region accounts for 39% of governmental consumption, and the Eastern Region accounts for 83% of industrial consumption. Electricity is an energy source for 99% of energy end users, 80% of whom consume less than 4,000 kWh per month with 1.4% consuming more than 10,000 kWh per month. Petroleum is an energy source for 21% of customers and gas for 7%.

The majority of survey respondents feel that energy bills are the least controllable of their operational costs, as a result, they rarely budget for energy efficiency measures. Where energy efficiency projects are undertaken, they appear to be entirely self-financed. The surveys did not identify, directly or anecdotally, any examples of financial institution involvement.

Ninety percent of energy service-industry respondents were unaware of local financing options and 80% were not prepared to self-finance energy audits or implement recommendations. Among energy end-user respondents, 58% were not prepared to self-finance energy audits or implement recommendations. Among incentive options for implementing energy efficiency programs, 40% preferred free studies, 23% preferred policies and regulations, reward schemes were preferred by 23%, and soft loans was preferred by 14%. Among common energy efficiency measures implemented in the residential sector, 90% turn off unused equipment, 34% have energy saving lamps, and 34% clean their air-conditioning filters. Self-financed energy-efficient purchases were reported by 30% of residential respondents; of these 40% were for lighting, 34% for air conditioning, 16% for refrigerators and freezers, and 14% for water heaters.

A compilation of key information about energy-efficient equipment and estimates of possible energy and monetary savings was made for dissemination by local organisations, the media, and the Internet. It is part of an on-going nation-wide public-awareness programme designed to educate consumers and to influence their buying decisions in favour of energy-efficient devices. A permanent energy-efficiency exhibition is planned for the newly constructed Prince Salman Science Oasis in Riyadh, and others may follow in other regions. The exhibits will include interactive displays and information about energy efficient air-conditioning systems, lighting, cooking, water heating, and insulation.

7.7 Energy efficient equipment leasing programme

This is an ongoing programme designed to eliminate financial barriers to the purchase of energy-efficient equipment by supplying energy-efficient equipment to consumers and recovering the cost through lease payments added to electricity bills. Energy savings are expected to cover the cost of the leases. Initially the programme will target air-conditioning units, and replace energy-inefficient units with energy-efficient units. Electricity demand is expected to fall by an amount proportionate to the number of units supplied through the programme. An important aspect of the programme is that units are traded-in. Without this provision, the old air-conditioning units would be retained and utilized in a location that wasn't previously air conditioned; instead of reducing energy consumption, this would add to consumption and increase demand on the grid. The trade-in condition also allows proper disposal of hazardous materials and correct recycling and disposal of the old units.

7.8 Technical and management training

In addition to the seminars and training mentioned above, training in various other energy efficiency disciplines was provided in the form of workshops and seminars. Most workshops were offered in several venues across the country but were only made available to suitably qualified professionals as a major consideration was training trainers. Over 400 personnel from all sectors attended the training sessions, many of whom have gone on to establish regular training sessions and energy efficiency programmes in their own institutions and companies. Sessions included:

- Quick savings programmes
- Detailed energy audits
- Energy-efficiency project financing
- Performance contracting and ESCO development
- Boiler and furnace efficiency
- Steam systems
- Motor efficiency
- Technical efficiency and energy management
- Demand-side management
- Energy efficiency technologies

8. Policy recommendations

During the course of the programme the need for additional government initiatives were identified; the most critical of which are:

8.1 Endorsement

Enforceable legislation is essential if energy efficiency measures are to have long-term national effect. For energy efficiency to become an integral and fundamental part of doing business, active and unequivocal support and endorsement by the Government of Saudi Arabia is essential. Leading institutions, such as KACST and SASO, ministries such as MOWE and MPMR, and major industries must be seen to take the lead and to support the concept and goals of the energy efficiency and conservation.

8.2 Transparency

Government needs to take the lead in transparency. All government agencies and ministries, as well as industry and commerce, must make information about their energy consumption patterns and energy efficiency activities available. Historically, commercial confidentiality has been cited for the reason very little data has been made available to public oversight. Without data, however, advisory and regulatory engineers and consultants are unable to provide detailed advice and feasibility studies to improve plant efficiency. Lack of data also degrades the ability to formulate public policy and to assure the security and economic growth of Saudi Arabia through efficient and wise use of its energy resources.

Public sector and large industrial plants and facilities should conduct mandatory energy audits and the verifiable non-confidential results need to be in the public domain. Ideally, information claimed to be confidential should be supported with documented reasons in support, and should, to the governing authority's satisfaction, adequately demonstrate that disclosure could seriously and prejudicially affect their commercial interests.

8.3 Training

Government needs to address the critical shortage of qualified engineers and technicians. There are at present insufficient suitably qualified and experienced engineers and technicians to perform the necessary work on a national scale. The provision of training needs to be accelerated.

8.4 Awareness

Government support is needed to increase public awareness of energy efficiency to a level substantially higher than its present level. High-level public figures need to endorse the effort to conserve the country's finite natural resources; in local and national legislation; in professionally produced programmes and campaigns of public awareness; but most importantly, through public, private, and professional example.

8.5 Financing

An initial programme of government intervention or subsidy, perhaps in the form of early adoption loan guarantees or a sliding scale of energy rebates, is required to encourage financial institutions to finance projects identified and endorsed as technically and economically feasible.

8.6 Inertia

Even when positive benefits are clearly demonstrated, resistance to change and a reluctance to adopt new technologies and methods extends from government procurement specifications to the adoption of energy efficiency standards for appliances and equipment. Decades of low energy prices and the consequent endemic wastefulness, requires the imposition and coordination of top-down energy efficiency policies and regulations to invoke changes in energy usage. Even with high level endorsement, public awareness campaigns, and financial inducements, public acceptance is likely to lag behind awareness. For this reason, the government should consider a staged implementation of policies, starting with those with the greatest public need or support.

8.7 National centre

Responsibility for energy efficiency rests on everyone, yet to ensure that measures are applied consistently and efforts are not duplicated, central coordination is required. This could be provided by a national centre for energy efficiency. In addition to continuing and extending the work initiated by NEEP, the centre should be able to advise at all levels and on all aspects of energy efficiency and conservation, and should be responsible for ensuring that they inform all aspects of governmental policy.

9. Initiatives for reducing energy consumption on the supply side

9.1 Modernisation of existing power plants

Most power generation capacity in Saudi Arabia comes from open cycle gas turbines and from steam turbines with efficiencies around 30%. Based on current mix of power generation capacity in Saudi Arabia, the nominal efficiency for the current power generation mix is around 29.5%. This relatively low energy level can be improved by converting existing open cycle gas turbines (OCGT) and steam turbine (ST)-based power plants to combined cycle gas turbines CCGT which can achieve efficiencies as high as 50% in Saudi Arabia. The value of constructing CCGTs rather than OCGTs has been recognised in Saudi Arabia. For, example the new power plant constructed at Qurayyah in the East part of SA was originally intended to be OCGT (15 x 127 MWe). The OCGT plant with a total capacity of ~ 1.9GW was completed in May 2010 with an investment cost of $ 570 Million. However, recently a $300 Million contract was awarded to GE to supply five STs for a major expansion of the Qurayyah OCGT plant and converting it to CCGT. This will increase the output of the power plant from 1.9 GW to 3.1 GW.

Considering OCGT, there are currently more than 3 GWe of power plants which can be potentially converted to CCGT. Our calculations show that this conversion can potentially provide an opportunity for oil saving of 14-15 million barrels/year which equate to additional revenue of $Bn1.2/year at current oil prices. The conversion to CCGT will produce more electricity for the same amount of fuel. However, the operation of CCGT will require additional infrastructure and so additional investment will be required for the conversion. The techno-economic feasibility of converting to CCGT then needs to be investigated on a case-by-case basis.

9.2 Minimum efficiency standard for new power plants

The overall nominal efficiency of power generation in Saudi Arabia can be improved by setting a minimum efficiency standard for new power plants. This will ultimately lead to a reduction in the additional capacity required by 2030 (currently at 35 GWe) and will consequently lead to primary energy/oil savings. These minimum efficiency standards are dependent on climatic conditions and so they need to be estimated specifically for Saudi Arabia and may differ from one region to another. For example, higher ambient air temperature will increase the compression power requirement thus reducing efficiency. In addition, higher condenser water temperatures will reduce the efficiency of the steam cycles. Thus, these need to be taken into consideration in setting minimum efficiency standards for Saudi Arabia. We recommend a minimum efficiency standard of 36% for new gas turbines and 52% for CCGT.

OCGT power plants will still be required to meet peak demand. However, it is obvious that significant improvements in the overall efficiency level and reduction in energy consumption can be achieved by building more CCGT power plants. Considering a situation where all newly-built power plants to 2030 are CCGT, oil savings of 140 million barrels of oil/year can be achieved. This corresponds to revenue of $Bn11/year at current oil prices.

9.3 Minimum overall efficiency standard for co-generation desalination plants

Desalinated water demand in Saudi Arabia is expected to increase over the next two decades with an additional 20-30 desalination plants required by 2030. Desalination plants are energy-intensive with heat being the major requirement for multistage flash vaporisation. Many desalination plants in Saudi Arabia use heat from co-generation units where electricity is also generated and exported to the grid. In 2008, 19 TWh of electricity and about 1.2 Billion m^3 of water were delivered from co-generation desalination plants.

A minimum overall efficiency (heat + electricity) of 80% is recommended for Saudi co-generation plants. This will ensure higher utilisation of available heat and consequently potentially more desalinated water for the same amount of electricity generated. Analysis of available data from Saudi co-generation desalination plants shows that power-to-water ratio (PWR) is on average 1600 MWe per million m^3 of desalinated water. Depending on the prime mover used in co-generation, PWR can be as high as 4000 MWe per million m^3 of desalinated water. This shows that there is a scope for improving power output from Saudi Desalination plants.

9.4 Implementation of a renewable energy programme to displace fossil fuels

Analysis for the current report shows that there is a great potential for solar energy in Saudi Arabia. A photovoltaic area of 22 - 40 km^2 in Saudi Arabia can produce as much electricity as a 1000 MWe oil-fired power station. We have shown that if subsidy on oil is removed then, at current oil prices (assuming $80/barrel), electricity from PV in Saudi Arabia can be more competitive than electricity from oil-fired power generation. Our estimates show that in order to stimulate PV in Saudi Arabia, an incentive equivalent to approximately 64 Halalah/kWh needs to be introduced.

Our review of renewable energy options for Saudi Arabia also showed a great potential for solar thermal applications. Solar electricity generating systems (solar thermal concentrators) can achieve conversion efficiencies of 18-30%. Solar absorption cooling where thermal energy from the sun is used for re-generation in absorption chillers is another possible application. Despite its great potential, the wide scale deployment of solar energy in Saudi Arabia faces several barriers. These include lack of incentives, high costs (considering current subsidies on oil), lack of public knowledge, and lack of professional training.

10. Initiatives for reducing energy consumption on the demand side

10.1 Introduction and implementation of a new energy conservation law

An energy conservation law in Saudi Arabia will assist in facilitating energy savings throughout society improving efficiency and economic benefits of energy use. In this context, energy conservation refers to the reduction of loss and waste in various energy stages from energy production to energy consumption and the use of energy more efficiently and rationally by strengthening management of energy use by adopting technologies which are technically and economically feasible and environmentally friendly KSA can formulate energy conservation policies and plans which can then be incorporated into the country's economic and social development program. KSA can also develop programmes to enhance and encourage research in the science of energy conservation. An energy conservation authority can be created to become in charge of the supervision and management of the energy conservation work throughout the country. In addition, the construction of new industrial projects exploiting technologies with excessive energy consumption and severe waste of energy should be prohibited and monitored by this authority.

National standards for energy conservation should be established and published. For sectors which are not covered by the national standards, relevant sectional standards of energy conservation should be established and filed with the standardization authority.

The following policies on energy conservation technologies should be particularly encouraged

- General energy conservation technologies which have demonstrated their maturity and economic benefits.
- Co-generation of thermal power and district cooling and co-supply of heat from co-generation to industrial processes.
- Economic operation for electrical motors, industrial fans and pumping equipment in the industrial sector by developing speed adjustable motors for energy savings and electrical and electronic energy-saving technologies for electric power.

10.2 Introduction and implementation of building regulations

About two thirds of the electric energy generated in Saudi Arabia is used for operating buildings. More than 60% of this energy is used in air conditioning. This high consumption of electricity combined with escalating energy prices has created a need for the introduction of energy conservation measures in buildings in KSA. While a Saudi building code exists, this needs to be enforced so that actual savings can be achieved. Since the main purpose of building regulations in KSA would be to reduce cooling demand, they should consider establishing building designs which encourage the implementation of energy-saving measures. This should cover the types of construction structures, materials, facilities and products which improve building designs and at the same time reduce cooling energy consumption and lighting.

Building regulations in Saudi Arabia should

- Include energy efficiency and energy conservation as a main component of the design, construction and operation of the building
- Provide the tools for enforcing the Saudi Building Code
- Provide the means for inspection and monitoring of energy efficient measures during the construction phase and afterwards during the operation of the building
- Introduce a licensing regime where energy conservation are an important requirement for approval

11. Scenario analysis and potential oil savings

This section gives brief analysis of two scenarios for reducing primary energy consumption and consequently oil consumption based on the policy initiatives discussed above. Two different scenarios are considered to illustrate how the growth in primary energy demand might be constrained. Both scenarios are compared with the Business As Usual (BAU) scenario presented in Section 5 which assumes the current predicted rate of growth will be reached by 2030. The two scenarios represent a lower, conservative estimate, Low Primary Energy Conservation (LPEC), and an upper estimate, High Primary Energy Conservation (HPEC). The upper estimate assumes that a more ambitious reduction can be achieved. Both scenarios include a combination of energy supply, energy demand and some renewable energy.

In the power generation sector, several options are considered for reducing energy consumption. These include the following

1. Improving the power generation efficiency by increasing the share of CCGT capacity is considered. Currently CCGT makes only 7% of the total generation capacity. CCGT is the most effective natural gas-based power generation technology with efficiency of at least 50%. For CCGT in KSA, an additional benefit can be recognised. Currently, 45% of power generation is natural-gas-based and if more natural gas-based power plants are built, more oil can be released for export. Two levels of CCGT are considered here. An additional 35 GW of power generation capacity is required in the next two decades. In the LPEC scenario, an additional 5 GW of CCGT is considered while in the HPEC scenario, an additional 30 GW is considered.
2. Reducing electricity consumption by (a) improving appliance efficiency and (b) utilising electricity conservation measures in order to reduce cooling/heating load in the residential, commercial, and other sectors. The LPEC scenario considers a reduction of 5% in electricity consumption which can then be translated into primary energy savings. The HPEC scenario considers 10% reduction in electricity consumption.
3. Using renewable energy for generating electricity. This paper focuses on PV as a means for generating electricity. The LPEC scenario considers 1 GW of PV. At current PV module prices, this corresponds to investment costs of $4-6 billion, a saving of 5 million barrel of oil per year which can be directed for export and CO_2 savings of 2.1 Mt/year.

The HPEC scenario considers a 5 GW of PV capacity which corresponds to investment costs of $20-30 billion. These costs, however, are expected to decrease as the rate of production increases. A PV capacity of 5 GW would displace 5 million barrels of oil per year (worth $300 - $400 million based on current international oil prices) for export and save 5 Mt/year of CO_2.The desalination sector in KSA uses about 11% of the total primary energy consumption. As shown in Section 4.3, this is also expected to be the case in 2030 with primary energy consumption reaching 325 TWh in 2030 in order to satisfy the increasing demand for water.

In KSA, currently 153 TWh are consumed in the desalination sector producing 19 TWhe of electricity and 1135 m^3 of desalinated water/year. The LPEC scenario assumes a reduction of 5% in primary energy in the desalination sector while the HPEC scenario assumes a reduction of 10%. Reducing energy consumption in the desalination sector can be achieved in several ways:

- Utilisation of solar energy as a source of the electrical and thermal energy required by desalination plants. In membrane desalination plants (such as reverse osmosis) electrical energy is required to operate the pumps and desalination units. In thermal distillation plants, both electrical and thermal energy are required and so both solar PV and solar thermal applications can be utilised.

- Lowering the energy consumption per m^3 of desalinated water.
- Reducing the demand for desalinated water will consequently lead to lower primary energy consumption in the desalination sector.

In the BAU scenario, primary energy consumption in the industrial sector amounts to 379 TWh. The LPEC scenario assumes a modest reduction of 5% while the HPEC scenario assumes that a 10% reduction could be achieved.A summary of the assumptions used for these two scenarios is shown in Table 5. Figure 10 compares the LPEC and HPEC scenarios with the BAU scenario.

Sector	Reduction measure by 2030	LPEC scenario	HPEC scenario
Power generation	Additional CCGT, GW	5	30
	Reduction in Electricity consumption	5%	10%
	Solar PV, GW	1	5
Desalination	Reduction in primary energy consumption, %	5%	10%
Industry	Reduction in primary energy consumption, %	5%	10%

Table 5. Assumptions for the low and high energy conservation scenarios.

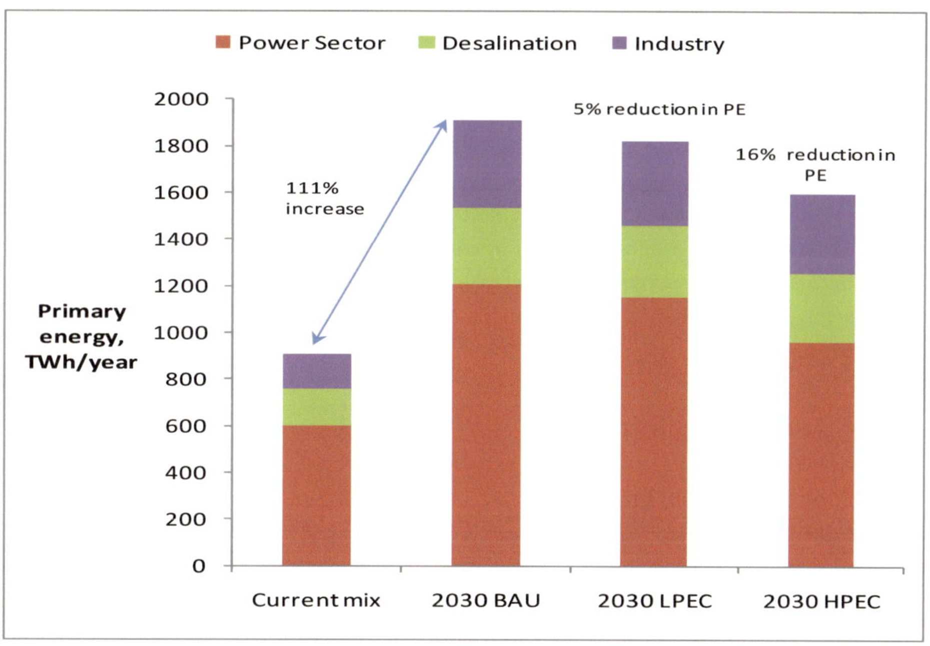

Fig. 10. Reduction in total primary energy according to scenario 1 (LPEC) and scenario 2 (HPEC).

Under the LPEC scenario, a total primary energy saving of only 5% from all the proposed measures would be achieved compared with BAU. Under the more ambitious HPEC scenario, a reduction of 16% could be achieved. A summary of the results is given in Tables 6 and 7. Table 6 shows that significant oil savings and additional revenue from export can be achieved from even the "low primary energy conservation" scenario. Also Table 6 shows the potential CO_2 savings associated with energy savings in each of the scenarios. The CO_2 savings associated with oil savings in the HPEC scenario are 13% of the total emissions in 2030. These figures account for the difference in CO_2 emissions between oil and natural gas-fired generation.

Sector	BAU	LPEC	HPEC
Primary energy consumption, TWh/year	1912	1820	1597
% power	63	63	60
% desalination	17	17	18
% industry	20	20	22
% Reduction from BAU	-	5%	16%
Oil savings compared to BAU, M barrels/y	-	54	185
Additional revenue at today's oil prices, $Billion/year	-	4.3	14.8
CO_2 savings associated with oil savings, Mt CO_2/y	-	92	126

Table 6. Comparing LPEC and HPEC scenarios to BAU scenario in KSA.

Base case assumption	Potential PE savings TWh/year	PE savings compared to BAU, %
5 GW of CCGT	12	0.6%
30 GW of CCGT	123	6.4%
1 GW of PV	11	0.6%
5 GW of PV	55	2.8%
5% reduction in electricity consumption	34	1.8%
10% reduction in electricity consumption	67	3.5%
Improving co-generation efficiency from 60% to 65%	16	0.8%
Improving desalination efficiency by 12% (from 60% to 67%)	32	1.7%
10% reduction in industrial primary energy consumption	38	2%

Table 7. The effect of changing each of the parameters in comparison to BAU scenario.

It is clear from Table 7that most savings can be achieved by changing the mix of the power generation sector. The construction of CCGT power plants in the future can lead to significant savings in primary energy and to also additional oil exports. Electricity efficiency improvements in the residential, commercial and industrial sectors offer the next most significant savings

12. Conclusion

This paper presented different affordable initiatives for reducing energy consumption on both sides of energy supply and demand. Also, the paper proposed two scenarios which have been considered to demonstrate the effectiveness of combination of measures including fuel switching, energy efficiency and solar energy. Each scenario compared the savings that could be achieved relative to Business As Usual (BAU), which assumes contained unconstrained growth until 2030. The first scenario includes 5 GW of new Combined Cycle Gas Turbines(CCGT), 1 GW of Photo Voltaic (PV) and only a 5 % improvement in energy efficiency. The total primary energy would be reduced by 5 % compared with BAU. A second ambitious scenario which includes 30 GW of CCGT, 5 GW of PV and a 10 % improvement from energy efficiency measures would lead to an estimated 16 % reduction compared with BAU.

13. References

Abdel-Jawad,Mahmoud. (2001). *Energy sources for coupling with desalination plants in the GCC countries*, report. ESCWA.

Alowaidh, M., and Alnutifi, A., (2010). *Technology-Based Improvement for KSA Energy Intensity*, Internal Report, Saudi Aramco.

Electricity and Cogeneration Regulatory Authority. (2008).*Updated Generation Planning for Saudi Electricity Sector*. Report. Riyadh.

Electricity and Cogeneration Regulatory Authority. (2009). *General Directorate, Economy and Tariff Affairs*, Annual Statistical Booklet on Electricity Industry. Riyadh.

Electricity Cogeneration Regulatory Authority.(2006). *Updated Generation Planning for Saudi Electricity Sector*. Saudi Arabia.

Elhadj,Elie. (2004). *Household water and sanitation services in Saudi Arabia: an analysis of economic, political and ecological issues*, SOAS Water Research Group, Jeddah.

International Energy Agency. (2006a). *Energy Balances of OECD Countrie*s. Statistics Division, International Energy Agency.

International Energy Agency. (2006b). *Economic Indicators and Energy Balances of Non-OECD Countries.* Statistics Division.

International Energy Agency. (2009). *Energy Balances of OECD Countries. Statistics Division.*

Liebendorfer, K.M., Andrepont, J.S. (2005). *Cooling the Hot Desert Wind: Turbine Inlet Cooling with Thermal Energy Storage (TES) Increases Net Power Plant Output 30%*, Proceedings of the ASHRAE Annual Meeting, ASHRAE Transactions, Volume 111, Part 2.

Ministry of Water and Electricity. (2006). *Electricity Growth and Development in the Kingdom of Saudi Arabia,* Annual Report, Saudi Arabia.

Ministry Of Water and Electricity. (2009).*Electricity Growth and Development in the Kingdom of Saudi Arabia*, Annual Report, Riyadh.

National Energy Efficiency Program. (2006). *Discovering Business Opportunities in the Energy Services Industry*, Second Workshop, Riyadh, Saudi Arabia.

National Energy Efficiency Program. (2010). *Energy Efficiency Opportunities in the Kingdom of Saudi Arabia*, Report , Saudi Arabia.

Said,S., ElAmin,I. and AlShehri,A. (2205). *Renewable energy potentials in Saudi Arabia.* Report. Saudi Arabia.

Saudi Arabia Standards Organization. (2005). *Energy Labelling and Minimum Energy Performance Requirements for Air-Conditioners,* Standard No. 3459, Saudi Arabia.

Saudi Arabia Standards Organization. (2006). *Energy Labelling Requirements of Household Electrical Clothes Washing Machines.* Standard No. 3569, Saudi Arabia.

Saudi Arabia Standards Organization.(2007). *Energy Performance Capacity and Labelling of Household Refrigerators, Refrigerator-Freezers, and Freezers,* Standard No. 3620, Saudi Arabia.

Saudi Arabian Standards Organization. (2005). *Energy Labelling and Minimum Energy Performance Requirements for Air-Conditioners.* Standard 3459, Saudi Arabia.

Saudi Electricity Company. (2009). Annual Report, Riyadh.

Saudi Electricity Company.(2010). Annual Report, Riyadh.

World Bank. (2007). *World Development Indicators,* Report, Washington, D.C.

14

A Comparison of Electricity Generation Reference Costs for Different Technologies of Renewable Energy Sources

Alenka Kavkler[1], Sebastijan Repina[2] and Mejra Festić[3]
[1]*Faculty of Business and Economics, University of Maribor and EIPF Economic Institute Ljubljana*
[2]*EIPF Economic Institute, Ljubljana*
[3]*Faculty of Business and Economics, University of Maribor and EIPF Economic Institute Ljubljana*
Slovenia

1. Introduction

The target value for electricity production from renewable resources in Slovenia till 2020 has amounted to additional 3.146 GWh. The highest share of this additional electricity will be obtained in majority from hydro power plants: additional 1.299 GWh will be obtained from big hydro power plants (that requires 618,3 mio € for investment); wind power plants with additional 567 GWh (that requires 345,6 mio € investment); photovoltaic with 469 GWh (which requires 1.641,5 mio € additional investment); biomass with additional 267 GWh (that requires 11,3 mio € of investment); small hydro power plants with additional 194 GWh and the needed investment amount of 148,8 mio €; natural gas with additional 191 GWh (that requires the investment amount of 95,5 mio €); and geothermal power plants with 150 GWh (that required investment amount of 93,8 mio € investicij). (Kavkler, Festić, Repina 2009).

Considering the economy and the contribution of particular technologies of renewable energy sources (RES) to the national economy macro indicators an adequate system for state incentives and abatements of RES investments in energy industry had to be made. The key eligibility condition for support of investments in renewable energy sources is the electricity cost price of particular RES technologies. Financial aid for electric power generated from renewable energy sources is defined in "The decree regulating subsidies for electricity generated from renewable energy sources", published in the Official Gazette of the Republic of Slovenia", nr. 37/09. Among other things the decree also defines (see art. 1):

- types of RES production facilities eligible for subsidies;
- allocation of RES production facilities into size classes;
- a more detailed definition of subsidies;
- the method of reference costs determination for electricity generated from renewable energy sources;
- the method of subsidy rate fixing and the eligibility conditions for support.

Production facilities exploiting the following renewable energy sources meet the eligibility criteria for subsidies (art. 3):

i. energy potential of watercourses;
ii. wind energy exploited in land-based production facilities;
iii. solar energy exploited in production facilities using photovoltaics;
iv. geothermal energy;
v. energy generated from biomass;
vi. energy generated from biogas originating from the treatment of biomass and biologically degradable waste
vii. energy generated from landfill gas
viii. energy generated from biogas originating from sludges from the treatment of industrial waste water;
ix. energy generated from biologically degradable waste.

Size classes of RES production facilities defined in "The decree regulating subsidies for electricity generated from renewable energy sources" are listed in Table 1.

Size classes of RES production facilities	Potential
1. Micro	less than 50 kW
2. Small	less than 1.000 kW
3. Middle	from 1 to 10 MW
4. Big	over 10 to including 125 MW

Source: "Methodology of reference costs determination for electricity generated from renewable energy sources", IJS, p. 9, table 1.

Table 1. Size classes of RES production facilities.

Subsidies are defined as potentially eligible financial aid to the generation of electricity by particular RES technologies if their production costs of electricity top the market price. The fifth article of "The decree regulating subsidies for electricity generated from renewable energy sources" defines two types of subsidies for RES production facilities:

1. compulsory purchase of electricity;
   The subsidiary centre purchases all net generated electricity at prices defined by this decree.
2. financial aid for current operations;
   These subsidies are granted to the net generated electricity if production costs top the market price of electricity.

Since reference costs of electricity generation (RCEG) are the starting point for the calculation of subsidy amounts for RES production facilities in the continuation of the paper a short description of RCEG methodology and RCEG calculations for different RES technologies will be introduced. A sensitivity analysis taking into account the financial volume of the investment and the interest rate of the loan will also be displayed.

2. A short description of the employed methodology

Reference costs of electricity were calculated in accordance with the "Methodology for the reference costs calculation of RES generated electricity" instructions that had been prepared at the Institute Jozef Stefan Energy Efficiency Centre. Investment risk was accessed by means of sensitivity analysis.

2.1 Reference costs of electricity

In the "Methodology for the reference costs calculation of RES generated electricity" reference costs of electricity generation (RCEG) are defined on page 7 as follows:

RCEP represent the total annual operation costs of a typical RES production facility reduced for all revenues and benefits of operations (sale of heat, etc.) and can be formulated in Eur/MWhel by means of the following equation:

RSEG = (COSTS - REVENUES) / ELECTRICITY

COSTS	= annual investments (annual instalment) + operating expenses (Eur) + cost of fuel (Eur)
REVENUES	= sale of heat (Eur) + other benefits (Eur)
ELECTRICITY	= annually generated electricity (MWh)
	= installed power (MWel) * annual operating hours (h)

The method of RCEG calculation is based on an annuity method of investment cost evaluation that also takes into account the cost of capital and the required return on invested capital respectively.

The calculation of RCEG for RES production facilities based on cogeneration of heat and electricity (CHE) and those that use different fuels is also affected by the following two parameters ("Methodology for the reference costs calculation of RES generated electricity", p. 10):

1. electricity efficiency (EffEl),

i.e. the ratio between the CHE production facility potential and fuel input potential;

2. thermal efficiency (EffT),

i.e. the ratio between the output calorific power (useful heat) of the RES production facility and fuel input potential.

Fuel consumption and the generation of useful heat can be calculated by means of the following equations (p. 10):

Fuel consumption (MWh) = Electricity potential (MWEl) * Operating hours (h)

Useful heat (MWh) = Electricity potential (MWEl) * EffT/ EffEl * Operating hours (h)

= Generated electricity (MWh) * EffT/ EffEl

2.2 Sensitivity analysis

We do not know the exact net cash flows of the investment since they are exposed to numerous risks and can only be estimated. By means of sensitivity analysis we get to know how the changes of certain variables influence the volume of cash flows and consecutively the investment effectiveness indicators. Each time only one of the variables is varied assuming that the values of all other variables remain unchanged. It is of crucial importance that critical variables whose changes have a substantial influence on electricity reference costs are chosen (Brigham and Houston, 2001).

3. Assumptions and data

Assumptions are recapitulated from the "Methodology for the reference costs calculation of RES generated electricity":

- Depreciation period: 15 years

Datum corresponds to an average depreciation period for RES production facilities with regard to the existing practice.

- Share of own resources: 40 %
- Loan: 60 % of the investment.
- Required yield on own invested resources: 20 %

The required yield on own invested resources in Slovenia is relatively high because of the possible production transfer abroad.

- Cost of loan: 6.5 %

Calculation of loan costs is made on the basis of EURIBOR for 2008 (4.7 %) with an extra payment of 1.8 %.

- Discount rate: 12 %

Discount rate is defined as a weighed average of capital costs (WACC). For solar power stations (in accordance with the guidelines from abroad) because of the most expensive technology a lower discount rate (8 %) was used.

- Annual cost of labour: 25.000 Eur/person.
- Basic price of wooden biomass in 2008: 23 Eur/MWh
- Average price of substrata mixture for biogas plants in 2009: 14,98 Eur/MWh
- Value of useful heat for all sizes of RES production facilities in 2009: 26,74 Eur/MWh

Tables containing basic data for different types of power plants are taken from the "Methodology for the reference costs calculation of RES generated electricity". For waste incinerators unfortunately there are no available data. Data used include the potential of power plants (MW), number of annual operating hours, amount of investment (Eur/kW), costs of maintenance, operation and insurance (as a % of investment) and the cost of labour (number of persons employed) and are stated in Tables 2 to 7.

Electrical efficiency for a small production facility using biomass that exceeds the 90 % share of fuel energy is 12 % while for a middle sized it is 17 %; a minimum 70 % operating efficiency is required.

Size class	Size	Operating hours	Spec. investment	Maintenance	Operation	Insurance	Labour
	MWe	h/year	Eur/kWel	% inv.	% inv.	% inv.	nr. of persons
up to 50kW	0,05	4.000	2.300	0,9 %	0,6 %	1,5 %	0,03
up to 1MW	1	3.500	1.700	1,5 %	0,6 %	1,7 %	0,4
up to 10MW	5	3.500	1.500	1,5 %	0,6 %	1,8 %	1,8
up to 125MW	30	3.500	1.400	1,5 %	0,6 %	1,8 %	9

Source: "Methodology for the reference costs calculation of RES generated electricity", IJS, p. 20, table 5

Table 2. Hydroelectric power plants - basic data.

Size class	Size	Operating hours	Spec. investment	Maintenance	Operation	Insurance	Labour
	MWe	h/year	Eur/kWel	% inv.	% inv.	% inv.	nr. of persons
up to 50kW up to 1MW up to 10MW	5	2.100	1.200	0,3 %	0,2 %	1,3 %	0,5
up to 125MW	50	2.100	1.100	0,3 %	0,2 %	1,3 %	5

Source: "Methodology for the reference costs calculation of RES generated electricity", IJS, p. 21, table 6.

Table 3. Wind farms - basic data.

Size class	Size	Operating hours	Spec. investment	Maintenance	Operation	Insurance	Labour
	MWe	h/annum	Eur/kWel	% inv.	% inv.	% inv.	nr. of persons
up to 50kW	0,05	1.050	3.620	0,1 %	0,05 %	0,4 %	0,015
up to 1MW	0,5	1.050	3.330	0,1 %	0,05 %	0,4 %	0,15
up to 10MW	2	1.050	2.685	0,1 %	0,04 %	0,4 %	0,5
up to 125MW	10	1.050	2.455	0,1 %	0,04 %	0,4 %	4

Source: "Methodology for the reference costs calculation of RES generated electricity", IJS, p. 24, table 8.

Table 4. Solar power plants (as independent objects) - basic data.

Size class	Size	Operating hours	Spec. investment	Maintenance	Operation	Insurance	Labour
	MWe	h/annum	Eur/kWel	% inv.	% inv.	% inv.	nr. of persons
up to 50kW up to 1MW up to 10MW	5	6.000	4.600	2,0 %	0,7 %	1,2%	12
up to 125MW	(1)	(1)	(1)	(1)	(1)	(1)	(1)

(1) Individual treatment of production facilities.

Source: "Methodology for the reference costs calculation of RES generated electricity", IJS, p. 25, table 10.

Table 5. Geothermal power plants - basic data.

Size class	Size	Operating hours	Spec. investment	Maintenance	Operation	Insurance	Labour
	MWe	h/annum	Eur/kWel	% inv.	% inv.	% inv.	nr. of persons
up to 50kW							
up to 1MW	0,5	5.500	4.500	2,0 %	0,8 %	1,2%	1
up to 10MW	2	5.500	3.200	2,0 %	0,8 %	1,2 %	3
up to 125MW							

Source: "Methodology for the reference costs calculation of RES generated electricity", IJS, p. 28, table 11.

Table 6. Production facilities using biomass that exceeds the 90 % share of fuel energy.

Size class	Size	Operating hours	Spec. investment	Maintenance	Operation	Insurance	Labour
	MWe	h/annum	Eur/kWel	% inv.	% inv.	% inv.	nr. of persons
up to 50kW	0,05	6.800	4.000	2,0 %	0,8 %	1,2 %	0,12
up to 1MW	0,5	6.800	3.800	2,0 %	0,8 %	1,2%	1
up to 10MW	2	6.800	3.300	2,0 %	0,8 %	1,2 %	3
up to 125MW	(1)	(1)	(1)	(1)	(1)	(1)	(1)

(1) RCEG are not defined.
Source: "Methodology for the reference costs calculation of RES generated electricity", IJS, p. 31, table 13.
For biogas plants electrical efficiency of 34 % was used.

Table 7. Biogas plants operating on biogas produced from biomass - basic data.

4. Results

4.1 Reference costs of electricity

Reference costs of electricity generation (RCEG) for 5 types of power plants were calculated on the basis of methodology developed at the Institute Jozef Stefan Energy Efficiency Centre. Calculations for small and middle sized hydro power plants were done separately while for other types of RES production facilities the greatest power for which relevant data existed was used. For waste incinerators unfortunately there were no data available. Results stated in Table 8 are in accordance with the RCEG from "The decree regulating subsidies for electricity generated from renewable energy sources", Official Gazette of the Republic of Slovenia", nr. 37/09 and from the "Methodology for the reference costs calculation of RES generated electricity".

Type of power plant	Power (MW)	Investment Spec. (Eur/kW)	RCEG (Eur/MWh)
Hydro power plant (small)	1	1.700	92,16
Hydro power plant (big)	30	1.400	76,57
Wind	50	1.100	86,74
Solar	10	2.455	269,22
Geothermal	5	4.600	152,47
Biomass	2	3.200	167,43
Biogas	2	3.300	140,77

Table 8.

As already mentioned solar power stations have the highest while hydroelectric power plants and wind farms have the lowest reference costs. Results are displayed graphically in Picture 1.

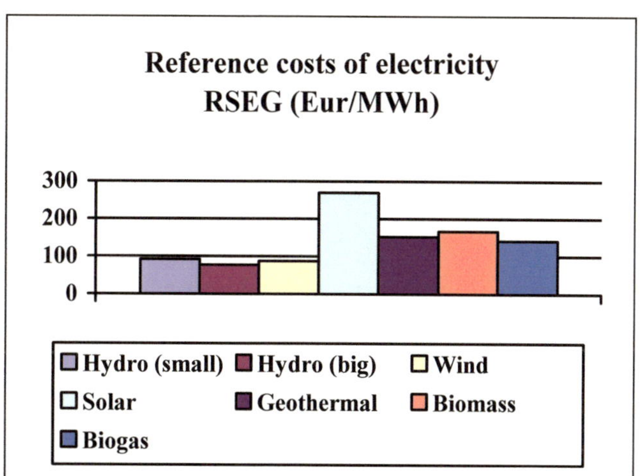

Fig. 1. Graphical display of electricity reference costs.

4.2 Sensitivity analysis

Taking into account the available data it was reasonable to perform a sensitivity analysis in view of the amount of investment and the interest rate of the loan. With investment costs a reduction of 10% and 20 % and an increase of 10% and 20 % was taken into account. Interest rate of the loan was varied as follows: 4,5 %, 5,5 %, 7,5 % and 9,5 %. The results are displayed in Tables 9 to 15. Investments are more sensitive to the investment input modification.

RATE OF INVESTMENT					
	- 20 %	- 10 %		+10 %	+ 20 %
RCEG(Eur/MWh)	78,19	85,17	92,16	99,14	106,13
INTEREST RATE OF THE LOAN					
	4,5 %	5,5 %		7,5 %	8,5 %
RCEG(Eur/MWh	88,30	90,20	92,16	94,18	96,26

Table 9. Results of the RCEG sensitivity analysis for small hydroelectric power plants.

RATE OF INVESTMENT					
	- 20 %	- 10 %		+10 %	+ 20 %
RCEG(Eur/MWh)	65,06	70,82	76,57	82,32	88,07
INTEREST RATE OF THE LOAN					
	4,5 %	5,5 %		7,5 %	8,5 %
RCEG(Eur/MWh	73,39	74,95	76,57	76,23	79,94

Table 10. Results of the RCEG sensitivity analysis for big hydroelectric power plants.

RATE OF INVESTMENT					
	- 20 %	- 10 %		+10 %	+ 20 %
RCEG(Eur/MWh)	71,67	79,21	86,74	94,27	101,8
INTEREST RATE OF THE LOAN					
	4,5 %	5,5 %		7,5 %	8,5 %
RCEG(Eur/MWh	82,58	84,62	86,74	88,92	91,16

Table 11. Results of the RCEG sensitivity analysis for wind farms.

RATE OF INVESTMENT					
	- 20 %	- 10 %		+10 %	+ 20 %
RCEG(Eur/MWh)	319,93	244,58	269,22	293,67	318,51
INTEREST RATE OF THE LOAN					
	4,5 %	5,5 %		7,5 %	8,5 %
RCEG(Eur/MWh	264,29	266,76	269,22	271,69	274,15

Table 12. Results of the RCEG sensitivity analysis for solar power plants.

RATE OF INVESTMENT					
	- 20 %	- 10 %		+10 %	+ 20 %
RCEG(Eur/MWh)	130,41	141,44	152,47	163,49	174,52
INTEREST RATE OF THE LOAN					
	4,5 %	5,5 %		7,5 %	8,5 %
RCEG(Eur/MWh	146,38	149,37	152,47	155,66	158,94

Table 13. Results of the RCEG sensitivity analysis for geothermal power plants.

RATE OF INVESTMENT					
	- 20 %	- 10 %		+10 %	+20 %
RCEG(Eur/MWh)	150,69	159,06	167,43	175,79	184,16
INTEREST RATE OF THE LOAN					
	4,5 %	5,5 %		7,5 %	8,5 %
RCEG(Eur/MWh	162,81	165,08	167,43	169,85	172,34

Table 14. Results of the RCEG sensitivity analysis for production facilities using biomass that exceeds the 90 % share of fuel energy.

RATE OF INVESTMENT					
	- 20 %	- 10 %		+10 %	+20 %
RCEG(Eur/MWh)	126,81	133,79	140,77	147,75	154,72
INTEREST RATE OF THE LOAN					
	4,5 %	5,5 %		7,5 %	8,5 %
RCEG(Eur/MWh	136,91	138,81	140,77	142,79	144,86

Table 15. Results of the RCEG sensitivity analysis for biogas plants operating on biogas produced from biomass.

The results of the sensitivity analysis of electricity reference costs for small sized hydro power plants are displayed graphically in Picture 2. A comparison of sensitivity analysis based on the modifications of observed input parameters for the remaining types of RES technologies leads to similar conclusions.

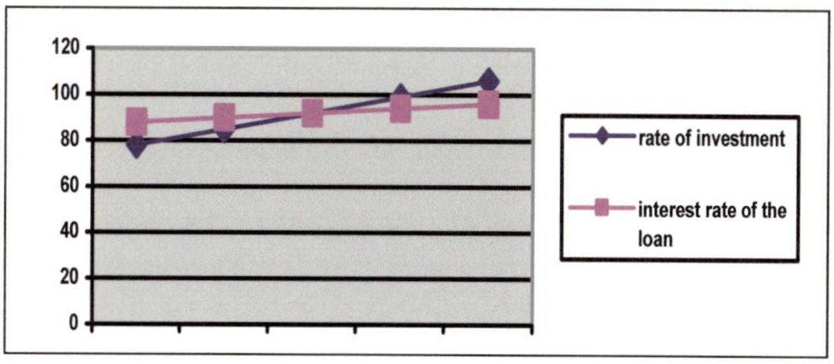

Fig. 2. Graphical display of sensitivity analysis for small sized hydro power plants.

The inclination of the pink line indicates that reference costs are more sensitive to the modification of the investment rate than to the interest rate of the loan. The cross-section point of both lines represents the starting value of electricity reference costs for small-sized hydro power plants (92,16 Eur) reached in accordance with basic assumptions regarding the rate of investment and interest rate of the loan that are 1.700 Eur/kW (Table 8) and 6,5 %.

Modification of the loan interest rate obviously does not represent a significant economic risk because an increase of the effective interest rate for 1 percentage point leads to an increase of electricity reference costs from approximately 1,5 Eur for big hydro power plants to approximately 3 Eur for geothermal hydro power plants. If the investment input is increased for 10 % electricity reference costs will increase from 5 to 9 %. A 5 % increase is noticed with biogas plants and production facilities operating on the basis of wooden biomass while a 9 % increase of RCEG is noticed with solar power plants. Results depend on the share of investment costs in total costs. When comparing changes of RCEG it is also necessary to take into account the cost of fuel for biogas plants and production facilities operating on the basis of wooden biomass while for the remaining five types of RES technology this is not the case.

5. Conclusions

In accordance with the proposal of the European Parliament directive Slovenia should by 2020 reach a 20 % share of RES energy in the total energy consumption. The results of our analysis display that reference costs for RES generated electricity are substantial and for most technologies even higher than the market price of electricity. In the years to come energy policy will play a crucial role as it will have to see to the decrease of capital costs and at the same time establish market conditions and incentive schemes which meet the technically conditioned effectiveness and the service life of RES production facilities.

At present in Slovenia a "Plan of action to reach target shares of RES generated electricity final consumption by 2020" is in the process of preparation. The latter will also form the basis for the adoption of the Slovenian national action plan. The target of the national action plan is to generate additional 3.000 GWh of RES generated electricity in 2020 in comparison with the currently generated volume. This plan of action makes several proposals for different possibilities and scenarios in order to reach this target.

6. References

[1] Berk, I. Lončarski, P. Zajc: Business Finance (*Poslovne finance)*, Ljubljana: Ekonomska fakulteta, 2004.

[2] Festić, Mejra, Kavkler, Alenka, Repina, Sebastijan. 2009. Investment in the renewable resources. Ljubljana: EIPF Economic Institute.

[3] E. Brigham, J.F. Houston: *Fundamentals of financial management*, Harcourt College Publishers: Orlando, 2001.

[4] Institut Jozef Stefan, Center za energetsko učinkovitost: Methodologies for estimation of reference prices of electricity production costs production from renewable resources (*Metodologija določanja referenčnih stroškov električne energije proizvedene iz obnovljivih virov energije*, Ljubljana: IJS, 2009.

[5] Inštitut za raziskave v energetiki, ekologiji in tehnologiji (IREET): Action plan for obtaining the desired level of final electricity production from renewable resources till 2020; summaries of reports (*Akcijski načrt za doseganje ciljnih deležev končne porabe električne energije iz obnovljivih virov do leta 2020, Povzetek treh faznih poročil).* Ljubljana: IREET, 2009.

[6] *A Decree for supporting electricity production from renewable resources (Uredba o podporah električni energiji, proizvedeni iz obnovljivih virov)*, UL RS, št. 37/09.

15

Recycling Hierarchical Control Strategy of Conventional Grids for Decentralized Power Supply Systems

Egon Ortjohann, Worpong Sinsukthavorn, Max Lingemann, Nedzad Hamsic, Marius Hoppe, Paramet Wirasanti, Andreas Schmelter, Samer Jaloudi and Danny Morton
Department of Power Engineering, University of Applied Sciences South Westphalia / Division Soest, Soest, Germany

1. Introduction

The objective is to develop an efficient control strategy, which is adaptable and flexible for power electronic inverter based Distributed Generation (DG) to interconnect to each other and to existing power systems. Since the proposed control strategy will be developed based on the hierarchical control structure of conventional power systems in [1, 2], it is able to handle not only modern DG sources, but also conventional sources. The general overview of the hierarchical control levels through inverter based DG are structured as shown in Fig. 1. These hierarchical control levels are the primary control at unit level, the secondary control at local level and the tertiary control at supervisory level. Moreover, as mentioned, the active controlled region of the grid is covered by higher voltage levels. With the proposed strategy, this controlled region can be expanded to medium voltage and low voltage distribution networks by active grid integration of Distributed Energy Resource (DER) based Energy Conversion Systems (ECSs) through inverters. The future inverters must be operating as intelligent and multi-functional interfaces between any ECS and grid in [3, 4].

2. Control strategy of distributed generation based on conventional power systems

Future power distribution requires extra expandability and flexibility in the integration of DG. The inverter which is used for interfacing DERs to the grids is an important part of a DG system. Therefore, the control strategy in the interconnected grids should be combined with the control methodology of inverters (grid forming, grid supporting and grid parallel modes). Load management, synchronization and load sharing with respect to generation rating, meteorological forecasting and user settings, is required in order to implement a control methodology of inverters into an interconnected system. Moreover, due to the flexibility and expandability of an inverters' control strategy, inverters in different feeding modes can be implemented into interconnected grids. In the following sections, the description and function of hierarchical control strategies are clarified.

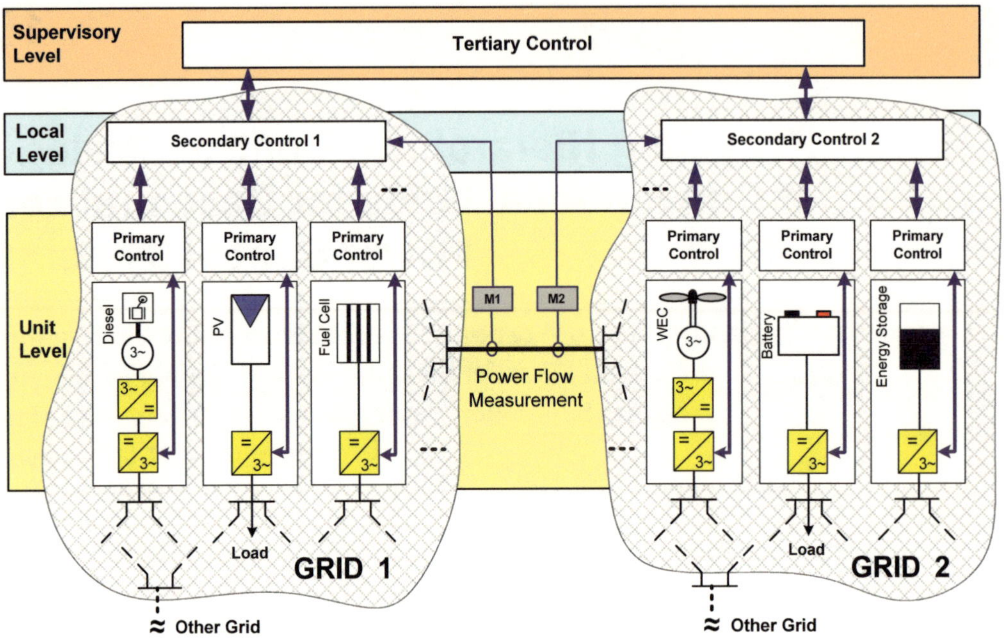

Fig. 1. Overview of hierarchy control strategy in interconnected grids.

2.1 Primary control

According to the future requirements of power supply systems, DER based DG should be actively integrated into the grid control. The power electronics like inverters are the primary interfaces that could fulfill this purpose. This inverter is needed to match the requirements of decentralized electric power systems [12], which must be enabled to be actively integrated into the control of the power system's state variables (frequency and voltage). Normally, the power produced by ECSs is DC power. This is fed to the grid through the inverter that produces an AC output of a specific voltage magnitude and frequency. This means that inverters provide decoupling between the voltages across the terminals of the ECSs from one side and the grid voltage from the other side. It also provides a decoupling between the frequency of the ECSs from one side and the grid frequency from the other side. The philosophy of inverter topologies is categorized as shown in Fig. 2. The inverter topologies are based on power flow from an ECS into a grid, which may be driven by a grid or by ECS itself.

Feeding modes of the inverter can be separated into two types, which are ECS driven feeding and grid driven feeding. A grid can be designed via several inverters with different operating functions and power ratings as discussed in [13].

An ECS driven feeding mode may be realized through a grid parallel inverter. In a grid driven feeding mode, the power flow from the ECS is controlled regarding the power requirements of the grid while in an ECS driven feeding mode, the power flow is controlled according to the requirement of the ECS itself. A grid feeding mode can be realized through two different cases, which are grid forming and grid supporting modes. Moreover, coupling mode such as droop control is defined also as a control function in primary control, which is needed for power sharing purpose when inverters are operated in parallel. The basic control function of primary control for inverters is summarized as shown in Table 1. In the following part of this chapter, only the structure of grid feeding mode is clarified, since it provides the active grid integration control function.

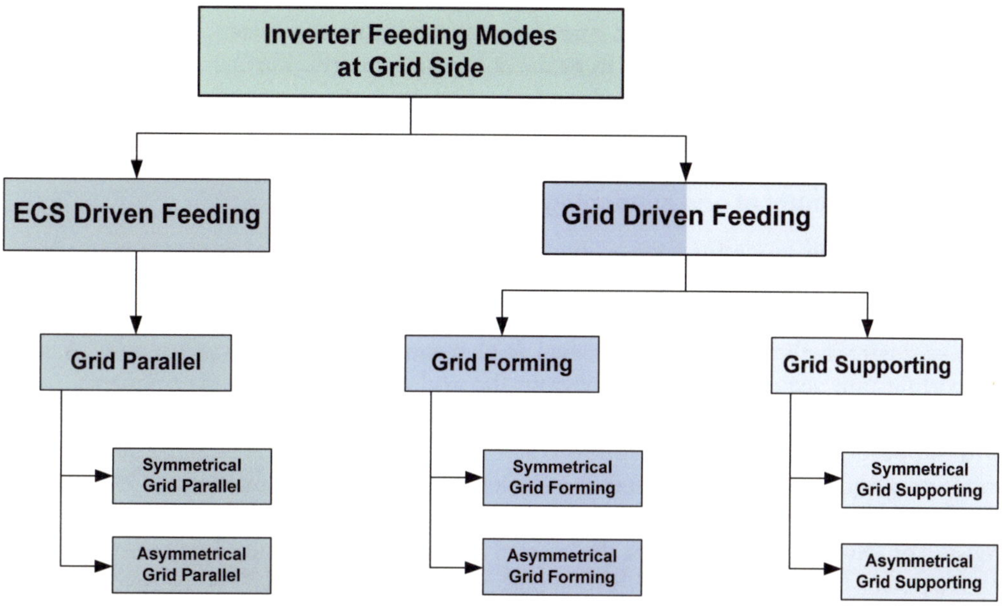

Fig. 2. Feeding modes related to the grid side [13].

Mode	Functionalities
Grid Forming	f- and V-control with nominal reference values (grid side driven)
Grid Supporting	P- and Q-control with external reference values from load dispatcher (grid side driven)
Grid Supporting	P- and V-control with external reference values from load dispatcher (grid side driven)
Coupling	$\Delta f/\Delta P$ and $\Delta V/\Delta Q$ droop (grid side driven) $\Delta P/\Delta f$ and $\Delta Q/\Delta V$ droop (grid side driven)
Grid Parallel	P- (and Q-) control with reference values from source (unit side driven)

Table 1. Basic control functions of primary control.

2.1.1 Primary control of grid forming mode inverter

A grid forming mode inverter is responsible for establishing and maintaining the state variables (voltage and frequency) of the grid. This is done by adjusting its power production to keep the power balance in the system. It also has to feed as much current into the grid as necessary. In the control scheme of the grid forming mode inverter, the voltage is controlled by the d-component, while the frequency is controlled by the q-component. The power injection in the connection point of the inverter is related to active and reactive power controllers. In addition, as inverters are parallel operated, synchronization and load sharing are also required in the control systems. A grid forming mode inverter includes conventional droop control functions at the primary control in the unit level as shown in Fig. 3.a. The load sharing is handled, using the $\Delta\omega/\Delta P$ and $\Delta V/\Delta Q$ conventional droop control functions. The voltage droop ($\Delta V/\Delta Q$) is related to the reactive power variation of the grid and the frequency droop ($\Delta\omega/\Delta P$) is related to the active power. Moreover, at the summation points of the active and reactive power controls, the offset power of the active power and reactive power are fed from the secondary control. Further information of the grid forming mode inverter included control structure is discussed in detail in [4], [6-13]. Therefore, the svm, dq transformation and PLL blocks will be not described.

2.1.2 Primary control of grid supporting mode inverter

A grid supporting mode inverter produces a predefined amount of power (active and/or reactive powers), which is normally specified by a management control unit. These predefined amounts of power can be adjusted according to the system requirements and user settings via the higher control level (e.g. secondary and tertiary controls). A grid-supporting unit acquires its frequency from the grid as there is only one frequency in the grid. Therefore, if the grid frequency changes due to any disturbance, the frequency of the grid supporting mode inverter follows that change. The control strategy for the grid supporting mode inverter included conventional droop control functions at the primary control in the unit level is shown in Fig. 3.b. The grid supporting mode inverter generally consists of four controllers, two for the current (*id* and *iq*), and two for the power (*P* and *Q*). Active power (*P*) is controlled by the real part of the grid current "*id*", while reactive power (*Q*) is controlled by the imaginary part "*iq*". The offset power from the secondary control is fed into the summation points of the active and reactive power controls. Further information of the grid supporting mode inverter including control structure is discussed in [4].

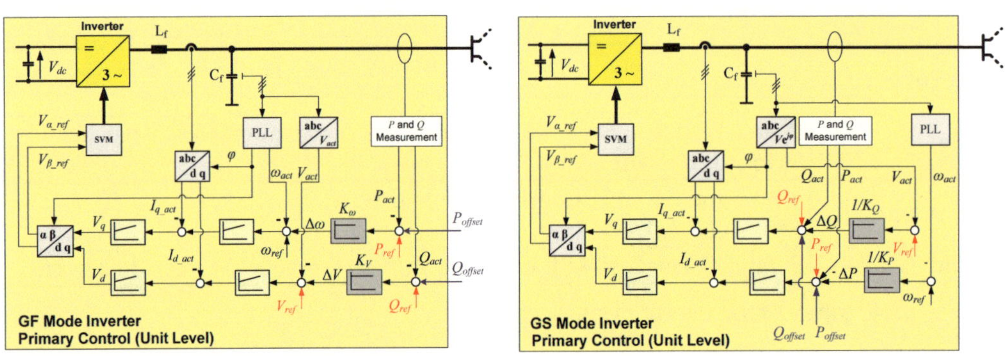

(a) (b)

Fig. 3. (a) Grid forming mode inverter with traditional droop controls [4]; (b) Grid forming mode inverter with traditional droop controls [4].

2.2 Secondary control

The secondary control is a centralized automatic function to regulate the generation in a control area based on secondary control reserve in order to balance power within that certain control area or control block and maintain interchange power flow, as well as the stabilization of the system frequency within the control area. In future power systems, the secondary control is also needed for local control in specific areas. This is possible leading to the cluster strategy using the secondary control, which is proposed and discussed in detail in [5]. As mentioned, the control functions of the interconnected DG power system are adapted from the conventional system. All the basic control functions of secondary control at the local level also follow from the control strategy of conventional system as follows:

- Maintain and control the state variables voltages and frequency to the nominal rated values
- Control the power exchange of interconnected grids
- Detect and maintain local power unbalances such as tie lines and interconnection points

In general, the secondary control has two main controls. The first one is the control that is responsible for frequency and active power. The second one is responsible for voltage and reactive power. Both controls are explained in detail as follows:

2.2.1 Frequency and active power control

The secondary control of DG power systems implemented through inverters is adapted from the secondary control of conventional power systems. The secondary control functionalities of DG power systems are similar to the conventional one, which is responsible for maintaining frequency, balancing power in the grid and transferring power between the grids. Therefore, the technical principles and fundamental backgrounds are not again described in detail. However, there are important issues of secondary control related to frequency and active power control in DG that should be taken into consideration. The operation control of secondary control should temporarily disconnect any certain responsible area from the main grids whenever a risk occurs such as unintentional islanding, grid instability, etc. In this section, the control structure of frequency and active power control is focused on. Fig. 4. shows the frequency and active power control for interconnected grids. The system frequency fact needs to be fed to the droop control. The system frequency (f_{act}) needs to be fed to the droop control. The droop factor is equal to the droop factors summation of the inverters in the grid. In addition, the active power flows are measured at the connection points of the grids. Both power signals are fed to the summation point with the active power transfer reference, which is managed from tertiary control at supervisory level. Later on, the summed signal will go through the PI controller and be sent to each primary control at unit level with the sharing factors for each one.

2.2.2 Voltage and reactive power control

As mentioned, the new control strategy should be adapted based on conventional power systems. Therefore, in case of the voltage and reactive power control, the secondary control of the DG power system is also implemented through the inverter, which is adapted from the secondary control of conventional power systems. The load flow optimization is required to control voltage and reactive power, as the voltages in power system are more sensitive than frequency. The control structure of voltage and reactive power control in interconnected grids is shown in Fig. 5. In the interconnected point, reactive power at the bus between the grids is measured. These measured values (Q_{12} and Q_{21}) will be adjusted to the amount of the reactive power transfer between the grid ($Q_{ref,12}$ and $Q_{ref,21}$), which are

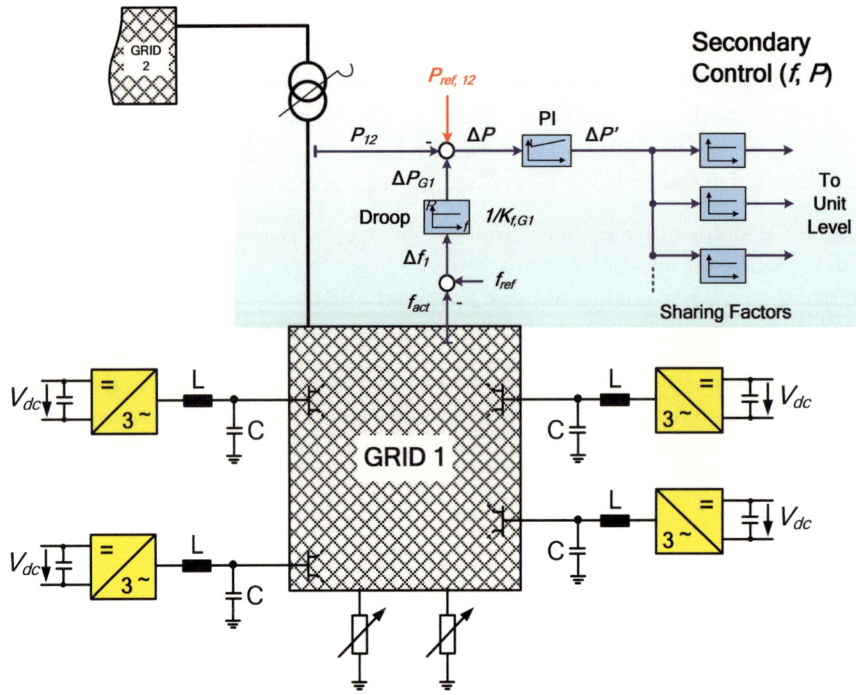

Fig. 4. Frequency and active power control with sharing factor in interconnected grids.

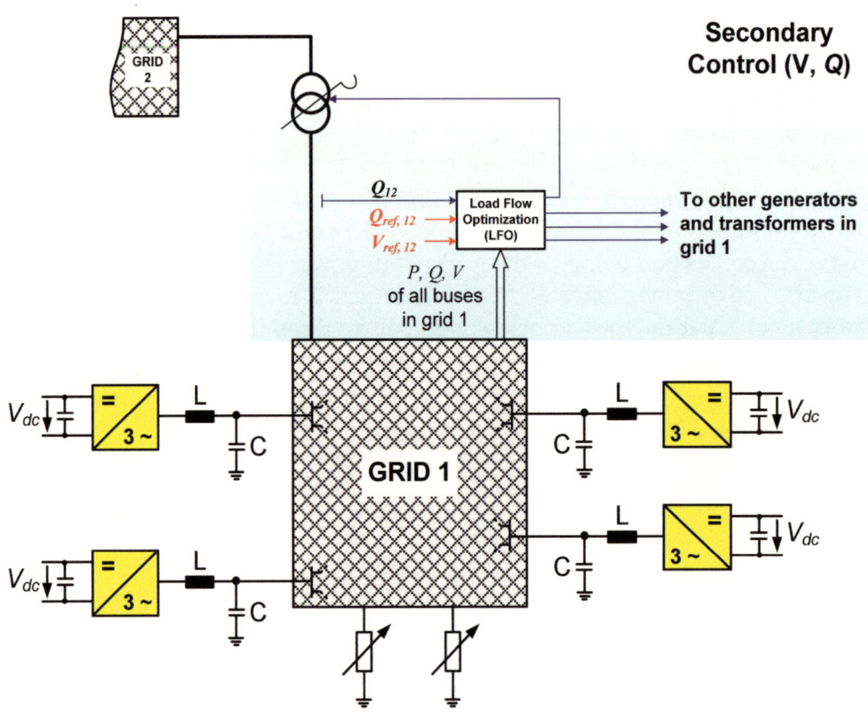

Fig. 5. Voltage and reactive power control with sharing factor for interconnected grids.

calculated from the higher control level such as the tertiary control at supervisory level (red arrows). The reactive power bus (Q_{12}), reference value of reactive power transfer ($Q_{ref,12}$) and reference voltage of interconnection point ($V_{ref,12}$) will be fed into the load flow optimization (LFO) block [15]. All data of bus voltage, active and reactive power in the grid are required for the calculation process of optimization. After the optimization, new offset of voltage and reactive power will be fed into primary controls as well as transformers in the grid.

3. Control strategy of interconnected distributed power systems

The hierarchical control strategy of the interconnected distributed power systems can be adapted based on the control structure of the conventional power systems. There are also generally three main control levels to manage the entire power system which are unit level, local level and supervisory level. This section introduces an example control structure of interconnected grids including the combination of grid forming and grid supporting inverters as shown in Fig. 6. This sample layout of control strategy in interconnected grids is controlled by the control strategy adapted from the conventional power system. The control strategy focuses directly on the flexibility of the inverter control structure that can be implemented along with the control strategy for grid interconnection.

Having a look at a grid forming mode inverter, its role is to establish and maintain the state variables (voltages and grid frequency). Therefore, in the control scheme of the grid forming mode inverter, the voltage is controlled by the d-component. In contrast, the frequency is controlled by the q-component. The power injection in the connection point of the inverter is related to active and reactive power controllers. Synchronization and load sharing are also required in the control systems. The load sharing will be handled using the voltage and frequency droop control functions for the primary control at unit level. The voltage droop is related to the reactive power variation of the grid and the frequency is related to active power. The secondary control at local level is included to bring system frequency back to nominal value, as well exchanged active and reactive power can be transferred between the grids. In addition, voltage and reactive power control can be possibly done by load flow optimization, which is adapted from conventional one in [9]. For optimization purpose, tertiary control will calculate and manage the reference values in the controllers. Note that, the optimization for voltage and reactive power is not a main focus in this work. For, the grid supporting mode inverter, it is used for power balancing and produces predefined amounts of power. These predefined amounts of power can be adjusted according to the system requirements and user settings via the secondary and tertiary controls. The grid supporting mode inverter feeds the grid with a specified amount of power, which is active and reactive power, or a combination of both. The control strategy for the grid supporting mode inverter using active and reactive power consists of four controllers, two for current (i_d and i_q), and two for active and reactive power (P and Q). Active power is controlled by the real part of the grid current "i_d", while reactive power is controlled by the imaginary part "i_q". For the grid parallel mode inverter, there is no need to have a secondary control, since it is a power production unit that is not controlled according to the requirements of the electrical system. However, it can be possible to implement the active and reactive power control into the grid parallel mode inverter.

This tertiary control is related to the supervisory level, which organizes energy management of the overall power system (i.e. system optimization, dispatch control strategy, load flow

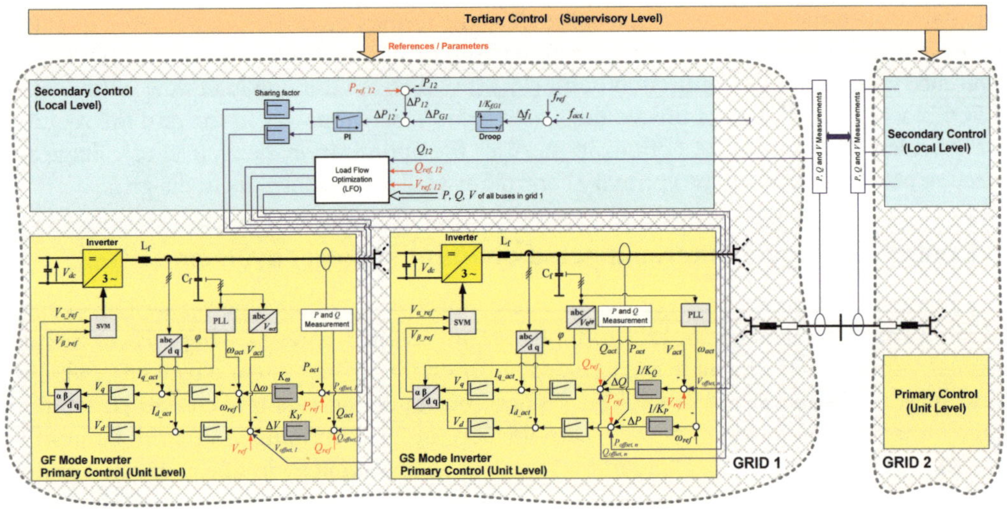

Fig. 6. Example of the control strategy in interconnected grids including grid forming and grid supporting mode inverters [6].

management, meteorological forecasting, network management and communication management). The tertiary control collects information of the interconnected grids such as forecasting data, power profile, load data, etc. The optimization is processed in this level to get the reference values fed to the local and unit levels. This also includes the optimized power dispatch between grids. However, the optimization process can be computed in other additional level depending on strategies of the control structure.

4. Case study

To verify the proposed control strategy, the model is tested by simulation of two grids including grid forming mode inverters as shown in Fig. 7. The first grid is supplied by grid forming mode inverters (GF1 and GF2). The Second grid is supplied by two grid forming mode inverters (GF3 and GF4). The power system operates at the rated frequency f_{rated} = 50 Hz and the reference voltage line to line $V_{L\text{-}L}$ = 400 V_{rms}. Rated apparent power of the grid forming mode inverters is S_r = 125 kVA. Both grids are linked via a tie line (NAYY 4×50 SE: R = 0.772 Ω/km and X = 0.083 Ω/km). The secondary control is included in the simulation to control power in each grid as well as power exchange between the grids. The active power and reactive power loads of the first grid are the same as those of the second grid which starts at 16 kW and 7.3 kvar. The total active power and reactive power are 32 kW and 14.6 kvar respectively. At t = 15 s, in the first grid, the active power steps up to 20.2 kW and reactive power load steps up to 7.37 kvar. Therefore, the total active and reactive powers after the load step are 36.2 kW and 14.67 kvar respectively. At t = 40 s, the exchanged active and reactive powers of 2 kW and 1 kvar respectively are transferred from the first grid to the second grid.

Active power of the inverters is shown in Figs. 8.a. At the beginning, the inverters supply active power of approximately 32 kW; around 8 kW is supplied by each grid forming mode inverters. At t = 15 s, the step load of 4.2 kW is added to the first grid. All the inverters of the system directly supply to compensate the additional load step. After the step, the secondary

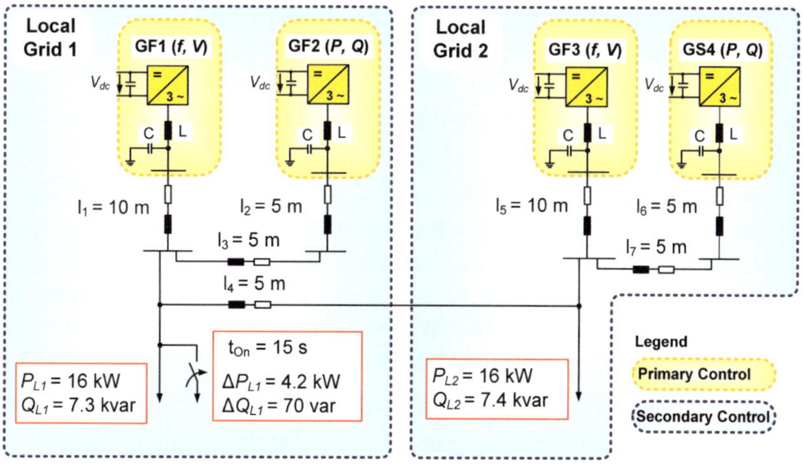

Fig. 7. Two grids including two grid forming mode inverters each.

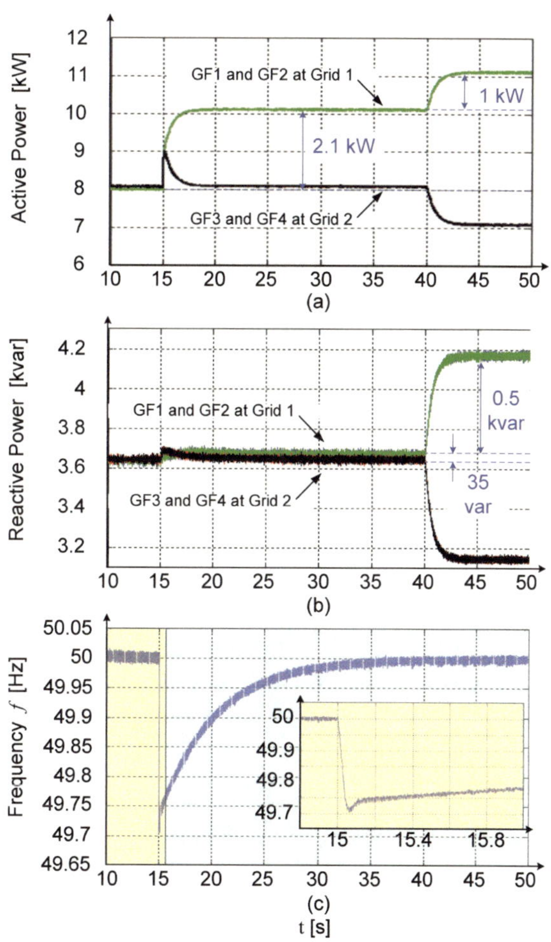

Fig. 8. (a) Active power, (b) Reactive power, (c) Frequency

control manages the generating units at the first grid to response to the disturbance to the grid by itself. Therefore, the active power of GF1 and GF2, which are located close to the first grid, are steadily increased, while active power of inverters at the second grid are steadily decreased. At t = 40 s, active power is transferred from the first grid to the second grid of 2 kW.

Reactive power of the grid forming mode inverters is shown in Fig. 8.b. At the beginning, all inverters supply reactive power of approximately 14.6 kvar. At t = 15 s, the step load of 70 var is added to the first grid. All the inverters of the system directly supply to compensate the additional load step. Later on, the inverters in the first grid will maintain themselves due to the secondary control which provides the same behavior response as the active power. At t = 40 s, the reactive power is transferred from the first grid to the second grid is 1 kvar.

System frequency is shown in Fig. 8.c. The primary and secondary controls have direct impact on the frequency behavior. Due to the droop control function of the primary control, at the load step t = 15 s, the frequency drops from the nominal frequency. This frequency drop can be brought back to the nominal value by secondary control.

5. Conclusion

As the penetration of DG systems in the grid is increasing, the challenge of combining large numbers of DERs in the power systems has to be carefully clarified and managed. The control strategy and management concept of the interconnected systems should be flexible and reliable to handle the various types of DG. The chapter introduces a control strategy for DG interconnected grids based on the control strategy of conventional power systems. The proposed strategy is integrating DERs and managing interconnected grids to operate in parallel. The power dispatch, exchanged power, frequency control and voltage control can be automatically managed by the proposed strategy. The simulation results illustrate that the strategy can be implemented into the power systems. The grid integration is additionally supported by adaptability, flexibility and efficiency of the proposed strategy. Mini-grids can be widely interconnected to each other and existing conventional systems to form huge power systems. Moreover, this proposed strategy is compatible with the concept of multi-level clustering power systems. The multi-level clustered secondary control strategy is the consequent development to be followed in the stepwise evolution from the historical centralized power system towards a more decentralized structure. It allows stepwise implementation and improvement of the power system control. The success of the proposed control philosophy will finally clear the way for future smart gird applications by creating a basic, technical power system control approach, based on control theory and the physical behavior of the power system. This is the fundamental basis for all other optimization, management and economic functions of the "smart grid" vision.

6. References

[1] UCTE: "Operation Handbook - Introduction", Final v2.5 E, 24.06.2004, July 2004.

[2] UCTE: "Operational Handbook - Policy 1: Load-Frequency Control", Final Version, March 2009.

[3] O. Omari: "Conceptual Development of a General Supply Philosophy for Isolated Electrical Power Systems", Dissertation, Joint-PhD Program between The

University of Bolton, Bolton, UK, in cooperation with South Westphalia University of Applied Sciences - Soest, Soest, Germany, January 2005.

[4] Mohd: "Development of Modular Grid Architecture for Decentralized Generators in Electrical Power Supply System with Flexible Power Electronics", Dissertation, Joint-PhD Program between The University of Bolton, Bolton, UK, in cooperation with South Westphalia University of Applied Sciences - Soest, Soest, Germany, January 2010.

[5] M. Lingemann: "Advanced Implementation Strategies for Distributed Energy Converters in Electrical Grids" Dissertation, Joint-PhD Program between The University of Bolton, Bolton, UK, in cooperation with South Westphalia University of Applied Sciences - Soest, Soest, Germany, June 2011.

[6] W. Sinsukthavorn: "Development of Strategies and Methodologies for Dynamic Control of Distributed Generation in Conventional Grids and Mini-Grids" Dissertation, Joint-PhD Program between The University of Bolton, Bolton, UK, in cooperation with South Westphalia University of Applied Sciences - Soest, Soest, Germany, June 2011.

[7] O. Omari, E. Ortjohann, D. Morton S. Mekhilef: "Active Integration of Decentralized PV Systems in Conventional Electrical Grids", PV Technology to Energy Solutions Conference and Exhibition, Barcelona, Spain, May 2005.

[8] E. Ortjohann, A. Mohd, N. Hamsic, D. Morton, O. Omari: "Advanced Control Strategy for Three-Phase Grid Inverters with Unbalanced Loads for PV/Hybrid Power Systems", 21th European PV Solar Energy Conference, Dresden, Sep 2006.

[9] E. Ortjohann, M. Lingemann, A. Mohd, W. Sinsukthavorn, A. Schmelter, N. Hamsic, D. Morton: "A General Architecture for Modular Smart Inverters", IEEE ISIE'08, Cambridge, July 2008.

[10] Hans-Helmut Graf: " Einfluss einer sekundären Spannungs- und Blindleistungsregelung auf das Verhalten elektrischer Energieversorgungsnetze", Dissertation, 1984.

[11] E. Ortjohann, W. Sinsukthavorn, M. Lingemann, A. Mohd, N. Hamsic, D. Morton: "Multifunctional Grid Front-End for Dispersed Energy Resources", International Symposium on Industrial Electronics, Bari, Italy, July 2010.

[12] Mohd, E. Ortjohann, N. Hamsic, M. Lingemann, W. Sinsukthavorn, D. Morton: "Isochronous Load Sharing and Control for Inverter-based Distributed Generation", International Conference on Clean Electrical Power Renewable Energy Resources Impact. Capri, Italy, June, 2009.

[13] E. Ortjohann, A. Mohd, N. Hamsic, M. Lingemann, W. Sinsukthavorn, D. Morton, "Inverterbased Distributed Generation Control using Droop/Isochronous Load Sharing," PP&PS IFAC Symposium on Power Plants and Power System Control, Tampere Hall, Finland, July 2009.

[14] E. Ortjohann, A. Arias, D. Morton, A. Mohd, N. Hamsic, O. Omari. "Grid-Forming Three-Phase Inverters for Unbalanced Loads in Hybrid Power Systems", IEEE 4th World Conference on Photovoltaic Energy Conversion, Waikoloa, Hawaii, May 2006.

[15] M. Hoppe: "Betriebsführung elektrischer Netze mit dezentralen Stromerzeugern" Diplomarbeit, South Westphalia University of Applied Sciences - Soest, Soest, Germany, September 2009.

16

Energy Efficiency and Electrical Power Generation

Hisham Khatib
World Energy Council
Jordan

1. Introduction

Reducing demand for energy, through energy efficiency measures and conservation, is the cheapest and most effective way for enhancing energy security, providing a stable basis for economic planning as well as cutting carbon emissions.

Energy efficiency is the most influential means of achieving sustainable energy future. It will significantly economize on energy use, reduce investments in energy infrastructure, reduce costs and maximize consumer welfare; as important it cuts off on emissions. In its World Energy Outlook 2010, the International Energy Agency (IEA), adopted the BLUE Map scenario which sets a goal of halving global energy-related CO2 emissions by 2050, compared to 2005 levels, and sets out the least-cost pathway to achieve that goal; through the deployment of existing low-carbon technologies (IEA-WEO 2010).

In 2050, with business as usual, global CO_2 emissions are expected to reach 57 Gt of CO_2. The Blue Map emissions aim at reducing them to only 14 Gt. Although this has become a rather unrealistic goal, however, what of interest to us here is that the most effective means of achieving these emissions reduction is through energy efficiency in end-use fuel and electricity generation. In the IEA 450 Scenario, energy efficiency is capable, by 2035, of achieving up to 48% of the abatement goal, more than twice that of renewables or Carbon Capture and Storage (CCS). The total reductions in CO_2 emissions as a direct result of energy efficiency measures are almost equal to the combined containment by all other means - CCS, renewables, fuel switching, nuclear, etc. (see Figure One).

This is a survey Chapter that deals with many aspects pertaining to energy efficiency, particularly in the electricity generation sector also industry. The electrical power generation sector is chosen for special emphasis due to its present low efficiency and the growing importance of electrification in the global economy and final energy use, as we are going to see later. It is not intended in this Chapter to go into the intimate details of technologies of electricity generation and how they affect and improve efficiency and reduce carbon emission in a carbon-constrained world. Such issues are covered in the literature (Sioshansi, 2010). What we are doing in here is outlining the important subject of energy efficiency and how the power generation activity fits into this, the present status and future prospects.

At the end of the Chapter there is a short section that refers to Energy Governance, which is important in attaining energy efficiency in all energy utilization activities including that of electricity generation. This section also deals with energy subsidies that proliferate in the electricity sector, particularly in oil exporting countries and some developing countries, and are detrimental to achieving energy efficiency including that of electricity generation.

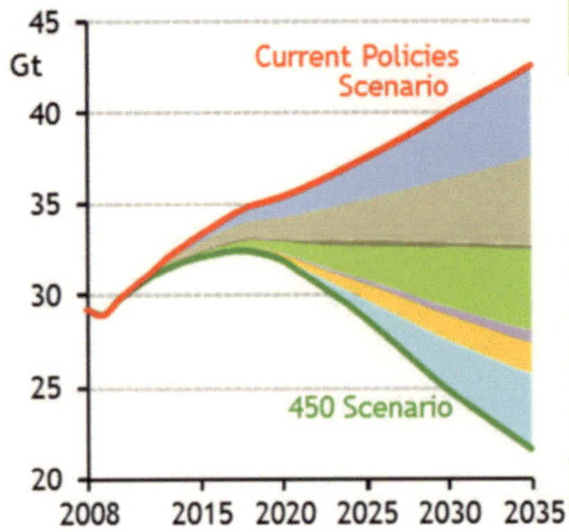

	Abatement		
	2020	2030	2035
Efficiency	71%	49%	48%
- *End-use (direct)*	*34%*	*24%*	*24%*
- *End-use (indirect)*	*33%*	*23%*	*23%*
- *Power plants*	*3%*	*2%*	*1%*
Renewables	18%	21%	21%
Biofuels	1%	3%	3%
Nuclear	7%	9%	8%
CCS	2%	17%	19%
Total (Gt CO_2)	3.5	15.1	20.9

Fig. 1. IEA 450 Scenarios for Reducing Global CO2 Emissions. Source: (IEA -WEO 2010)

2. Energy efficiency

Let us first of all start by defining what we mean by "energy efficiency". According to the IEA - something is more efficient if it contributes more services for the same energy input, or the same services for less energy input. For example when a compact florescent light (CFL) bulb uses less energy than an incandescent bulb to produce the same amount of light, the CFL is considered to be more energy efficient (IEA- Governance). Energy efficiency is also defined as "percentage of total energy input to a machine or equipment that is consumed in useful work and not wasted as useless heat" (ABB 2010). The EU Directives define efficiency as "a ratio between an output of performance, service, goods or energy and an input of energy" (ESD Article 3 b).

The exact value of global energy efficiency is a debatable figure. Of the 12,500 million tons of oil equivalent (Mtoe) of primary energy consumed in 2010, 8,600 Mtoe are delivered as total final energy to users and consumers. There are estimated 7,500 Mtoe lost, mostly as low and medium-temperature heat (WEA). This means that globally the energy efficiency ratio does not exceed 40%, which is extremely low and demonstrates the future strides that can be achieved in this quarter. A large proportion of the energy losses in delivering final energy to consumers are in the electricity generation sector, and that is why this sector receives emphasis in this paper. Losses in converting delivered energy into useful energy in final use also occur in all sectors of the economy, mainly in industry, transport and building.

Energy efficiency has to be differentiated from energy conservation. Energy conservation refers to efforts made to reduce energy consumption. Energy conservation can be achieved through increased efficient energy use, in conjunction with decreased energy consumption and/or reduced consumption from conventional energy sources (Wikipedia 2011). Whereas conservation can be achieved through behavioral changes, education and pricing, energy efficiency can be mainly improved by technology, investments, codes and practices also energy pricing. Energy efficiency also needs to be differentiated from energy management which means timing energy use in a manner that minimizes cost and maximizes useful utilization of efficient facilities.

Energy is consumed for production, services and comfort. Its extent of use varies from one country and region to another depending on technology achievement, prices of energy products, economic output and weather. Primary energy intensity of a country is its total primary energy consumption per unit of the country's gross national product (GDP). Primary energy intensity measures the total amount of energy required to generate one unit of GDP. GDP is expressed at constant exchange rate or purchasing power parity (ppp) to remove the impact of inflation and changes in currency rates. Using purchasing power parity rates instead of exchange rates to convert GDP in the same currency (e.g., $) makes it possible to account for differences in general price levels: it increases the value of GDP in regions with low cost of living (case of developing countries) and, therefore, decreases their energy intensities.

Prior to the age of increasing energy prices, i.e. pre - 1973, the growth of energy consumption used almost to mirror that of global economic growth, i.e. elasticity of energy demand versus that of economic growth was almost 1:1, with little improvement in global energy intensity. However, in recent years, and due to rising energy prices, environmental awareness and resource depletion, as well as technology oriented efficiency programs, significant trends took place in reducing energy intensity. Global energy intensity decreased by 1.4% per year over the last two decades and is continuing to do so (see Figure Two). Over the last twenty years the world economy grew at an average of 3.2% annually, while primary energy consumption grew from 8,800 Mtoe to 12,500 Mtoe i.e. an annual average of 1.9% only. This lead to these significant improvements is energy intensity. (See - Global Energy Efficiency Trends - ABB 2010).

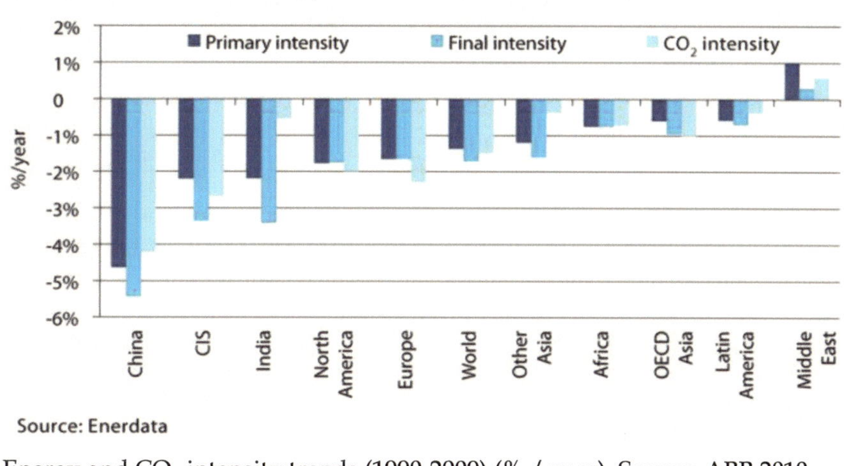

Fig. 2. Energy and CO_2 intensity trends (1990-2009) (% / year). Source: ABB 2010.

Most reductions in energy intensity occurred in OECD countries. One of the reasons of the decrease of the energy intensity in OECD countries is that the heavy and high energy intensity industries (steel and cement industries, aluminum smelters, etc.) moved out of these countries into places where energy is still cheap (like China and India) and where there is less national emphasis on environmental and emission issues. OECD economies are emphasizing the services sector which is less energy intensive, while many developing countries are building and expanding their industrial base with emphasis on heavy industries and utilization of local natural resources while exporting industrial products to OECD countries. Correspondingly the true country figure of energy intensity should be modified to reflect imports to OECD countries of industrial material that carry a high energy bill (more of that later in the Industry sector).

3. Efficiency in the electricity generation sector

The efficiency of electricity generation is a complex phenomenon. It depends on many factors and varies from one country to another. National primary energy endowments plays a prominent role; countries with viable hydro electric sites try to exploit these as a priority, so they will also exploit local coal resources in spite of their environmental impacts. The endowment of national (or imported) resources of natural gas presents opportunities for high efficiency and relatively inexpensive electricity generation. Nuclear is limited to few countries with relatively advanced financial and human resources and its efficiency is below average, also below average is the efficiency of new renewable generation - wind and solar.

This Chapter will cover these issues with a global view pointing out to global averages but also to individual technologies. It will cover the existing generation technologies and tries to predict future world averages depending on trends and furthering new technologies particularly in replacing vintage plant with more efficient CCGT and high efficiency coal plant with advanced parameters.

Energy losses occur in every activity in the electricity sector, not only in the utilization of electricity, but mainly in the electricity supply industry – electricity production (generation), its transmission to consumer centers as well as its distribution to users and consumers. It is mainly in electricity generation where the majority of losses occur. Till today the average net efficiency of electricity generation worldwide is around 34%, with almost two thirds of calorific content of primary fuel input into electricity generators is lost as waste heat (Khatib 2003). It has to be emphasized that there are no exact dedicated data on global net electricity generation and its exact consumption of fuels. So we have to rely on published data in the global annual energy surveys, the latest of which is the "International Energy Outlook 2011" of the US Energy Information Administration (IEO, 2011). Efficiency of electricity generation, which was as low as 25% in the 1950s, significantly improved to around 34% - 35% today, and is continuously but slowly improving mainly through the increasing use of combined cycle facilities firing natural gas and improved technologies in material use. In this Chapter when we are referring to net electricity generation efficiencies we are only accounting for net power sent out from the station, thus ignoring the energy use by the plant auxiliaries which can be as high as 5-6 % of generated power in case of steam plants and 1-2 % in gas turbine plants.

Figure Three (IEA- ETP 2010) shows how the efficiency of coal firing power stations was held steady at almost 34% over the last twenty years, while that of gas firing facilities improved from 34% in 1990 to an average of 43% in 2010 through the wide introduction of combined cycle gas turbine plant (CCGT) facilities firing natural gas.

Electrical efficiency is most important because the world is electrifying. Energy services are now being increasingly offered in the form of electricity, rather than in any other form -

mechanical or human. Whereas energy demand is expected to grow at an annual rate of 1.6 % over the next 25 years, that of electricity is expected to grow at a rate of 2.3 %, reflecting this growing electrification (IEO 2011); that is why improving efficiency of electricity production figures so high in any effort to improve global energy efficiency - see Figure 3.

There are also major losses in the electricity grid - the transmission network as well as distribution lines. The transmission network that delivers bulk electricity from major power stations into bulk supply substations, can incur losses as high as 1 - 3 % of transmitted energy (depending in the length and voltage of the network). Other 6 - 10% losses occur in the distribution networks (also depending on the network configuration - length and conductors). Therefore, world average transmission and distribution (T&D) losses are around 8 - 9 % of energy sent-out from the power station (IEA-ETP 2010). When this is taken into account, of the primary energy input into the power plant only 30 - 32% of this energy reaches consumers, i.e. less than one third. This demonstrates the efficiency dilemma of the electrical power system.

3.1 Efficiency of electricity generation

Efficiency of the electricity generation depends on the mode of generation. Hydroelectric production has an efficiency of over 90%; whereas vintage thermal plants firing coal have efficiencies of no more than 25%. Modern steam plant, firing coal, is becoming the preferred means of generation, particularly in China and India, where thermal plants represent 80% of production due to availability of cheap coal. Steam generation, mainly firing coal, presently represents half of world electricity generation and future improvements in efficiency of electricity production depend on technology advancement in this quarter. Major losses mainly occur in the thermal plant itself, in condensing steam; also in auxiliaries and generators and these can be as high as 5-6% of the generated electricity.

3.2 Improving efficiency of electricity generation

There are two approaches to improve steam power generation efficiency: one is through increasing live steam parameters (pressure and temperature) to develop supercritical (SC) and ultra-supercritical (USC) technologies; another is by system integration, a typical example is Integrated Gasification Combined Cycle (IGCC). In the next 10 years, SC and USC will be built in significant numbers, and these new plants are likely to remain in use until 2050 for electricity production from coal because of their flexibility and general advantages of lower cost, reliability, high availability, maintainability, and operability. IGCC currently looks promising in its ability to produce deep CO_2 reductions at least cost while maintaining high generation efficiency.

Pulverized coal combustion (PCC) is currently the predominant technology for generating electricity from coal and represents 43% of fuels used in electricity generation. It also accounts for more than 97% of the world's coal-fired electrical capacity. Most existing plants operated at less than SC steam conditions, i.e. less than 34% efficiency, with best examples reaching 39% efficiency.

New pulverized coal power plants - utilizing SC and USC - operate at increasingly higher temperatures and pressures and therefore achieve higher efficiencies than conventional units. Supercritical power generation has become the dominant technology for new plants in industrialized countries. Now considerable efforts are underway in the United States, Europe, Japan and also China, to develop 700 C–class Advanced Ultra-Supercritical (A-USC) steam turbines. If successful, this will raise the efficiency of the A-USC units to about 50% by 2020 (Lui - 2010).

Natural gas (NG) presently accounts for 20% of electricity production. The increasing availability of NG and its prices are favoring the utilization of NG in high efficiency combined cycle gas turbines (CCGT) where efficiencies of over 50% are becoming common and as high as 60% are increasingly possible. This increasing CCGT trend and improvements in the efficiency of thermal plant through the increasing use of critical and super critical steam turbines means that the efficiency of electricity generation is destined to continue to improve, although slowly, in the future.

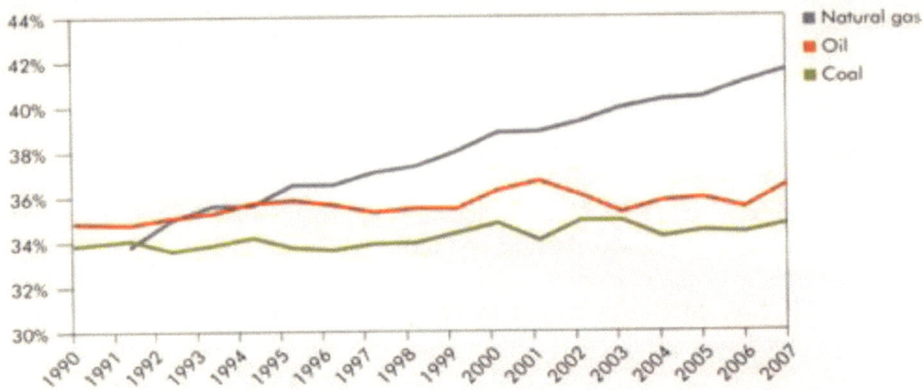

Fig. 3. Development of Efficiency of Thermal Generating Plant. Source: ETP 2010.

3.3 Future trends in efficiency of electricity generation

As already indicated the efficiency of electricity generation is destined to continue to improve year after another. Such improvements are prompted by the rising cost of primary fuels, but also and as important by environmental considerations. Mitigating carbon emissions from generating plants is a global environmental concern. Carbon taxation, which is becoming common in most OECD countries and few other countries, will enhance already mentioned technological trends to improve efficiency of coal plants and switch to cleaner renewable resources, particularly hydro and wind, also the more efficient CCGT plant firing natural gas.

But such global improvements are going to be very slow due to the existence of a vast inventory of vintage low efficiency plants which have long lives (sometimes even extended), the high investment cost of introducing new plants and also the efficiency penalty of mitigating emissions through the introduction of clean technologies, like that of carbon capture and storage (CCS) which tend to significantly reduce efficiency of new plants fitted with such facilities.

Table 1 below gives a prediction of the future development of electricity generation in coming next twenty five years and its rising proportion of primary energy sources.

Year	Expected Net Generation (TWh)	Fuel consumption (Quad BTU)	% of global use
2005	19 125	194	38.4
2015	22 652	227	39.4
2025	28 665	281	41.8
2035	35 175	337	43.8

Source: IEO 2011

Table 1. Future Electricity Generation Trends

The fact is that the annual amount of fuels destined to electricity generation will grow at a rate of almost 2.1% annually, while net electricity generation will grow at a higher rate of 2.3%, indication an annual improvement of efficiency of 0.2% percentage points, Figure 4. Over a period of 25 years (2011-2035) this means that the efficiency of electricity generation is likely to improve by 5% as indicated in Table 1 above.

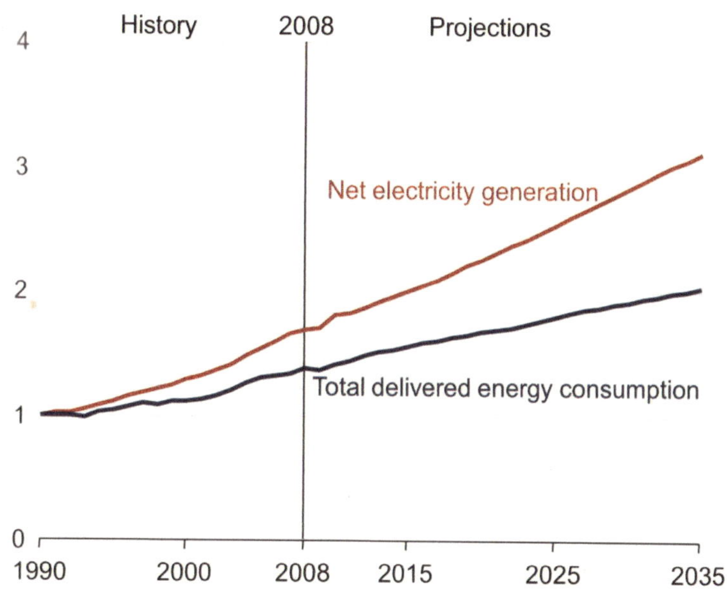

Fig. 4. Future trends in energy consumption and electricity generation. Source: EEA, 2011.

3.4 Carbon emissions from electricity generation

Thermal power stations firing coal are major emitters of CO_2 with around 750 – 1000 grams CO_2/kWh in 2010. With net electricity generation of 20,500 TWh in that year, total CO_2 emissions from the electricity sector amount to around 12,000 million tons, this is 40% of global carbon emissions. With each one percentage point improvement in efficiency of electricity generation, CO_2 emissions can be reduced by as high as 350Mt annually (Creyts, 2007).

It has to be realized that slow but continuous progress has been achieved in reducing emissions from generation plant during recent years. Past vintage plants of the late twentieth century burning coal operated at an efficiency of 25% while emitting 1.30 tons CO_2

/MWh. New thermal generation operating at an efficiency of 45% and burning coal is emitting only 0.720 tons CO_2 /MWH, while a new CCGT burning NG can has an efficiency of 60 % and emits only 0.320 tons CO_2 /MWH. This is only one quarter of the emissions per unit of electricity one quarter of a century ago.

Nuclear plants, due to security reasons have to operate at relatively moderate temperatures and pressures. Correspondingly its efficiency tends to be at 35% or lower. This is a low efficiency value compared to modern thermal plants. However a nuclear plant has the advantage of no emissions. The same applies to new renewable resources (wind and solar) which tend to have low or no emissions but operate at low efficiency in comparison to hydro plants which have high efficiency.

3.5 The smart grid and energy efficiency

The introduction of the Smart Grid is the latest innovation in the electricity supply industry. It allows for two way communication between suppliers and consumers and also the connection of the distributed generation to the national grid, correspondingly it enhances the integrity of the supply and its management. As important it will contribute to improve the efficiency of the public electricity generation. It will enable both the suppliers and consumers to manage demand as to reduce peaks, improve load factors and thus enhance loading of efficient generators and reduce the contribution of the inefficient peaking plant. But it will take many years until the beneficial effects of smart grids begin to significantly show in improved global efficiency of electricity generation (Khatib, 2011).

3.6 Factors affecting the efficiency of generating plant

Power plant efficiencies are typically defined as the amount of heat content in (BTU) per the amount of electric energy out (kWh), commonly called a heat rate (BTU/kWh). Such efficiency is affected by (NPC, 2007):

- Design Choices that present a tradeoff between capital cost, efficiency, operational flexibility and availability
- Operational practices that aims at full load, avoiding steam leakages and utilizing integration systems, etc.
- Fuel, particularly utilizing hard dry coal that possesses less water and ash.
- Environmental Control, to reduce emissions (NO_x and SO_x); represent *parasitic loads* that decrease efficiency. Similar penalties are introduced by applying CCS technologies that significantly increase cost and reduce plant efficiency.
- Ambient temperature, colder cooling improve efficiency of generating power, however high altitude negatively affect output and efficiency of gas turbines.
- Method of Cooling- methods for cooling steam turbine effluent can be through once-through cooling, wet cooling tower and indirect dry cooling. There are penalties to utilize wet cooling towers which can range 0.8-1.5% and as high as 4.2-8.8% with a dry cooling tower.

To this must be added penalties brought about by aging, normal deterioration and bad operation, low maintenance and management practices. Deterioration can be addressed by: refurbishment, replacement in kind, upgrade with advanced design, modify original design, repowering and retirement with replacement by new construction.

3.7 Improving efficiency of the generating cycle

The main consideration in improving the efficiency of the generating cycle is through ustilising the heart content of the exhaust gases and cooling water. This is done mainly through combined cycle gas turbines (CCGT) in which the exhaust gases of the GT are fed into a steam boiler and steam turbine generator- thus increasing efficiency of the generating cycle by 50%. A typical CCGT plant now operates in the range of 45-55%, but efficiencies as high as 60% are already achievable.

The other means to improve the efficiency is by combined electricity and heat production where the condensed steam from the steam turbine outlet is ustilied as a heat source mainly in industry but also various applications including district heating (if possible). This also considerably improves efficiency of utilising fuels. An average efficiency of around 50% can be reached in the EU by conventional thermal electricity and heat production. Tremendous strides have been achieved in this regard during the last twenty years as demonstrated in the following Figure 5.

Improvements in efficiency of the utilization of generation fuels can be attained through the combined production of power and water in which the steam from the generating process is utilized for desalination by the flashing process, thus significantly improving the utilization of the generation fuels. Further improvements in output and efficiency are obtained by many new technologies like cooling the input air to gas turbines and Direct Steam Injection Heaters.

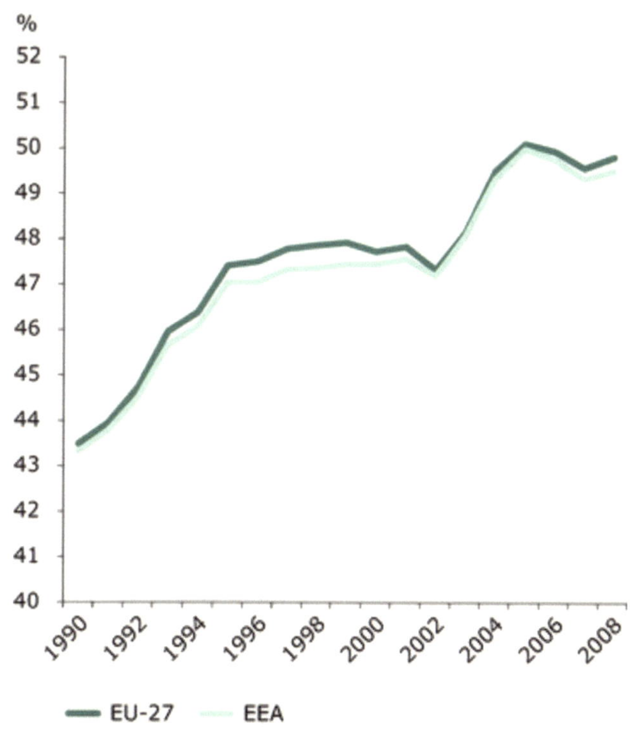

Fig. 5. Development of efficiency of conventional thermal electricity and heat production. Source: EEA 2011.

4. Energy efficiency in industry

Industry is the major energy user by sector, slightly ahead of power generation, buildings and transport. It is also a major user of electricity. In 2011 it is expected to use one quarter of global electricity production compared to one fifth in 1990. Due to advancement in technology major improvements took place in the efficiency of energy use in the industrial sector during recent years. The energy required per unit of value added (industrial energy intensity) has decreased in all regions since 1990. As a result of the globalization of industrial activities, energy intensity levels are converging. At world level, it fell by 1.6 percent / year between 1990 and 2009 (1 percent / year between 2000 and 2009).

Industrial energy intensity is lowest in Europe and OECD Asia, where it stands 40 percent below the world average since 1990. Due to the migration of heavy industries to the developing world, the European industrial energy intensity has decreased by 2.5 percent / year, while North America and OECD Asia show reductions of around 1.5 percent / year over the period. Energy intensity is slightly higher in North America than in Europe and OECD Asia, it stands just 13 percent below the world average and 50 percent above the EU average. Industrial cogeneration (CHP) is aiding the improvement in energy use in industry.

As mentioned above, one of the main reasons for improvements in the industrial energy intensity of Europe and North America is the shift of much of the heavy industrial production from these countries into China, India and few other developing countries which are rich in cheap energy resources.

Energy intensive industries are: the steel industry (20% of global energy use in industry), chemical industry (14%), cement industry (11%), paper industry (6%) and the aluminum industry. These heavy industries account for over than 55% of global energy use in industry. (ABB 2010)

Advancement in technology in heavy industries (like: increasing use of electric process in steel, change from heavy into light chemicals, the dry process in cement production and recycling in the aluminum industry, etc.), was the main reason for improvements in efficiency in energy use in industry.

The energy required per unit of industrial value added has been decreasing in all regions; as a result of the globalization of industrial activities, energy intensity levels are converging. The global economic crisis in 2009 had a strong impact on industrial trends, especially in developed countries, where energy-intensive industries were severely hit.

In the steel industry, which is the main industrial energy consumer, significant progress has been made over the last 20 years thanks to the spread of the electric process. However, the main producers, in cheap energy regions, still use outdated processes like open-hearth furnaces (Russia and Ukraine) or small-sized plants and low quality ore (China).

In the chemical industry, the energy consumption per unit of value added decreased in all main producing countries except in the United States, which is the world's main producer of chemical products and also the country with the highest energy intensity. The spread of dry processes and kilns using pre-heaters and pre-calciners let to reductions in the average energy consumption per ton of cement (specific energy consumption) in several large producing countries. The sharpest drop in specific energy consumption was achieved in China, thanks to the replacement of small cement plants by larger facilities.

Nevertheless, a great deal still remains to be done, as shown by the significant energy savings potentials identified in the references. The energy efficiency improvement potential can be as high as 40 percent in the case of steel and 20 percent in other industries.

5. Energy governance

To achieve the benefits of energy efficiency calls not only on technological advancement but also on institutional, legal, public awareness and similar coordination arrangements. This is called "energy governance" and is defined as "combination of legislative frame works, funding mechanisms, institutional arrangements and coordination mechanism, which work together to support implementation of energy efficiency strategies, policies and programs". Such energy governance is needed to overcome the many barriers to energy efficiency. These barriers are market, financial, institutional and technological, as well as lack of awareness and information.

To these must be added another major barrier which is "energy subsidies". The existence of subsidies is a major barrier to the attainment of the goals of energy efficiency, including that of generation and this is dealt with in detail below.

5.1 Dealing with barriers

It is not intended here to go into full details of how to deal with barriers to energy efficiency. These are detailed elsewhere (IEA-Energy Governance).

It is enough to mention here the importance of regulatory, pricing, fiscal and financial measures in overcoming such barriers. The most important is setting minimum energy performance standards (MEPS) for equipment, apparatus, buildings and vehicles with obligatory energy audits to ensure attainment of MEPS as well as appliances labeling and enforce building certification. Pricing has to aim at avoiding unnecessary subsidies with an increasing electricity tariff so that higher consumption is penalized by higher unit tariff that curbs overuse. Fiscal and financial measures need to provide incentive to energy efficiency investments, by providing revolving funds and contingent financing facilities.

Awareness and dissemination of information is most important in facilitating adoption of energy measures, particularly among the public. This is lacking in many cases and requires public information campaigns and promotions, enforcing the already mentioned appliances labeling, also demonstration and training. The energy service companies (ESCOs) can play a major role in this regard. They can provide the services demanded by industry as well as the public in certification, audits, training and promotion as well as disseminating public information. They also can be intermediaries in facilitating funding. Such ESCOs are now common in industrialized countries but are missing in most developing economies where their services are most badly needed.

5.2 Energy subsidies as a major impediment to energy efficiency

A subsidy implies selling an energy product (refined fuel or electricity) at prices lower than the product's opportunity cost. The opportunity cost is the price which the product would have obtained had it been exported instead of direct burning or utilization in local electricity production. We have to note that energy cost includes, beside the fuel's opportunity cost, the amortization of investment as well as operational and maintenance costs.

If we apply this definition, subsides proliferate energy use in most of the developing world. Electricity tariff subsidies are more widespread than subsides in selling refined products like petroleum for cars. This is due to the fact that electricity is a basic commodity that is widely used in all aspects of the economy as well as the society, particularly among limited income groups. This forces governments to provide it (like water) at prices lower than actual cost.

The dangers of subsidies manifest themselves in over use of the cheap resources more than is required by actual needs. Instead of spending the country's limited resources on human development, subsidies lead to waste, inefficiency and over use. This is besides the harmful environmental implications of this over use.

There is always the call for the need to provide energy in subsidized manner to certain limited income consumer groups as well as industries. If there is real need for this, then the subsidy should be made in cash and not in energy bills. Electricity subsidies lead to inefficient use and waste. Correspondingly it is essential to charge fair electricity prices that cover actual cost to avoid waste. Subsidizing the industry and economy can be in cash payment or other fiscal means, if need be.

Energy subsidies proliferates mostly in oil/energy exporting countries where refined products are provided at only a fraction of what they command in the export market and electricity tariffs are so low thus encouraging waste and over use. In few instances, in oil and gas rich countries, electricity is also provided free!!

The IEA (IEA-WEO 2011) estimates that the value of fossil-fuel consumption subsidies, including electricity generation, in 2010, amounted to $400 billion. Most of these subsidies encouraged wasteful over use with social, economical and environmental damage. Most important they were an impediment to achieving the goals of energy efficiency. The value of the energy products market now approaches $5 trillion annually. Subsidies account to almost 8% of this market value. If market pricing can reduce this by a half almost 4% of world energy consumption can be saved by phasing out subsidies.

The following Table 2 refers to the World's top energy subsidized economies. Mostly they are oil exporting countries.

Iran	82
Saudi Arabia	44
Russia	39
China	25
India	23
Egypt	20
Venezuela	20

(Source: IEA - World Energy Outlook 2011)

Table 2. Energy Subsidies in Billion $ (2010).

6. Conclusions

This chapter dealt with many aspects pertaining to energy efficiency, particularly in the electricity generation sector. The electrical power generation sector is chosen for special emphasis due to its present low efficiency and the growing importance of electricity as the major energy carrier.

Energy efficiency can be defined as percentage of total energy input to a machine, equipment or a facility that is consumed in useful work and not wasted as useless heat. The exact value of the global energy efficiency is debatable. Presently, however, it does not exceed 40%; that of the average net electricity generation is even lower at 34%. These are extremely low figures and demonstrate the strides that can be achieved in this quarter. Most of the energy losses in delivering final energy to consumers are in the power system, and that is why this sector is central to this discussion. Significant losses in converting delivered primary energy into useful energy in final use also occur in all sectors of the economy, mainly in industry, transport and buildings.

Generation efficiency has to be differentiated from electricity conservation or demand side management. Whereas conservation can be achieved through behavioral changes, education and pricing, power generation efficiency can be mainly improved by technology, investment, codes and practices also energy (and carbon) pricing, introducing edits and enforcing regulations. Power generation efficiency varies from one country and region to another depending on economic output, technology achievements, price of electricity, and weather. Many barriers exist that are delaying the deployment of generation efficient technologies.

In the global quest for curbing emissions, generation efficiency and electricity conservation figures out to be the most effective and cheapest means to reduce carbon emissions. In many cases it is a win-win situation where benefits can be attained at minimum (or no) cost, with short pay-back periods. However this means overcoming many barriers, one of which is the proliferation of subsidies in the developing world. Any environmental strategy that does not have power generation efficiency and electricity conservation as a center of interest will be missing the target.

Due to increasing cost of primary energy and its products, also environmental awareness, the efficiency of utilizing energy is improving worldwide year after another. Major efficiency improvements happened in OECD countries in recent years through lowering energy intensity in their economies. However this was assisted by the migration of heavy industries to developing economies. In spite of all technological advances and regulatory arrangements, energy efficiency worldwide does not exceed 40%. This demonstrates the still wide scope for improvements in this regard.

The efficiency of the electricity generating sector demands particular attention. Of the primary energy input to power stations less than one third of the calorific input reaches consumers, in many countries only one quarter. The wider introduction of combined cycle plant and higher parameters in PC plants are gradually improving the electricity sector efficiency. This is particularly important because the world economy is increasingly electrifying.

In the world-wide pursuit for improving energy security and reducing environmental impacts, improvements in energy efficiency are the cheapest and easiest means to attain results. Any global sustainable energy strategy which does not have improvements in energy efficiency as its center of interest will be missing the correct emphasis. No more this is evident than in the case of electrical power generation.

7. References

ABB - Global Energy Efficiency Trends – ABB product 2010

Creyts, J., A. Derkach, S. Nyquist, K. Ostrowski, and J. Stephenson. 2007. "Reducing U.S. Greenhouse Gas Emissions: How Much at What Cost?" U.S. Greenhouse Gas Abatement Mapping Initiative Executive Report. December 2007. McKinsey & Company. Available online at http://www.mckinsey.com/clientservice/ccsi/pdf/US_ghg_final_report.pdf

EEA 2011, "Efficiency of conventional thermal electricity generation (ENER 019) – Assessment published Aug 2011", European Environmental Agency.

Enerdata - Global Energy Intelligence. More information about Enerdata available www.enerdata.net

ESD – Energy Services Directive, Article 3b, 2006

ETP 2010 – IEA Energy Technology Perspectives, 2010

Lui, L. & G. S. Gallagher, "Catalyzing Strategic transformation to a low-carbon economy- A CCS roadmap for China", Energy Policy Journal, Vol. 38, Jan 2010, pp 59-74

IEA-Energy Governance: IEA Energy Efficiency Governance Handbook - Second Edition-2010

IEA-WEO 2011: The International Energy Agency – World energy Outlook 2011

Khatib, H., "Economic Evaluation of Projects in the Electricity Supply Industry", I.E.E., London 2003.

Khatib, H., "Smart Grids", International Seminar on Nuclear War and Planetary Emergencies – 43rd Session", World scientific Publishing Co., 2011.

NPC – National Petroleum Council, "Topic Paper # 4 – Electric Generation Efficiency", 2007

Sioshansi, F., "Generating Electricity in a Carbon- Constrained World", Elsevier, 2010.

WEA – World Energy Assessment "Energy End-Use Efficiency" Eberhard Joehen, UNDP 2000.

Wikipedia - "Energy Conservation", 2011